D0735573

TECHNICAL EDITOR'S HANDBOOK

TECHNICAL EDITOR'S HANDBOOK

A Desk Guide for All Processors of Scientific or Engineering Copy

GEORGE FREEDMAN
and
DEBORAH A. FREEDMAN

DOVER PUBLICATIONS, INC.
New York

for Ruth Freedman

Our wife and mother—and a former technical editor—who could always tell when we were going astray as we were writing this book, and who always set us straight.

Copyright

Copyright © 1985, 1994 by George Freedman and Deborah A. Freedman. All rights reserved under Pan American and International Copyright Conventions.

Published in Canada by General Publishing Company, Ltd., 30 Lesmill Road, Don Mills, Toronto, Ontario.

Published in the United Kingdom by Constable and Company, Ltd., 3 The Lanchesters, 162–164 Fulham Palace Road, London W6 9ER.

Bibliographical Note

This Dover edition, first published in 1994, is an unabridged republication of *The Technical Editor's & Secretary's Desk Guide*, first published by McGraw-Hill Book Company, in 1985. A Foreword to the Dover Edition has been specially written, and biographical matter specially augmented, for this edition.

Library of Congress Cataloging-in-Publication Data

Freedman, George.
[Technical editor's and secretary's desk guide]
Technical editor's handbook : a desk guide for all processors of scientific or engineering copy / George Freedman and Deborah A. Freedman.
p. cm.
Originally published: The technical editor's and secretary's desk guide, New York : McGraw-Hill, 1985.
ISBN 0-486-28009-8
1. Technical editing—handbooks, manuals, etc. I. Freedman, Deborah A. II. Title.
T11.4.F74 1994
808′.0666—dc20 94-3172
CIP

Manufactured in the United States of America
Dover Publications, Inc., 31 East 2nd Street, Mineola, N.Y. 11501

Foreword to the Dover Edition

When a book goes into a reprint edition, what is implied is that years after it was originally written it is still relevant. Its content is still essentially correct, so that its value to the reader is not significantly diminished by the passage of time. We feel that *Technical Editor's Handbook* satisfies these criteria. But we wrote it 1984 (it was published by McGraw-Hill Book Company in 1985 as *The Technical Editor's & Secretary's Desk Guide*). How much has the passage of a decade "diminished" it? Inevitably, somewhat.

In that sense this book stands as a testimony to the ever-accelerating speed of the accumulation of new knowledge, new technology and new practices, not only in our daily lives but in our professional lives as well—and, specifically, in the ways we reproduce technical and scientific text, which itself has advanced in content. In fact these rates of change are now so fast that even were the book entirely rewritten today it would still—in a sense—be out-of-date the moment it appeared on the shelves of bookstores and in catalogs, since it is always about a year from the time a book is written to the time it appears in print. And because a lot can happen in science and technology in a year, a book such as this one has no choice but to achieve its relevance and the correctness of its content "on the fly," and with the hope that, in spite of progress, it retains its value, since it is all there is of its kind.

One of us has been through a similar reprint experience before with the precursor of this book, which dealt with the same subject. (The other author did not participate in the creation of that book, being of but tender years at the time.) *A Handbook for the Technical and Scientific Secretary* by George Freedman was published by Barnes and Noble in 1967 and reprinted—also by Dover Publications—in 1974. While the sixties and early seventies were a time of great technical advance, they did not hold a candle in this regard to the eighties and the early nineties. For example, while computers were making their presence known almost thirty years ago, when the original *Handbook* was written, there was little awareness of the fact that someday word processors would become the tool of choice for both authors and writers (see Chapter 1 for the distinction between the two)—then relegating most typewriting machines to the trash heap. Nor did any epoch-making trend appear eight years later when the reprint version came out. Even as recently as 1974, most secretaries still spent most of their working time pecking at the keys of typewriters.

But their role was diminishing fast and in our "new" book of 1984 we predicted that "even though not every secretary has to have a word processor," their roles would multiply—as indeed they have. But just being aware of this fact—and how fast change has come upon us—is all the reader needs to derive value from the content of Chapter 1 on general aspects of reproducing technical material in what we called then the "automated office."

Interestingly, the problem is far less an issue in the chapters on mathematics, chemistry and physics. In those fields, while new worlds have been conquered in spectacular fashion in the decades since these chapters were written, there is little new for the text reproducer that will give him or her trouble when he/she makes reference to the pertinent pages. Most of the notations and practices described therein are timeless. And since these three disciplines underlie most kinds of engineering, the same statement pertains to them. The same goes for the chapter on electricity and electronics, but not for the sections on vacuum tubes. Today, while they are not quite as dead as the proverbial doornail (we still use x-ray tubes in hospitals, traveling wave tubes in military missiles and magnetrons in our microwave ovens—and engineers and scientists produce thousands of documents yearly on those subjects), had we it to do over today, the subject of "tubes" would have received less emphasis.

As for the Glossaries—all 18 of them—they still provide the fastest "fixes" in the book. As with any such listings of terms, new ones should be added yearly, but even without the latest possible additions, they should remain useful for 98% of the cases for which the writer chooses to consult them.

It is the computer chapter—the one which deals with the newest science and technology—that is most impacted by the advances of recent years. We must acknowledge that many new terms and reproduction practices necessarily do not appear in that chapter, but what it says about the field is still valid and still useful.

GEORGE FREEDMAN AND DEBORAH A. FREEDMAN

Contents

PART II REFERENCE SECTION

Preface

This book is addressed to those who transcribe or type scientific or technical material in the form of memos, letters, reports, proposals, or manuscripts intended for distribution or publication. Typically, the user of this book will be a personal secretary or a technical typist in the publications department of a technological firm; but the book will also serve as a guide for others in such a department—technical editors, technical writers, and graphics specialists.

When an author's draft is rendered into printed form, the result is most successful when there is a background of team effort between the author and those who process that draft. Additionally, the quality of such documents is highest when the processors—the writers, transcribers, typists—have acquired an understanding of the particular formats and conventions pertinent to the disciplines being dealt with. Based on these ingredients of good collaboration between author and writer and the processors' understanding of how to write what the authors create, what is rendered into printed form becomes what it must be—an effective instrument of communication between scientific and technical professionals.

No printed material is a good communication instrument unless it is easy to read. It is inherent in technical and scientific text that the content is never "easy." Nor is the format, since it is so replete with symbols, foreign alphabets, and graphics, and often even lacks recognizable words. Thus it behooves those who have the responsibility for putting such text into printed form not to add to the complexity. Rather, by clear rendering of each symbol, by correct placement of each expression, by common-sense planning of each page, the writer must make the task of the scientific or technical reader easier (for a task it is). The result can be very satisfying. When a page of complex mathematical or chemical text, for example, is well done—which means well-crafted and therefore understandable to its audience—it will also be attractive and tasteful. It is the goal of this book to serve as a guide to attaining these qualities in this very special and very important field of scientific and technical communication.

This handbook presents the basic rules that apply in virtually every field of technical writing. It goes on to describe fine points that occur infrequently but which may cause trouble when encountered. There are many examples and illustrations which will serve as models for accepted usage. The book also provides the scientific rationale that underlies the rules. Thus the reader is provided with a wealth of material.

Recognizing that each writer, secretary, and typist will have individual needs—and desires—with regard to how deeply to master this material, instruction is therefore presented in two ways—in terms of just the mechanics of processing text and in terms of the meanings (in very basic terms) that underlie those mechanics. Since each chapter (except the first) presents these two different treatments of the same material in two separate sections, the user of this book has the option of choosing how much to learn about any subject.

Each scientific discipline has its own sets of rules. There is a uniqueness to each such that one would never mistake a page of chemical text for a page of electrical engineering text. At the same time, there is much overlap, for there is much that is similar too. For example, both chemistry and electrical engineering make use of other disciplines such as mathematics.

Chapter 1 concentrates on what is common to all technical writing, regardless of the discipline. Thus it contains instructions for presenting foreign alphabets and scientific symbols on the printed page. It describes the use of prefixes and suffixes common to all fields of science and engineering and how they are linked to root words. Abbreviations are similarly treated, as are physical units. Simple techniques for creating graphics are given. The chapter also serves as a guide to references of many sorts. Accepted procedures for preparation of manuscripts for publication are provided. The chapter concludes with a description of what has come to be known as "the automated office," where many of the readers of this book work.

Chapters 2 through 6 treat the major disciplines of science and engineering, showing how they differ from each other, how they are unique. These disciplines are mathematics, computer science, physics, electricity and electronics, and chemistry. Although some fields of science and technology have been left out, very little goes on in American factories and laboratories that is not included under one or more of these headings.

These six chapters are supplemented by Part 2, which consists of glossaries containing charts and lists of reference material arranged for quick referral. Part 2 also includes a "Dictionary of Technical Terms" containing about 3000 of the most commonly encountered words and expressions. These are presented with procedures for word division and with simple "stripped down" definitions.

THE AUTHOR-WRITER TEAM

Let us draw a distinction between *authoring* and *writing*. Authoring creates the text. Writing, by transcribing it to a printed page, renders it easily readable and easily accessible to a wide audience. The result of author-writer collaboration is the finished document. Why should the authoring and writing functions be separate? Because each function requires its own expertise. But both are required. Thus a precondition for achieving success (the readable, accessible, etc., document) is that there be teamwork.

What the Author Expects

The scientist or engineer who submits a sheaf of handwritten or single-finger-typewritten pages to a secretary or a publications department is offering a piece of his or her lifeblood. That is the way people always feel about something they have created. The author is its parent, as much as though the product were a wriggling newborn infant. When that creation is handed over to someone who had nothing to do with that creative process, the author-parent nevertheless expects that it will be treated with the same love—and sweat—that he or she has lavished on it.

But in addition to that, the author expects that person to possess a certain special competence. The author does not expect the secretary or the technical writer to be an engineer or a scientist—the author already is that. What is expected is that these people will be *literate* in the field. Therefore, a major goal of this book is to impart at least a measure of literacy to those of its readers who are deficient in that regard.

Scientific and Technical Literacy Required by the Writer

If anyone is going to be a writer in a particular language (and there is a particular language of science and engineering), a first necessary condition is that the person be at least somewhat literate in that language.

For our purposes *literacy* in science and technology has a number of implications not necessarily found in a dictionary. Scientific and technical literacy means

- Understanding special notations and how to reproduce them
- Understanding the meaning of the crucial terms in the text
- Knowing how to present supporting material in the form of graphics and other illustrations
- Knowing how to present data in tables and graphs
- Knowing how to present supporting material in the form of literature references
- Knowing how to organize a technical paper for publication, or a proposal to the government, or a document in the form of a manual for a piece of equipment

It is interesting that the idea of literacy has become popular in the now frequently used term *computer literacy.* Everyone agrees we must all have it, whatever our field of business or interest. To function in the world as it is today—or at least as it is becoming—it is necessary for all of us to become computer literate. Thus teaching of this subject now begins in the elementary schools. Indeed, Chapter 3 of this book, "Computers," essentially provides the information required for computer literacy.

What is intended is that each of the chapters of this book provide enough information for the reader to become literate in that subject. Thus Chapter 2 provides "mathematical literacy"—enough for a technical secretary and technical writer—and Chapter 6 provides "chemical literacy," and so on. Where the meaning of a particular expression that has not been explained in these chapters is required, the "Dictionary of Technical Terms"—with very short, easily understood definitions—will help (Part 2).

Literacy in the English Language

None of these requirements for technical literacy diminishes the requirement for the most important literacy of all. Like any other writer in any field in this country, the technical writer must be literate in the English language. Mistakes in spelling, punctuation, grammar, and rhetoric are as much sins in scientific documents as in any documents. High standards of English must be maintained at all times. That goes without saying. Thus no more will be said about English literacy in this book, except for a comment on authors.

Alas, the authors often present a problem here. Many, not all, maintain higher standards of technical literacy than English literacy. The writer must play a role here—if the document is to be up to the required quality of literacy in English. This is another reason that teamwork between the author and the writer is needed.

THE CHALLENGE OF TECHNICAL AND SCIENTIFIC COMMUNICATION

Let us appreciate the context of this book. We live in a scientific and technological world. Implicit in that fact is that the world never stands still. The world of the 1960s was very technological too, but it was primitive compared with the 1970s. Now we are in the 1980s, and everything is new again. What do the 1990s hold?

One essential ingredient of technical advance is that the technologists be able to talk to each other efficiently. The best way to achieve that—and to keep a record of the talk—is to get it down on paper. Our task is to do that competently and professionally. Only then can engineers and scientists do their jobs effectively.

Engineers and Scientists—Clientele for Text Processors

There are more than 3 million engineers and scientists in the United States. That means an astonishing 1.5 percent of the population consists of people who have undergone the rigorous years of higher education—at least 4, but often 5, 6, or 7—to qualify them as professionals fit to practice in their various fields. Furthermore, scare talk to the contrary (we are producing too many attorneys and business school graduates and not enough technical people), they are replacing themselves faster than they retire. Every year hundreds of thousands of young men and women enter the marketplace, armed with their new degrees in civil engineering, mechanical engineering, materials engineering, electrical engineering, chemical engineering, bioengineering, chemistry, physics, mathematics, computer science—and many others; the list is long.

These engineers and scientists and their disciplines are extraordinarily diverse. Yet they share certain basic scientific ways of approaching problems. They share the language of mathematics. And in all cases, their work means absolutely nothing unless they can communicate it. Everything accomplished by engineers and scientists, to have value, must be written down. This means that they all require the services of people as skilled in "writing down" as the authors may be in authoring the original copy.

This gives us the reason for this book, which is to provide a guide through the complexities of reproducing scientific and technical material—data and text, notation and narration—which it is hoped will be of value for secretaries, technical editors, technical writers, technical typists, technical publications departments, technical illustrators, graphic artists, and anyone else engaged in producing clear, readable, neat, unambiguous copy in any discipline of science and technology. It is a worthy field to be in, and we should be gratified to be part of it.

Today's Office—Where the Writer Works

Everything that has been said about technological advance is as true for the writer and processor of text as it is for the scientific or technical author. Today's writing tools place the ones of a decade ago into a relative stone age.

It used to be that engineers and scientists simply handed their scribbled memos, reports, or technical papers to the secretaries in the office—the same ones who arranged conference rooms, sorted out travel schedules, credited expense vouchers, and simultaneously handled incoming calls with the tact and judgment needed to keep the people they served out of trouble with *their* bosses. It was then the secretaries' job to "type it up," somehow achieving that goal along with all the others. This is still the common practice for a large segment of the technical community. For that group, the personal secretary must still do "everything."

But with the coming and acceptance of computer-based office machinery in the last decade, a revolution has begun. We now have *office automation.* It is inevitable that all but the smallest organizations will be changed by it. The result will be, as it already is in thousands of offices, that the personal secretary will become more of an office administrator than a text processor, and the publications department will grow in importance as a way to produce technical documents faster, cheaper, and more professionally.

It might seem that this will get the personal secretary off the hook as far as understanding how to reproduce technical material is concerned, but that is not so. All that has happened is that long documents are sent to "pubs." Shorter ones, one to five pages, still are best done by the personal secretary.

We must conclude that the personal secretary of an engineer or scientist must possess much of the same craft as the technical typist and the technical writer in the publications department. All of them have to know the subject matter of this book.

Chapter 1 describes some features of the most modern of automated office systems and how these can be tailored to technical writing.

ACKNOWLEDGMENTS

The authors wish to thank the Raytheon Company of Lexington, Massachusetts, which was most cooperative in providing insight into the subject of processing technical text in many of its divisions and subsidiaries. This company was also generous in contributing examples of typewritten text and graphics. The assistance provided by Ms. Stephanie A. Rodrick, manager of publications of the Raytheon Research Division, Mr. John E. Smith, manager of office automation systems of the Raytheon Equipment Division, and Mr. Ross Giuliano, business manager of the Raytheon New Products Center, was especially valuable in these regards.

We also wish to thank Mr. Joseph R. Adamski and Dr. Peter L. Toch of the Raytheon New Products Center technical staff, who were a resource of essential expertise to which we turned frequently in the preparation of the computer and chemistry chapters, respectively.

George Freedman
Deborah A. Freedman

How to Use this Book

This is a reference book. To get the best value from it, you needn't read it from cover to cover.

USING A FAST-SKIM APPROACH

If your author is a chemist, by all means skim the chemistry chapter. If your author is a physicist working in the field of optics, by all means skim the section on optics. In effect, get the lay of the land. In any case skim the chapter on mathematics. All engineers and scientists use mathematics. Also, read Chapter 1. It describes basic techniques and practices common to all scientific disciplines.

RULES

The correct processing of scientific and technical material requires familiarity with particular rules of notation. Unless the rules are followed, with professionalism, the primary goal of the document—communication—will not be effectively attained. This book attempts to present these rules, along with some (pared-down) basic definitions of relevant scientific principles.

These explanations are usually presented in no more than one or two sentences, which we hope will provide "just enough" meaning so that the drift will come through and so that the text processor half of the author-writer team will gain at least a glimmer of what points the author is making. That is, understanding these rules is the same as attaining literacy in that subject.

It is the intent of these explanations that they be so phrased as to be able to be grasped even by those with no background, and—let's admit it—little interest, in science. For those who wish further explanation, beyond this minimum, there are further statements in the form of extended explanations and examples, given in smaller typeface.

Rules Differ

The rules for reproducing on paper the written creations of scientific and technical authors are different from the rules used by other kinds of authors such as novelists, journalists, or attorneys. Further, each discipline of science and engineering has its own rules (although there is a lot of sharing, as when a ceramist treats the subject mathematically or when a chemist describes the electrical properties of a chemical compound).

Where to Find Rules

The rules for writing about mathematics, chemistry, physics, materials engineering, electrical engineering and electronics, and computers are presented here. Necessarily, some disciplines are not dealt with, and some of the less used rules have been left out. But virtually every other field of science and technology is related to these basic ones. For example, the rules for presenting metallurgy text are based on the rules for presenting physical chemistry text (in the chemistry chapter); crystallography is part of crystal physics (in the physics chapter); and the rules for civil engineering are based on mechanics and hydraulics (both to be found in the physics chapter).

To find where the rules for any field are treated, use the index. Most of the index entries refer to rules by section number and page.

Since it is impossible in a book this size to give all the possible rules for all scientific notation and narration, inevitably some needed information will not be in this book. In such a case, the text processor must not hesitate to go to the author and request clarification. That is what is meant by author-writer teamwork, of which such a strong point has been made here. Failure to do this means that the product will be less effective than it could be.

FORMAT

Part 1

Each chapter is divided into two separate sections. In each case, the first is entitled "Basic Principles" and the second "Basic Practice—Techniques and References."

For the user of this book who desires simply to know techniques—how to draw a symbol, how to place subscripts and superscripts, how to construct an equation, how to draw a graph, etc.—the second section will suffice and indeed should serve as a quick, handy reference.

For those who are curious about the meanings of symbols, and the reasons for representing various things on the printed page, the first section provides more information. Furthermore, it is the first section of each chapter that deals with the *literacy* referred to here. Reference to the first section will usually be sufficient to provide an interested secretary or technical writer with an inkling of what the document is all about. Surely, many text processors could not care less. That is their privilege. Obviously, they are not employed to be scientists and engineers; the authors already have that assignment. But who can deny that knowing a little bit about the subject of the text will make a difficult and admittedly tedious and nit-picking job more interesting? Also, any particular group of authors will always tend to deal with the same disciplines. As months and years go by, the technical secretary will find that an occasional peek into the first section of a particular chapter will not be intimidating, and will pay off.

There are several other elements to be noted in the format of this book.

- Cross-references are generously provided so that related material can be looked up.
- Tables of words and terms (sometimes with descriptive sketches) that group them in terms of general subject matter are provided.
- Illustrations of important constructions, especially mathematical and chemical formulas and equations, are provided.
- Illustrations of simple graphics such as a secretary might be expected to reproduce are included.
- A number of sample pages, extracted from actual documents typed by secretaries and technical typists in a number of commercial companies and laboratories, have been reproduced (with permis-

sion of those establishments) to serve as real-world guides to good practice.

Part 2

Part 2 consists of two portions. The first is a set of glossaries containing hundreds of items—detailed data and references—primarily in the form of lists. It is an extension of the techniques sections of Chapters 2 through 6, providing more information on the mechanics of scientific and technical writing. Refer to the glossaries when detailed information not found in Part 1 is required. The second portion of Part 2 is the "Dictionary of Technical Terms." Several thousand of the most commonly encountered expressions are listed, shown in divided form for typing purposes, and defined in simple terms. There is a good chance that the crucial word needed to clarify the meaning of a paragraph or a page will be found there.

PART ONE

Technical Narration and Notation

CHAPTER ONE

Basic Techniques and Procedures

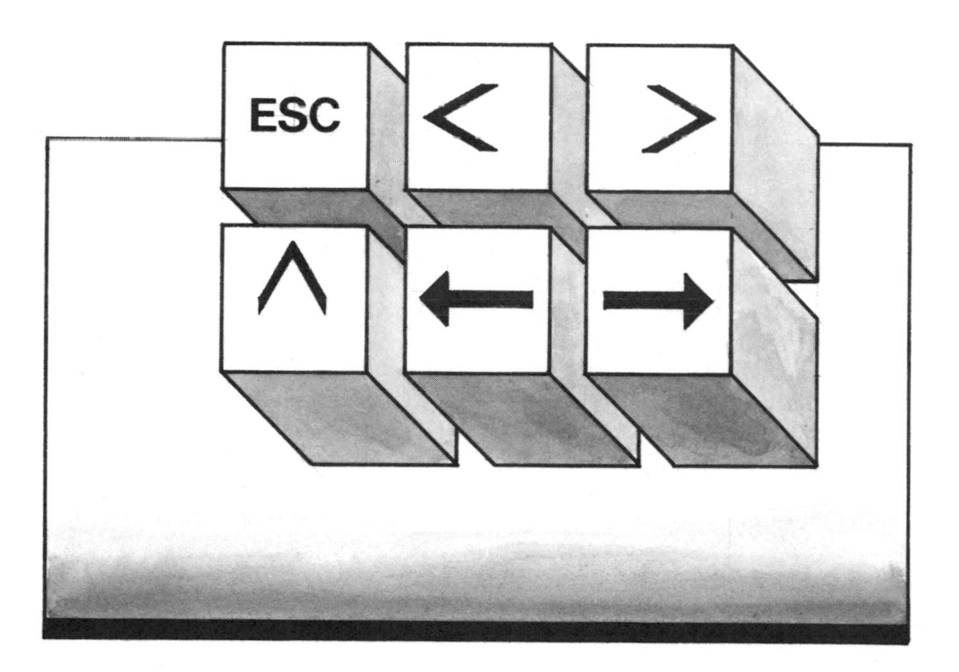

This chapter is basic in that the material it contains is pertinent to and common to all disciplines of science and technology. It forms a foundation upon which to build techniques for producing text in any scientific discipline, and the five chapters that follow may be considered as five distinct branches that grow from this foundation.

All engineers and scientists are familiar with the Greek alphabet, and all disciplines have developed *constructed* words, which employ special prefixes and suffixes, and other word parts based on acronyms and abbreviations. They also share conventions in naming and symbolizing units. In fact, different disciplines share many units and make use of the same mathematical signs and symbols. All these *language* matters are treated in the first section of this chapter.

The second section describes techniques for preparing *graphics*—sketches, graphs, etc.—and the third describes procedures for *preparing manuscripts for publication.*

Section four provides lists of *recommended reference books* to which the reader may resort should this book fail to provide sufficient information.

Finally, the *automated office* is described. It is the ultimate environment for the production of written scientific text, based as it is on the latest computerized equipment.

1.1 MATTERS OF LANGUAGE

1.1.1 Other Alphabets, Primarily Greek

We in the United States use the Latin alphabet. So do people in many other countries, including the British, Danish, Dutch, Finnish, French, German, Hungarian, Italian, Norwegian, Polish, Portuguese, Spanish, Swedish, and Turkish. That still leaves the rest of the world (which is by far the larger part), which uses Russian, Arabic, Chinese, Japanese, Hebrew, Greek, and many more alphabets.

This book is addressed to an American audience that writes with the Latin alphabet; but it is a special audience, one that deals with texts authored by scientists and engineers. People who work in technological and scientific fields find the Latin alphabet too limited. They find that it is often more convenient to use Greek as well as Latin letters when constructing new terms. This means that those who reproduce technical text must use (and develop a facility in writing) Greek letters as well.

1. Getting Greek Letters on Paper

How does one type Greek letters? With replacement keys and with special fonts such as may be found on special rotating balls or on special daisy wheels. (Refer to "Mathematical Signs and Symbols" in Section 1.1.4, where the same conditions pertain.) Failing that, they must be hand-drawn neatly with a fine black fiber-tipped pen.

Care should be taken that the same symbol always looks the same and that there be no chance for confusion on the part of the reader.

Do not confuse:

α (Greek letter alpha) with $\propto$ (proportionality sign)

κ (Greek letter kappa) with k (Latin letter)

η (Greek letter eta) with n (Latin letter)

ρ (Greek letter rho) with p (Latin letter)

ν (Greek letter nu) with v (Latin letter)

μ (Greek letter mu) with u (Latin letter)

There is one expedient which can be used for a number of Greek letters. They can be *constructed.* Table 1-1 lists a number of examples. You must use your judgment here. The appearance of the resulting Greek letters may be satisfying or not depending on the shape of the print on your machine. If they do not look convincing, go back to making them entirely by hand.

2. Greek Letters Become Words

Another important point about the Greek alphabet is that in some instances the name of a letter, through common practice, has become a word. And that word has acquired a technical meaning. Thus engineers will talk of *beta rays* or the *mu* of a vacuum tube (by which they mean its amplification factor). It is as though, were the Latin equivalent involved, they were using terms like "bee rays" or the "emm" of a vacuum tube.

TABLE 1-1
CONSTRUCTING GREEK LETTERS

Beta	Add a curl to an 8	β
Delta	Add a curl to an o	δ
Theta	Add a hyphen to an O	θ
Mu	Add a curl to a u	μ
Rho	Add a curl to an o	ρ
Sigma	Add a curl to an o	σ
Phi	Add a slant to an O	ϕ
Chi	Use a slant and add ⌒	χ
Psi	Use a slant and add ∪	ψ

3. The Greek Alphabet

Glossary A gives the Greek alphabet in both uppercase and lowercase letters (technical people use both). Note that there are a few variations in how to write five of the lowercase letters. Use the ones the author prefers.

4. The Aleph

There are a few other alphabets, such as archaic German, or Cyrillic (Russian), which scientists sometimes want to use. But these have never become popular and they are not treated in this book. One letter from one other alphabet is used, however. It is the Hebrew *aleph,* which stands for the letter a and is used in its uppercase form. The only thing to do with an aleph, if you ever encounter it, is to hand-draw it carefully. It looks like this: ℵ

5. The Latin Alphabet in Other Languages

There are some variations (differences in how to write certain letters) in the Latin alphabet that the technical writer or secretary should become familiar with just because the scientific and technical community knows no national borders. Glossary B lists these variations in the Latin alphabet. The author will often cite a reference by an authority in a foreign land and the typist has to be able to reproduce that reference in keyboarded text—with the alphabet variations. Often the variations will occur in the name of the foreign author. Here you must be especially careful, because you must *always get peoples' names right,* foreign or not. Examples of foreign names, some of which use these variations, may be found in Glossary P, "Names of Scientists and Engineers."

1.1.2 Plurals

Technical nouns, like other nouns, may be expressed in plural form. Although most of them become plural by the addition of the letter s, those based on Latin and Greek roots often do not and may thus cause confusion. Glossary C lists the ones that most commonly present difficulty in scientific texts, with the preferred usages.

1.1.3 Prefixes and Suffixes

Prefixes and suffixes enrich and enlarge words so that the new composite word conveys information that would otherwise require many individual ones to give the same message. In general, they add meanings of quantity and kind. Such added meanings are useful in ordinary language, but they are especially valuable in science and engineering, where variations in "how much" and "what sort of" are particularly complex and delicate.

1. Numerical Prefixes

The SI units of measure (Système International d'Unités) include a series of quantifying prefixes (also described in "SI Prefixes and Symbols" in Section 2.1.6). These are tabulated in Glossary D together with other prefixes that specify amounts.

2. Descriptive Prefixes

Glossary E is a selection of nonquantifying prefixes which designate various attributes including qualities, classes, relative positions ("supra-": "above"), or relative times ("post-": "after").

3. Suffixes

Suffixes function as prefixes do—to enrich and enlarge words. They too designate quantities (roughly) and qualities. Glossary F is a selection of the suffixes most commonly encountered in scientific and technical text.

1.1.4 Mathematical Signs and Symbols

Probably the most difficult task that confronts the processor of scientific and technical text is the reproduction of mathematical material. Before even beginning to construct a page of equations, one must decide how to represent each sign and symbol. Many approaches are possible depending on the available resources. Furthermore, the actions taken tend to be different for each symbol, again depending on the available resources.

1. Replacement Keys for "Old" Machines

Old-fashioned typewriters, which cannot be disregarded since millions are still in use, often can employ replaceable keys. Some manufacturers will supply a kit of Greek letters and mathematical symbols which can be used one at a time by manually inserting the required key as needed. This is a laborious process and its one redeeming feature is that the resulting text looks neat and professional. To find out whether such a system is available for your machine, contact your local repair service or the manufacturer.

2. Replacement Ball Fonts for Rotating- Ball-Type Machines

A better solution is available if you use a rotating-ball machine. You can usually purchase special balls, most of which have the required signs and symbols. Again, this is a laborious procedure—the typist must learn which key to strike to get a particular symbol—but it is more convenient than using individual replacement keys.

3. Daisy Wheels and Word Processors

Word processors can be fitted with special daisy wheels for mathematical work, and the typist will find this to be the most convenient procedure of all. But there is still the nuisance of changing from one wheel to the other and learning which key to strike for a particular symbol.

4. Hand Drawing—The Last Resort

In the end, the typist must resort to handwork. Occasionally templates will help, as with integral signs of different length. But when all else fails, the typist must resort to the most primitive procedure of all: the symbol must be drawn. When that is the case, let common sense be your guide. Do it neatly. Do it so that all similar symbols always look similar. And do it logically.

5. A List of Most-Used Mathematical Signs and Symbols

Glossary G summarizes most of the mathematical signs and symbols likely to be encountered.

1.1.5 Physical Units

Possibly the earliest unit which we still use today was the foot. And that was what it was—the length of someone's foot. Hence its name. Whose foot? The legend goes that it was a royal foot, belonging to an ancient king of England. But that is not important. What is important is that it illustrates the concept of the unit. A unit is a defined quantity used for measurement. Since everything physical can be measured (otherwise it would not be physical), all we need to proceed with the act of measuring is to define units—one for each physical item.

1. A List of Most Commonly Encountered Units

Glossary M lists the most common units with the abbreviations that are used to designate them. As with all aspects of the reproduction of technical text, you must be careful when you encounter these. *Get them exactly right*. For example, it makes a lot of difference whether you use uppercase or lowercase letters. As can be seen, T stands for the electrical unit *tesla,* while t stands for the weight unit *metric ton.* And G stands for *gauss,* but g for *gram.*

2. Notation—As Distinct from Units

Notation is a system of *symbols* used to describe scientific or technical phenomena in a particular field. Table 1-2 provides an example; it is a system of notation for the calculation of stress analysis as used by mechanical engineers. Comparison of this table with Glossary M, "Commonly Encountered Physical Units," will show why people tend to confuse systems of notation with systems of units. The two lists look very similar. That is, both consist of short definitions with letter symbols. But the distinction between them is this: *Notation* is quantitatively expressed *in units.*

Take the symbol l. It stands for *length.* The *amount* of length

TABLE 1-2
NOTATION FOR MECHANICAL STRESS ANALYSIS

Quantity	Symbol	Quantity	Symbol
Acceleration	g	Pressure per unit of area	p
Angle; total angle of twist	ϕ	Radius	r
Angle of twist per unit of length	θ	Resultant stress	S
Bending moment	M	Strain energy	U
Breadth, width	b	Temperature coefficient of expansion	α
Compressive force	C	Tensile force	T
Critical unit stress in column formulas	S_{cr}	Thickness	t
Deflection of beam (positive upward)	y	Torque, twisting moment	M_t
Density	ρ	Total force	F
Diameter	d	Total weight or load	W
Distance from neutral axis to extreme fiber	c	Ultimate stress	s_{ult}
Height	h	Unit compressive stress	s_c
Inside diameter	d_i	Unit normal stress	s_n
Length	l	Unit normal stresses in x-, y-, and z-directions, respectively	s_x, s_y, s_z
Modulus of elasticity in shear (modulus of rigidity)	G	Unit shearing strain	γ
Moment of inertia of area	I	Unit shearing stress	s_s
Normal force	N	Unit stress	s
Number	n	Unit tensile stress	s_t
Poisson's ratio	μ	Unit working stress	s_w
Polar moment of inertia of area	I_p	Velocity	v
		Vertical shearing force	V
		Weight	w
		Young's modulus of elasticity	E

which l represents is unknown until someone assigns l a quantity measured in *units,* such as inches or centimenters. Likewise, the symbol v stands for *velocity,* but *velocity* can only be expressed in numbers by using *units,* such as miles per hour or meters per second.

Because symbols like l and v are "variables" (that is, their quantitative value will change from one situation to the next), it is customary to set them off by italicizing them. In typescript this is shown by using the underscore. Usually notation symbols are letters of the alphabet (sometimes with subscripts or superscripts), and they are often like abbreviations of the concepts they stand for. For example, the concept *twisting moment* can be represented as M_t. In some cases, however, the concept is represented by a letter whose relationship to the concept is not obvious; for example, the symbol for *heat content* may be Q. Greek letters are often used, such as α for *angle;* these do not need to be italicized. As opposed to variables, units of measure are almost always abbreviated with letters taken from the Roman alphabet such as cm for centimeters. One of the few exceptions is ohms for which the Greek symbol omega (Ω) is used. See Glossary M for a fairly complete list of units.

Notation lists are not given in this book (except for the example in Table 1-2), because they are rarely consistent. There are different notation lists for every scientific or technical discipline, and each author within a field tends to invent new variations. Furthermore, notations in conventional use today will undoubtedly evolve into different forms with the passage of time.

1.1.6 Abbreviations

Certain expressions occur again and again in each scientific and technical discipline, and sometimes they are long and ungainly. Since technical authors are addressing their own professional colleagues, they quickly fall into the practice of designating such expressions with familiar accepted abbreviations. Just as baseball fans in Boston refer to the Boston Red Sox as Bosox, so a physicist may refer to infrared radiation as IR, and a computer engineer may refer to computer-aided design as CAD (pronouncing it "cad").

The most commonly used abbreviations are those that stand for units, such as cal for calorie or ft for foot.

The important thing to remember is that all these are the author's shorthand. As is true with all shorthand, small variations in how they are written may drastically change their meanings. But unlike shorthand, scientific abbreviations are devised with the aid of very few rules. The only thing that is consistent about them is that they are short, shorter than if the original term had been written out.

Sometimes the abbreviation ends in a period, more often not. Sometimes it begins with a capital letter, more often not. It may be an acronym consisting of all uppercase letters; then the letters may be set off by periods—but not necessarily. Since there are no rules, the processor of the text has no choice but to be careful. And the authors of this book have no choice but to admonish the readers to *get them right.*

There are times when the author will use an unfamiliar abbreviation or when abbreviations may resemble each other. It is then good practice for the author or editor to write out the term the first time it is used in the document, followed by the abbreviation, as in "revolutions per second (r/s)."

You should know that abbreviation usage is not cast in concrete. Sometimes an abbreviation will change with time: revolutions per second used to be rps, and that is still an accepted but not preferred usage.

Also, the same unit may be designated differently in different scientific or engineering specialities.

> NOTE: In this book you may see one symbol or abbreviation used in one part of the book and another one used in another part of the book or in a table. That is because no one system is accepted by everyone. The forms recommended in this book are generally accepted, but no attempt has been made to include all possible acceptable variations. If your authors use forms different from those you find in this book, don't change them; the author may have good reasons for using different terminology. But it never hurts to ask. The author may be using an old-fashioned abbreviation and may appreciate being brought "up-to-date."

Glossary N is a list of the most commonly encountered abbreviations.

1.1.7 Code Names

Scientific and technical text frequently makes use of code names. These are almost always acronyms made up of capital letters. The most commonly encountered are designations for technical societies and for government agencies and programs.

1. Scientific and Technical Organizations

Technical organizations are of great importance to scientists and engineers. Not only do they provide a means for setting standards of quality for work performed, but they also constitute a forum for people of the same technical discipline. It is by means of these organizations (societies, associations) that professionals keep up to date with developments in their fields and with each other.

It is the forum aspect that is of importance to those who read this book. The best way for a technical person to participate in such a forum is to present a paper. Such papers are delivered orally at technical conferences run by these organizations, and they are published in journals sponsored by them. Nowhere are the rules and techniques which this book attempts to describe held to a higher standard than in published papers. Procedures for preparing manuscripts for publication are presented in Section 1.3. *McGraw-Hill Dictionary of Scientific and Technical Terms* (see Glossary Q) lists over 300 scientific and technical societies. Glossary O is a selection of the most commonly encountered of these. *The Encyclopedia of Associations* (see Glossary R) provides the most exhaustive list available, its tabulation running into the thousands.

2. Government Agencies and Programs

Many thousands of code names are used to designate such governmental agencies as OSHA (Occupational Safety and Health Administration), ARPA (Advanced Research Project Agency), DoD (Department of Defense), and EPA (Environmental Protection Agency). The same kinds of letter combinations—all acronyms—are also the accepted names of radar surveillance systems such as PAVE PAWS and missiles or missile systems like SPRINT or MX.

It was to describe this naming procedure that someone coined the phrase *alphabet soup*. It has been used so much that is has become trite. Yet it is apt.

Both the agencies and the military systems tend to have high technical content, so the likelihood that you will process text sponsored by one or another agency or program is high. If you are ever in doubt about correct usage, consult the *International Code Name Handbook* or *Government Acronyms and Alphabetical Organizational Designations,* both referenced in Glossary Q.

1.1.8 Names of Scientists and Engineers

It can be argued that the greatest achievers in world history are those who have given their names to some thing or device or law or effect that they first conceived or perceived. We grow up with their names and cease to think of them as names; rather we think of them merely as words. Indeed they often pass the first test of words, for many of them can be found in the dictionary.

> We all know what a diesel engine is. It is what is under the hoods of most trucks. Not only do we forget that there once was a Rudolf Diesel, but we have so accepted the word quality of the term that it is usual not to capitalize it (although *Diesel engine* is also acceptable).
>
> We also talk of *degrees Celsius* and the *Bunsen burner* and the *dewar flask,* and hundreds more.

Say what you will about "the immortals of stage, screen, and radio"—or of any field—it is the people who have given their names to the laws, effects, and gadgets of science and engineering who will truly live "forever."

1. Get Them Right

Anyone who is in the business of reproducing scientific and technical text must expect to encounter these names. Many are foreign; many are constructed with strange spellings. Many incorporate the variations of the Latin alphabet described in Glossary B. We can't tell you much other than to *get them right.* Also, keep in mind that there are few rules. Sometimes the name has become a word, so it is not capitalized. Sometimes it remains a name and retains its capitalization. And sometimes it becomes an adjective with the addition of the suffix "-ian" or "-ean." Thus we have words like

Euclidean (from Euclid)
hamiltonian (from Hamilton)

Units are generally uncapitalized words, and like all words they can be subjected to all manner of prefixes and suffixes.

Thus we have units like

curie, millicurie, microcurie (from Madame Curie)

watt, wattage, megawatt, milliwatt (from Watt)

ampere, amperage, microampere (from Ampère)

farad, micromicrofarad (from Faraday)

and so on (many more, named after notables, may be found in Glossary M).

2. A Selection of Names

Glossary P is a selection of the most commonly encountered names of scientists and engineers. When the names are usually used as adjectives which specifically designate a particular device or physical law, the complete term is given in parentheses in Glossary P. Should this list prove inadequate, the *McGraw-Hill Dictionary of Scientific and Technical Terms* lists at least a thousand additional scientific and technical notables; their names have become (almost) legend, so you are sure to encounter many of them in your career as processor of technical text.

1.2 GRAPHICS

One dictionary definition of the term *graphics* is "drawings made in accordance with the rules of mathematics." It is a valid definition for many drawings that appear in technical text, but it is too narrow for our purposes. Technical documents often also include flowcharts for chemical processes or computer programs, photographs, graphs, and even simple lines with arrows at one end. We customarily add those to our definition of the term. In effect, anything that appears on the printed page that has not been keyboarded is graphics. And the majority of texts authored by scientists and engineers have at least one piece of graphics every few pages.

In most offices, the technical secretary is responsible for the graphics content of all documents—if not for the actual generation of the drawings, at least for their placement and captioning (but less so in large establishments such as the aerospace company example of Figure 1-7, where the publications department takes over).

1.2.1 The Secretary and Graphics

There is no question that graphics are best produced by people who make that skill and craft their life's work. Every publications department has illustrators and drafters. They are professionals in their disciplines, which means that their contribution to a technical document helps to make the document look professional. Does that mean that the technical secretary is off the hook? No. Graphics are part of the job of document preparation.

Some companies do not have a publications department, or even a single drafter. But even when such people are on the scene, arranging and scheduling an assignment to a publications department can be a nuisance. Sometimes the drawing is "simple." Then the author of the text is inclined to say, "Surely the secretary can do it . . . " So every technical secretary should have the ability to handle at least some of the graphics requirements of technical documents and in order to do this should have a number of rudimentary drafting tools.

1.2.2 Drafting Tools

The most important tools are a ruler with which to draw straight lines and a fine black fiber-tipped pen. In addition, the technical secretary should have the following items in a handy storage place next to the keyboard:

- A set of templates suitable for the technical content of the documents being processed
- A compass
- Triangles—45, 45, 90° and 30, 60, 90°
- Several french curves
- A small drafting board large enough to accommodate an 8½ by 11 inch sheet of paper

These are illustrated in Figure 1-1.

FIG. 1-1 Using drafting tools in the office. *(Drafting board courtesy of B. L. Makepeace Inc., Boston, Mass.)*

1.2.3 Basic Drafting Technique

No instruction is really required for using these tools—only an admonition: Be neat and be logical.

> This means all illustrations should be placed logically, centered on a page whenever possible. All lines that can be drawn with the aid of a straightedge or a french curve should be drawn that way rather than freehand. All circles should be made with the aid of a compass. All lines should have the same thickness and blackness—unless the author desires some lines to be more boldly drawn. All pointing lines connected to labels should be clear but unobtrusive and should be terminated, when so required, with neat hand-drawn arrowheads. In other words, the page on which the graphics will appear should be planned and the options thought through before pen is applied to paper. Probably the best scheme is to do everything with light pencil lines first. Then if it looks right, those lines can be drawn over with the pen.
>
> Detailed instructions for the special case of drawing graphs are given in "Graphs" and "Plotting Graphs" in Section 2.3.4. You are advised to refer to those sections.

Figures 1-2 through 1-5 show typical illustrations which a secretary should be able to generate satisfactorily. They are respectively a figure, a graph, a table, and a flowchart. Note that all include keyboard inputs.

1.3 PREPARATION OF MANUSCRIPTS FOR PUBLICATION

The word *manuscript* is used to describe a document intended for publication. It means literally "written by hand," and that is probably true of the author's first draft; but for our purposes—when it refers to copy

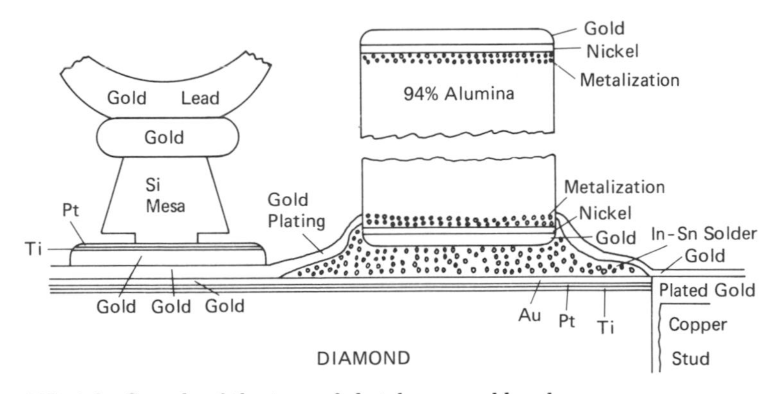

FIG. 1-2 Sample of the type of sketch you could make.

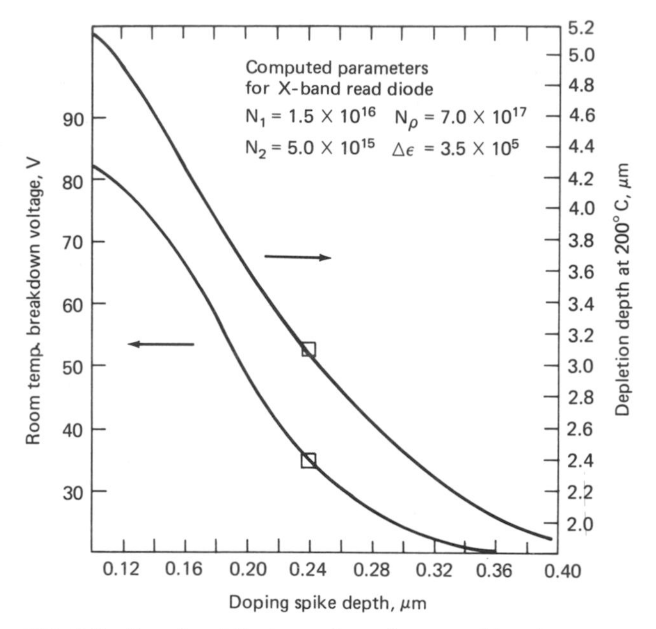

FIG. 1-3 Sample of the type of graph you could make.

submitted to a publisher—it has the primary meaning of "suitable for publication." What makes a manuscript suitable? Conformity with the publisher's requirements.

It is characteristic of most authors of technical and scientific texts to dislike conformity—to anything. But if they want their articles to be accepted for publication by professional journals, they have no choice. They must conform.

Any journal to which a manuscript will be submitted will gladly send a pamphlet describing its own specifications, its own rules. These must be read and understood first by the author and then by the processor of the author's text. Both are in the act, and both must abide by the list of rules imposed by the journal.

> For example, if the journal requires an abstract, the author must write one. And if the journal says, "Not to exceed 300 words," the manuscript must not exceed 300 words. And if the copy is requested to be "double-spaced with a 1½-inch margin on the right side," the typist had better generate double-spaced copy with a 1½-inch margin on the right side.

What follows is a set of procedures which are common to most scientific and technical journals.

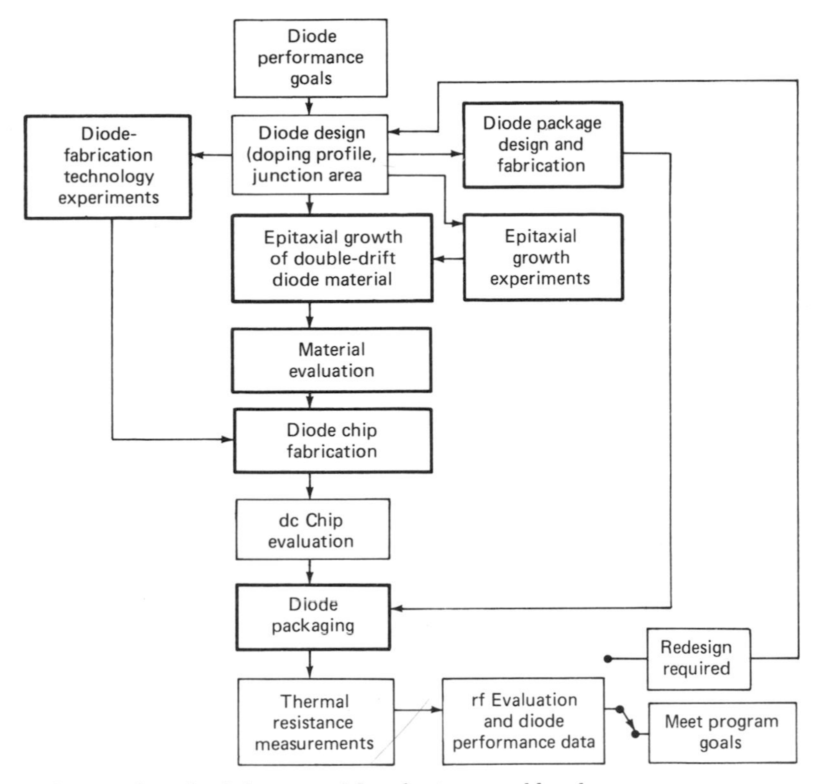

FIG. 1-4 Sample of the type of flowchart you could make.

1.3.1 Some General Rules

1. Format

Here are some rules of format that always pertain:

- Manuscripts should be typed double-spaced on standard 8½ by 11 inch white bond, one side only.
- Margins should be wide on the sides, top, and bottom.
- Pagination should begin at the abstract page.

Some journals do not give the manuscript to typsetters but photocopy it directly. Then the copy as submitted becomes the final copy, appearing in the publication *exactly* as submitted. In such cases the journal may make special demands as to typeface (font), margins, graphics, etc. Failure to conform may result in rejection of a manuscript, or at best, frantic revisions transmitted by overnight mail.

2. Content

At a minimum, the following should be included in the manuscript (if more is required, be assured that the journal will so advise):

- Title, fully capitalized
- Author's (authors') name(s), and affiliations
- Abstract
- Text
- Acknowledgments
- Appendixes, if necessary
- Footnotes
- Captioned tables
- Captioned figures

3. Check a Recent Issue of the Journal

This is a wise and prudent procedure for both the author and the writer. It should resolve most questions and doubts.

1.3.2 Abstract

The abstract is a summary of the subject treated in the article. It is generally presented as a single paragraph and on a separate page or pages from the main text. It normally appears in the journal in a different typeface (smaller, boldface, italics, etc.) from the main text; but as submitted in manuscript, it should have the same typeface, spacings, and margins as the main text. An abstract that exceeds 500 words is considered too long and should be rewritten.

1.3.3 Signs and Symbols

All directions by all journals contain a statement to the effect that all notations should be clear, unambiguous, uncluttered, and consistent with standard usage. They then go on to indicate the journal's preference for typed rather than hand-drawn signs and symbols and for similar appearance of the same sign or symbol as it reappears in the text. The glossaries and text in Part 2 were formulated with these requirements in mind, and if you utilize the information they provide, your copy should be acceptable in all but a very few special circumstances.

1.3.4 Subscripts and Superscripts

Another possible snare is subscripts and superscripts. If one distills all the instructions by journals on this topic, they come down to this: Write subscripts and superscripts so that it is clear what they are, where they start, where they end, and what base they apply to. There must never be any doubt about what part of an expression is a subscript or superscript. Chapter 2, "Mathematics," treats this subject in detail.

1.3.5 Graphics

Graphics, defined in this book as all portions of the text which are not primarily keyboarded—thus including drawings, graphs, photographs, flowcharts, etc.—are usually subject to different requirements by each journal. Section 1-2 suggests practices that are consistent with these requirements and should be helpful in meeting them.

Although graphics are not primarily constructed on a keyboard, most have keyboard ingredients. Instruction for captions, labels, numbering systems, etc. are commonly somewhat different from journal to journal and must be carefully studied and followed for each manuscript.

1.3.6 Tables

Here are some rules for tables:

- Tables should be placed on separate pages and not inserted into the text unless only four or five lines long.
- Tables should be typed double-spaced.
- Tables should be numbered sequentially. The author should take care that each table is referenced in the text.
- Each table should have a title. All titles should be set off from the body of the table, preferably with a single horizontal line.
- Column headings should include units where applicable. All column headings should be set off with a single horizontal line.

TABLE 2-1
MAGNETIC PROPERTIES OF Li- and Ni-FERRITES

Structure Formula	*$4\pi Ms$ at 20°C (G)*	*T_C (°C)*	*$2/K_1/M^{-1}$ (Oe)*
$(Fe)[NiFe]O_4$	3000	590	−580
$(Fe)[Li_{.5}Fe_{1.5}]O_4$	3400	640	−605
$(Fe_{.6}An_{.4})[NiFe]O_4$	5200	360	−200
$(Fe_{.7}Zn_{.3})[Li_{.5}Fe_{1.5}]O_4$	5000	430	−320

FIG. 1-5 Sample of the type of table you could make.

- Vertical lines may be used to set off columns, but this is optional. The main reason for such lines is to guarantee that there be no confusion with regard to the contents of each column.
- When the table requires footnotes, these should be designated by superscript letters. Footnotes should be placed sequentially below the table on the same page as the table.

Figure 1-5 on p. 1.19 illustrates most of the practices described.

1.3.7 Figures and Graphs

As pointed out in Section 1.3.5, the rules for preparing figures and graphs, like those for all graphics, are somewhat different for each journal. However, some general principles are rarely deviated from:

- Figures and graphs should be placed on separate pages and not inserted into the text unless very small (a few lines in height).
- Figures and graphs should be numbered sequentially. The author should take care that each figure and graph is referenced in the text.
- Each figure and graph should have a title. It is preferable that it be placed below the figure or graph.
- If possible, figures and graphs should be drawn so that they may be viewed with the paper in its normal aspect (long way up and down). However, if the material will not fit onto a page unless the paper is turned broadside (rotated 90 degrees), broadside presentation is acceptable.

Figures 1-2 and 1-3, showing a figure and a graph, respectively, conform to these practices and may be used as examples.

1.3.8 Abbreviations

Use standard abbreviations as listed in Glossary N. When the required abbreviation does not appear in this list, consult a scientific and technical dictionary such as those referenced in Glossary Q.

1.3.9 Footnotes and References

These appear at the end of the manuscript, not at the bottom of the page on which they are noted.

1. Numbering Systems

Some journals use superscript numerals, [1], [2], [3], etc. Others use letters [a], [b], [c], etc. A less common practice is the sequence *, †, ‡, §, ¶, etc.

2. Footnotes

The footnote may contain supplementary text or may cite references. In the latter case, various formats are required by different journals. Consult a recent issue of the journal for an appropriate format. A typical one, giving author, title (in quotation marks), journal name (abbreviated and underlined), volume number, page numbers, and date of journal issue or of publication, is shown in Figure 1-6.

Handley, J. A., and DeFuria, L. P., "Dependence of N_2 Adsorption on Particle Sizes of Various Alumina Desiccants," *J. Gas Technol.*, vol. 21, no. 2, 1979, pp. 341–359.

FIG. 1-6 A typical reference note.

1.3.10 Submittal of Manuscripts

Here is accepted practice for submittal of a manuscript to a journal or book publisher:

- Submit the manuscript as directed by the journal or publisher's instructions, or lacking these, directly to the editor.
- Include a letter of submittal, making clear the author's name and correct corresponding address.
- Pack the manuscript well for shipping, sandwiching it between cardboard stiffeners.

1.4 REFERENCE BOOKS AND MANUALS

A technical secretary, editor, or writer without reference books is like a castaway on a desert island. Of course, almost everyone has a dictionary—the most important book of all—and the typist probably has a guide to the operation of the word processor. But for the processing of technical and scientific text, more reference works are recommended.

1.4.1 Scientific and Technical Societies

Most professional societies in scientific and technical fields publish references to manuscript preparation for authors and technical writers. These references have become standards for the disciplines they represent. You will be well-advised to obtain copies of the references for those specialities of science and engineering treated in your office. In fact a good procedure is to learn which societies or associations the authors of the texts you will process belong to; then write these organizations for their references. All organizations are listed in *Encyclopedia of Associations* (see Glossary R). A new edition is published each year.

1. Some Important Societies

Table 1-3 is a selection of several of the most important technical societies in the United States with their addresses. Their scope is so wide that even though there are hundreds more with narrower specialties (see Section 1.1.7 and Glossary O), most engineers and scientists have membership in at least one of these.

TABLE 1-3
SOME SELECTED TECHNICAL SOCIETIES

American Association for the Advancement of Science (AAAS)
1515 Massachusetts Avenue, NW
Washington, D.C. 20005
(202) 467-4400

American Chemical Society (ACS)
1155 Sixteenth Street, NW
Washington, D.C. 20036
(202) 872-4600

American Institute of Aeronautics and Astronautics (AIAA)
1290 Avenue of the Americas
New York, NY 10019

American Physical Society (APS)
335 East Forty-Fifth Street
New York, NY 10017
(212) 682-7341

American Society of Mechanical Engineers (ASME)
345 East Forty-Fifth Street
New York, NY 10017
(212) 705-7722

Institute of Electrical and Electronics Engineers (IEEE)
345 East Forty-Seventh Street
New York, NY 10017
(212) 644-7558

Society for Technical Communication (STC)
815 Fifteenth Street, NW
Suite 506
Washington, D.C. 20005
(202) 737-0035

2. The Society for Technical Communication

The Society for Technical Communication (Table 1-3) differs from the others in that it is for technical writers as distinct from technical authors. It is the society of technical editors, writers, graphics specialists, and secretaries.

1.4.2 Recommended Scientific and Technical References

Glossary Q lists technical references which are considered useful. Not all are pertinent to all work, nor should this list be considered complete. But most answers to most questions in most fields can be found in them. And they are authoritative.

1.4.3 Recommended General References

Glossary R reminds us that after all the technology and science content has been rendered correctly, the finished document must still be written correctly in the English language. It lists some works on style which should be useful to both authors and technical writers.

1.5 THE AUTOMATED OFFICE

When we say something is *automated,* we mean that it runs without human direction. But that is never entirely the case. Consider the automatic shift in an automobile. Indeed it adjusts the gears from first to second to third without commands from the driver—but the driver still has a gearshift lever. Only the operator of the car can decide when to shift into reverse or when to go into first or second gear on a mountain road. In other words, the term *automatic* usually actually means "*partially* automatic," which merely means "more convenient."

This holds for the modern office as well. The word processor and the family of computer-controlled functions it brings with it tell us simply that the modern office is more convenient than it used to be, and we call that more convenient office the *automated office.*

1.5.1 The Extremes of Office Automation

As with all automated facilities, some go further than others. Probably the most sophisticated systems for the production of scientific and technical text are installed in today's aerospace companies whose primary business is with the Department of Defense and other departments of the U.S. government. Such companies normally employ thousands of engineers and scientists deployed in dozens of plants and laboratories around the world. These people are served in turn by hundreds of personal secretaries, technical typists, illustrators, drafters, and other graphics specialists. When you realize that such companies also have reproduction facilities of which most major publishers would be proud, you begin to understand how such an organization can turn out 1500-page proposals to the government *from scratch* in a 2-week period, and do such work routinely.

At the other end of the spectrum is the company that employs one engineer who is served by one secretary with one broken-down manual typewriter. Most technical reproduction in the nation lies between these extremes.

1.5.2 Office Automation in an Aerospace Company

Let us examine the typical aerospace company's system with the understanding that even though it represents today's ultimate in office automation, many of its features will inevitably extend in some similar form to smaller organizations, so that in the end even that manual typewriter in the one-engineer office will be replaced with some sort of electronic system.

Figure 1-7 diagrams the aerospace office system.

1. Input—Forms and Sources

The engineers and scientists provide authorship of copy, also known as *input.*

a. Many different forms Their input is in many forms which match their individual skills and the equipment available to them.

Thus the copy ranges from handwritten to typed to keyboarded on disks. The quality ranges from illegible to legible, from disorganized to beautifully organized, from hunted and pecked by two forefingers to 10-finger touch-typed.

The problems of organizing these highly various inputs are shouldered first by the program manager, the person who must get the proposal or report into the hands of the customer or contracting agency in the government. But in the end the organization of the proposal or report becomes the responsibility of the editor in the publications department.

b. Many different sources There can be at least as many different sources of inputs as the company has parts.

For example, several selections are contributed from a dozen different departments of the parent facility in Massachusetts. Some come from the Utah plant, some from the Iowa subsidiary, and some from a subcontractor in Florida; part of the introduction is prepared in the Brussels office in Belgium. Then there is a section that is retrieved from a tape library in

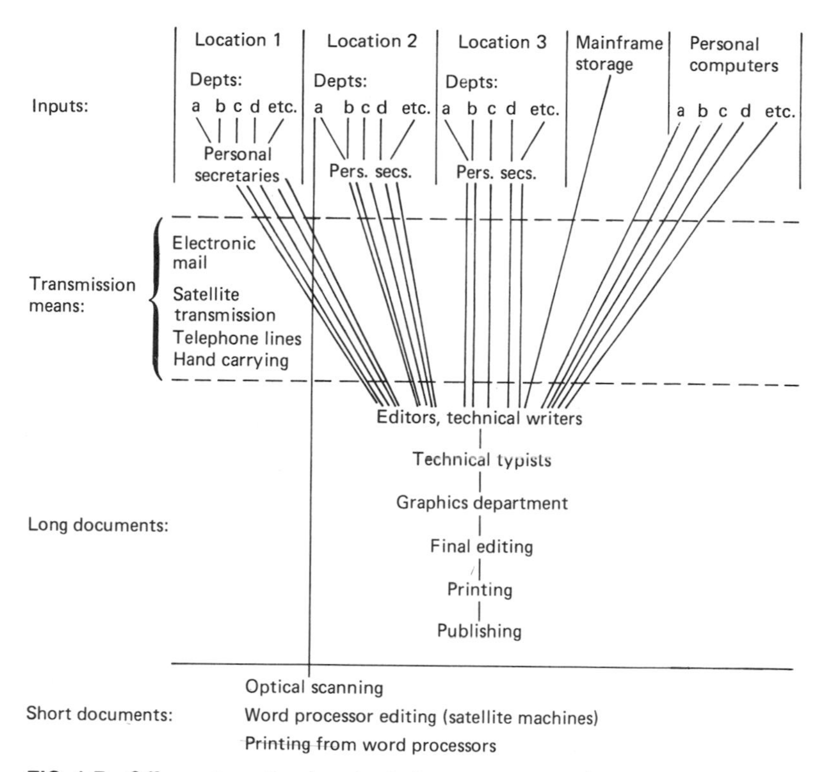

FIG. 1-7 Office automation in a typical aerospace organization.

the mainframe central computer in Texas, while a number of the engineers have generated their contributions to the text at home on their own personal computers. These are only some of the variations of input sources.

2. Transmission Means

For the most part, the authors know they have discharged their tasks when they hand their copy to the secretary in the office, who in turn hands it over to some *input center* in the company's publications department. This "handling" process is simple enough in the central location, but it can become very complex for the faraway locations.

In this example, the publications department is in the Massachusetts facility. Somehow, all the copy has to end up there, and the finished product has to come out of there.

By the use of telephone network lines and satellite communication links tied into electronic mail systems and with suitable modem connections (and with recourse even to an old-fashioned messenger service), all sources funnel copy to the publications department, where all text is actually processed (see Figure 1-7). The text is considered to have arrived when it has all been keyboarded to diskettes and when any segment can be called up on a word processor video screen in the publications department.

3. Short Documents

When the document is short, the personal secretary handles it. For such documents a good pragmatic policy is for each personal secretary to have a rotating-ball-type typewriter with a font size that is compatible with an optical scanner. (Indeed, the easiest policy is to use that font all the time, even when optical scanning is not used.) This means that not every personal secretary has to have a word processor. The typewritten pages prepared with this font are scanned and thus recorded automatically on floppy disks—the so-called diskettes. The secretary can then display the copy for editing on any one of a number of satellite word processors located throughout the office area. Thus short reports and letters are processed in a hybrid fashion—regular typing for the first copy, and word processing for editing and printing.

4. Long Documents

Longer documents—such as the elaborate proposal of the preceding example—are sent directly to the publications department, where, after preliminary organization and editing, they are given to technical typists, people whose function is to bang out copy. (Typically, there is one technical typist for every six personal secretaries.) Such typists become astonishingly skilled and are often so fast that no mechanical typing machine can keep up with them without jamming. So each of the technical typists has a word processor. That solves the jamming problem. Aerospace offices usually maximize equipment utilization and minimize capital investment by dividing the technical typing pool into two-shift and even three-shift operations.

5. Graphics

Adjacent to the technical typing location is the graphics section. Here illustrators and drafters do their work with drafting boards—and increasingly with electronic tools. With these electronic tools they can produce sketches on video screens from which printouts can be made. In either case, close collaboration is needed between the suppliers of graphics and the typists. Should the labels and captions be typed in first, or should the drawings be made first and the labels and captions typed in afterward? Each office decides the way that works best for it.

6. The Automated Office Must Still Follow the Rules

This section has described today's electronic mechanics of producing technical text—conveniently. What of the material described in the following chapters? Is it still valid? Does it still pertain? The answer to both questions is yes. Whether "ancient" or modern office technology is used for getting print onto paper, scientific principles are the same and the rules of notation are the same. Modern electronic equipment is no substitute for correct technique, understanding of the material, and good judgment, all of which are promoted in the following five chapters.

CHAPTER TWO

Math

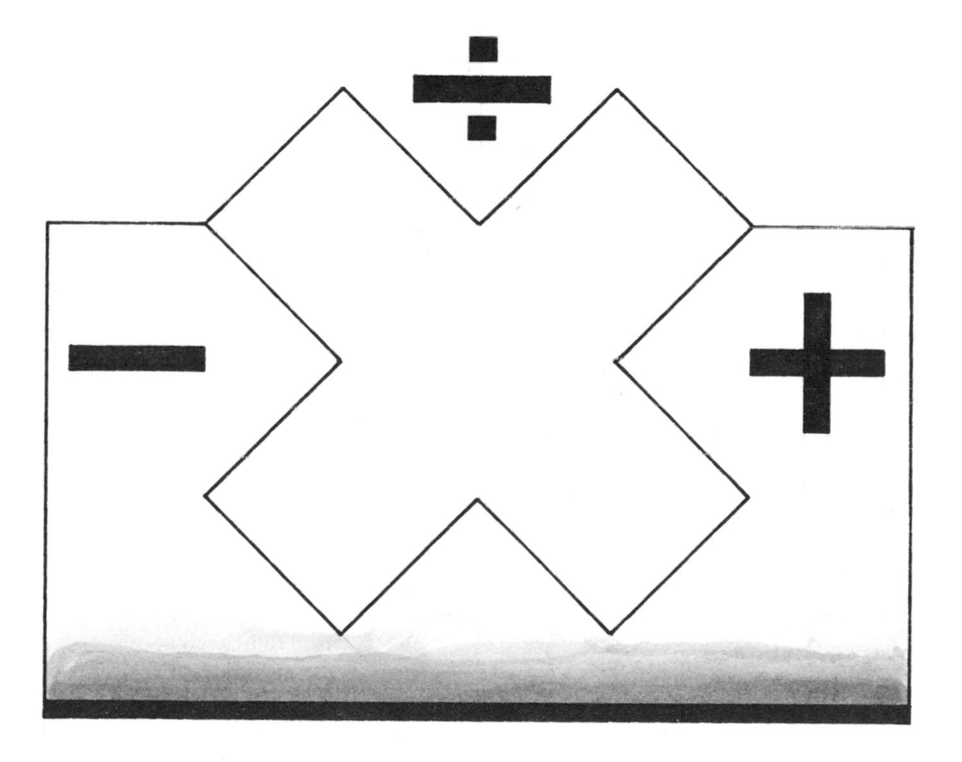

Basic Principles

Since every engineer or scientist uses mathematics, this is possibly the most important chapter in this book. And mathematical material is possibly the most complicated material to reproduce neatly and logically on a printed page. Yet, as you go through this chapter, section by section, you will see that it is not difficult to master. Let's start with numbers.

2.1 NUMBERS

A number is a symbol which stands for a quantity. Thus the symbol 5 stands for "five" and the two-part symbol 67 stands for "sixty-seven." Just as only 26 letters provide enough symbols to write all the literature of the English language, there are only 10 number symbols, which are enough to perform most of the calculations we will do in a lifetime. These are 1, 2, 3, 4, 5, 6, 7, 8, 9, 0.

NOTE: Such numbers, which are the ones used in counting and which are units in themselves, that is, *whole numbers* and not parts of numbers, are known as *integers*.

But engineers and scientists find these numbers insufficient for their purposes. They have had to devise additional symbols beyond the integers. These symbols stand for far more complex concepts, such as infinity, and for operations on numbers which provide new numbers, such as summation. Those that will be most frequently encountered will be described here, in terms of their meanings and how to write them.

2.1.1 Some Additional Symbols

Let us examine seven of the most commonly used symbols which technical people have added to the basic list of 10. The particular meaning of each should be understood. They are:

∞	Infinity	!	Factorial symbol
π	Pi	$\sqrt{\ }$	Root or radical sign
Δ	Increment sign	i	Imaginary number
Σ	Summation symbol		

1. ∞ (The Infinity Symbol)

The best way to understand what *infinity* means is to look at the word itself: *in-finite* means "not finite." Finite means having definable limits. So something that is infinite has no limits or boundaries or ends. When we refer to infinity as a number we mean a number that is larger than any we can imagine. A computer definition is that infinity is a number larger than any computer can store.

In science, something can be infinitely large or infinitely small. Or something can extend infinitely distant in time (and why not, time never ends) or imagination (that never ends either, at least for scientists). Obviously, the concept of infinity is "way out." But someone who has to deal with it frequently, such as a scientist or engineer, has to make it concrete. Thus infinity has been given a symbol so that it can be used in mathematical expressions.

Here is an example of the infinity symbol in a mathematical expression of the sort we will describe in Section 2.3.

$$2(x - 3) = \int_0^\infty 4x^2\,dx$$

2. π (pi)

This number stands for the ratio of the distance around a circle (circumference) to the distance across the circle at its middle (diameter). While it is more describable and definite than infinity, there is something infinite about it. The ratio is a little more than 3, which means that the circumference of a circle is a bit more than three times its diameter. More exactly it is

3.14169 . . .

The numbers after the decimal point go on forever. But it is enough to say that the ratio is "about" 3.14. That ratio is called *pi* and is designated by the lowercase Greek letter π.

Indeed, *pi* is the name of the Greek letter, π, the sixteenth character of the Greek alphabet. Its Latin equivalent is p. It may be found in Glossary A, which lists the Greek alphabet.

Even though π describes a geometric parameter, its use is not limited to geometry. Do not be surprised to find it often in mathematical text and in the special field of statistics.

Numbers like π, with places to the right of the decimal point that go on forever in a nonrepeating sequence, are called either *irrational* or *transcendental* numbers. Thus π is both an irrational and a transcendental number.

3. Δ (The Increment Symbol)

Engineers and scientists often deal with situations in which they need to know what would happen if they increased a number just a little. They show such increases with the symbol Δ, which is the Greek capital letter *delta*.

Suppose the text you are working with is about the rotation speed of a motor which is described as turning 1000 times every minute. Then the text asks the question, What would be the effect of increasing that speed to 1020 revolutions per minute (rpm)? Perhaps the increase of 20 rpm makes a crucial difference in the design of the motor. At the increased speed maybe the motor starts to vibrate uncontrollably, or maybe the engine begins to overheat, or maybe nothing happens. In any case, what is being discussed is the effect of a small change.

a. The increment Mathematical treatment of small changes employs the concept of the increment; a term that comes from a Latin word which means "to increase." So a small increase is called an *increment*.

In the motor example, the increment is 20 rpm.

NOTE: There should always be a space between a number and the unit of measure it designates. Thus, in this example

INCORRECT: 20rpm

CORRECT: 20 rpm

b. The negative increment, or decrement What if the speed goes down, not up? Is that still an increment? Yes. It is a negative increment. A negative increment is called a *decrement*.

If the motor were to slow down to a speed of 970 rpm from its original speed of 1000 rpm, its decrement would be 30 rpm.

c. Using the increment symbol How is the increment symbol used? Usually in equations.

In the motor example, the author might represent its speed by the symbol S. Thus

$$S = 1000 \text{ rpm}$$

Then, to represent the increment, the mathematical statement becomes

$$\Delta S = 20 \text{ rpm}$$

or for the decrement

$$\Delta S = -30 \text{ rpm}$$

Placing the increment symbol in front of the speed symbol shows that an increase or decrease in speed is meant.

NOTE: This should be written with no space between the Δ and the mathematical expression which follows it. Thus

INCORRECT: $\Delta\ S$

CORRECT: ΔS

d. Subscripts The author will often make use of subscripts (see "Subscripts in a Sequence" in Section 2.1.7).

Then a likely mathematical sentence expressed as an equation (Section 2.3.3) may say that the original speed was S_1 and the final speed was S_2. Then

$$S_1 = 1000 \text{ rpm} \qquad \text{and} \qquad S_2 = 1020 \text{ rpm}$$

and

$$S_1 + \Delta S = S_2 = 1020 \text{ rpm}$$

or for the decrement case

$$S_1 - \Delta S = S_3 = 970 \text{ rpm}$$

4. Σ (The Summation Sign)

We have just described the symbol Δ which indicates an increment that may be added to a base number. Another Greek letter shows a different kind of adding: Σ, the capital *sigma.* It is the Greek letter for S and it usually means "summation." Σ is thus the shorthand way to say "take the sum of." The symbol instructs the reader to add up a row of numbers or other mathematical expressions in sequence.

a. Limits of summation It is necessary to know where to begin and where to end the summation. That is known by saying "between this number and that number." You then write *this* number under the Σ and *that* number over it in little equations of the form $n =$.

Suppose you wanted to say, "Take the sum of the numbers between 8 and 14." In mathematical notation it would look like this:

$$\sum_{n=8}^{n=14} z$$

and it would mean:

$$\sum_{n=8}^{n=14} z = 8 + 9 + 10 + 11 + 12 + 13 + 14 = 77$$

The numbers written at the top and bottom of the sigma are known as the *limits of summation:* they are to be applied to the quantity that follows the summation symbol.

In this case the quantity being summed is z, but the limits of summation can be applied to any other mathematical expression as well.

b. Omitting *n* Very often the $n =$ is omitted, either from the top limit or from both limits, when writing the summation, because the reader is expected to understand that it is there even if it is left out.

Thus the preceding expression might be written

$$\sum_{n=8}^{14} z \quad \text{or} \quad \sum_{8}^{14} z$$

The choice is the author's.

c. Larger number goes on top Which quantity goes on top? Usually the larger one, as in the preceding examples. But sometimes the limits are symbols themselves. Then you have no way of knowing which is larger. If the text is written, you merely copy it, but if it is dictated, the transcriber must ask.

Here is one way of writing the summation of $x - 9$ between A and B:

$$\sum_{A}^{B} x - 9$$

5. ! (The Factorial Symbol)

The *factorial* symbol is an exclamation point !. When it appears following any number (with no space between), it changes that number into a new number in accordance with a prescribed set of multiplication operations.

The factorial symbol instructs the reader to multiply by each other all the *integers* (whole numbers) that can be counted from 1 up to that number.

Thus, factorial 8, or 8 factorial, is

$$8!$$

Which means that one should multiply all the integers from 1 to 8 by each other, like this:

$$8! = 8 \times 7 \times 6 \times 5 \times 4 \times 3 \times 2 \times 1$$

And if one goes through the actual multiplications, the answer is

$$8! = 40{,}320$$

NOTE: Verbally, the concept is expressed as either the word *factorial* followed by the number to which the factorial operations are to be applied or vice versa, as "factorial eight" or "factorial 8" (or "eight factorial" or "8 factorial") in the preceding example.

The factorial symbol is especially useful in statistics and in calculating probability, such as one's chances of winning a lottery. It is a neat shorthand way of indicating a somewhat involved mathematical operation.

NOTE: An interesting thing to keep in mind is that the factorial symbol is a sort of multiplication analogue of the summation symbol Σ. The summation symbol tells the reader that a sequence of numbers is to be added up between given limits. The factorial symbol tells the reader to multiply in a given way.

6. $\sqrt{}$ (The Root or Radical Sign)

Every time we multiply something by itself we are *squaring* it.

$2 \times 2 = 4$ (which is "2 squared")

$3 \times 3 = 9$ (which is "3 squared")

$4 \times 4 = 16$ (which is "4 squared")

If we want to do the same thing in reverse we *take the square root.*

Thus

The square root of 4 is 2.

The square root of 9 is 3.

The square root of 16 is 4.

To do this in mathematical notation with a symbol we use the root symbol $\sqrt{}$. Then the preceding root statements may be restated:

$$\sqrt{4} = 2$$
$$\sqrt{9} = 3$$
$$\sqrt{16} = 4$$

a. Long expressions Long root expressions require long root symbols.

The expression under the root sign can be *any* mathematical expression (of any length). It can be full of algebraic symbols, or it can be made up of simple numbers.

$$\sqrt{7843}$$
$$\sqrt{x + 4y - 66}$$

b. Tall expressions Tall root expressions require tall root symbols.

$$\sqrt{\frac{M^3 - N^2(x - 2)^{1/2}}{50(x + 2)[A/(2B + A)]}}$$

c. Higher-level roots The square root is also called the *second root.* That is, it tells of *two* identical numbers which, when multiplied together, will give the base number. All root signs mean square root unless otherwise indicated. That indication appears as a number (or other symbol) in the wedge space of the root symbol. One could take the seventh root or the third root or the *n*th root, that is, any root.

Here is an example of the third root:

Since

$$3 \times 3 \times 3 = 27$$

then

$$\sqrt[3]{27} = 3$$

NOTE: The third root is known as the *cube root* (just as the second root is known as the square root). And the reverse process—multiplying something by itself three times—is known as *cubing* (just as multiplying something by itself twice is known as *squaring*). All other roots are designated by the particular number of the root to be taken. This will be discussed in greater detail in the section on exponents (Section 2.1.4).

d. Radical Roots and root signs have another name. They are also known as *radicals*.

7. i (The Imaginary Number)

For the person writing in the language of mathematics, numbers can be real or they can be imaginary.

Here, the mathematician is saying that while certain concepts have no relation to ordinary experience, they may have real value—in spite of being "unreal." It never stops a scientist that something is "unimaginable."

Imaginary numbers occur when we take roots (which we have just discussed) of negative numbers. They always include the quantity $\sqrt{-1}$.

In mathematics, by the rules of multiplication, the multiplication of two positive numbers is positive, and the product of two negative numbers is also positive. This means that something multiplied by itself must always be positive. Thus

$$(+5) \times (+5) = +25 \quad \text{and} \quad (-5) \times (-5) = +25$$

Yet there is nothing to stop someone from writing

$$\sqrt{-25} \quad \text{for the square root of } -25$$

What multiplied by itself can be -25? We have already shown that it cannot be $+5$, not can it be -5. The only way for something to be multiplied by itself and still be a negative number is for it to be imaginary. It turns out that $\sqrt{-25}$ can be written as $\sqrt{25} \times \sqrt{-1}$. If we now take the square root of the 25, we can write

$$\sqrt{-25} = 5 \times \sqrt{-1}$$

Now we say that in this expression there is a real part and an imaginary part. The 5 is real and the $\sqrt{-1}$ is unreal, or *imaginary*.

a. The imaginary symbol The mathematician has invented a rather reasonable symbol for the concept of the imaginary number. It is the letter *i*, which is the first letter of the word *imaginary*. More specifically,

$$i = \sqrt{-1}$$

Thus $$\sqrt{-25} = 5\sqrt{-1} = 5i$$

b. Complex numbers Imaginary numbers (which contain an *i*) are also known as *complex numbers*.

8. Letters as Numbers

By now it should be clear that numbers can be shown as symbols other than the most commonly encountered 1, 2, 3, 4, etc. We have just described seven examples. But there are dozens more. Some of them are shapes—like the root sign—that are drawn, but most of them are letters of one sort or another. As you know, Roman numerals (I, II, IV, XX, etc.) are numbers constructed from the capital letters of the Latin alphabet. And we have just encountered the Greek π, Δ, and Σ. The fact is that scientists and engineers make numbers out of almost every letter of the Greek alphabet. Not only that, they use them in two forms, as we use the letters of our alphabet—that is, capital letters (uppercase) and small letters (lowercase). The Greek alphabet is shown in Glossary A.

NOTE: Other forms include script letters, resembling handwritten script; boldface roman letters, and German (Fraktur) type.

The use of letters from another alphabet to stand for certain numbers or concepts makes sense. If we were limited to using only our alphabet we would have two problems. First, there wouldn't be enough letters. Even multiplying the 26 letters by 2 to give 52 (that is, using both the upper- and lowercase forms) is limiting. As you will see, the scientists and engineers in all the different disciplines use hundreds of symbols. Also, putting a number or expression into a different alphabet distinguishes it and minimizes the chances for confusion.

Even the Greek alphabet is not enough. Certain symbols from the Hebrew, the archaic German, and the Cyrillic alphabets are occasionally used.

NOTE: Section 1.1.1 discusses some foreign alphabets. Use it as a reference.

2.1.2 Fractions

A fraction is a number too, because it describes a quantity. It indicates a particular mathematical operation—division. A fraction is written by placing one number or expression over another number or expression, with a line (the *fraction bar*) between them. The fraction says, "divide the top number by the bottom number."

The top number and the bottom number have their own names.

The expression above the fraction bar is the *numerator*.

The expression below the fraction bar is the *denominator*.

1. Simple Fractions

Simple fractions are simple enough. One number is placed over another number, with the *fraction bar* between them.

The fraction bar is usually horizontal, but it doesn't have to be. A slant bar may be used, and usually is when the fraction is included as part of single-spaced text. When a slant bar is used, the fraction is called a *shilling fraction*. Thus, in text, 4/5 may represent the fraction "four over five," which may also be said "four-fifths" (with 4 the numerator and 5 the denominator). But when there is room for the fraction to take up more than one line, it is better to write

$$\frac{4}{5} \quad \text{which means} \quad 4 \div 5$$

When the fraction is written with one number over the other, it is called a *built-up fraction*.

NOTE: Because the fraction indicates the operation of division, all the following expressions are correct and have the same meaning.

$$4/5 \quad \text{or} \quad \frac{4}{5} \quad \text{or} \quad 4 \div 5 \quad \text{or} \quad 5\overline{)4} \quad \text{or} \quad 0.8$$

2. Complex Fractions

Fractions may be of unlimited complexity. A complex fraction (as distinguished from a simple fraction) is one that has a fraction in either the top or bottom expression or both. But the same principles hold. There is always one number or expression on top (the numerator) and another number or expression on the bottom (the denominator). And a line (the fraction bar) separates them.

$$\frac{A + \dfrac{3}{4x}}{y}$$

or

$$\frac{x^2 + 2Ax + A^2}{\dfrac{x - A}{2A}}$$

3. Ratios

One important use of fractions is the *ratio*. One can usually recognize a ratio by the word *to* when two quantities are being compared.

The statement "3 to 5" is the same as the fraction 3/5. Also, the statement "3 is to 5 as 6 is to 10" (a proportion) is the same as 3/5 = 6/10.

2.1.3 Decimals

A *decimal* is just a shorthand way to write a fraction. It is a number whose denominator is always a multiple of 10, that is, a number that is being divided by 10, 100, or 1000, or any other number made up of 1 followed by 0s.

So 0.3 is the same as 3/10; both are three-tenths. In the same way 0.03 is 3/100 and is three-hundredths, 0.003 is 3/1000 and is three-thousandths, and so on.

Any fraction can be converted to a decimal by changing its denominator to some multiple of 10.

So 1/8 becomes 125/1000 when both numerator and denominator are multiplied by 125. This is true because when the top and bottom of a fraction are multiplied by the same number, the total value of the fraction is unchanged. Thus

$$\frac{1}{8} = \frac{125}{1000}$$

and by the shorthand of decimals, both can be expressed by the decimal number 0.125 (0.125 = 1/8 = 125/1000).

1. Writing a Decimal Expression

Decimals are numbers written with a decimal point followed by one or more digits.

Thus in the decimal expressions 107.31, 0.014, and 4.9, the expressions .31, .014, and .9, respectively, are the *decimal digits*. (See "Commas and Spaces" on p. 2.13 for more on writing decimal numbers.)

2. Ciphers

In the preceding section the examples 0.3, 0.03, and 0.003 were written with a *cipher* (a zero) before the decimal point. That is the best practice.

You may see the cipher left out in some circumstances, but that is usually an occasion where the author has been careless.

NOTE: There are a few times when leaving the cipher out is accepted practice, such as in writing probabilities.

3. The Importance of the Location of the Decimal Point

In a decimal expression, the *decimal point* is the most important symbol. Everything depends on its location.

a. Whole numbers to left of point All numbers to the left of the decimal point are whole numbers or integers.

b. Fractional numbers to right of point All numbers to the right of the decimal point are less than the smallest positive whole number, 1. This means that all numbers to the right of the decimal point are fractions of a whole number (no matter how many numbers appear).

c. Identification of decimal positions Each position to the right side of the decimal point has its own meaning and name. The first digit to the right of the decimal point is called a tenth, the second one is called a hundredth, the third one is called a thousandth, and so on, as shown in Figure 2-1.

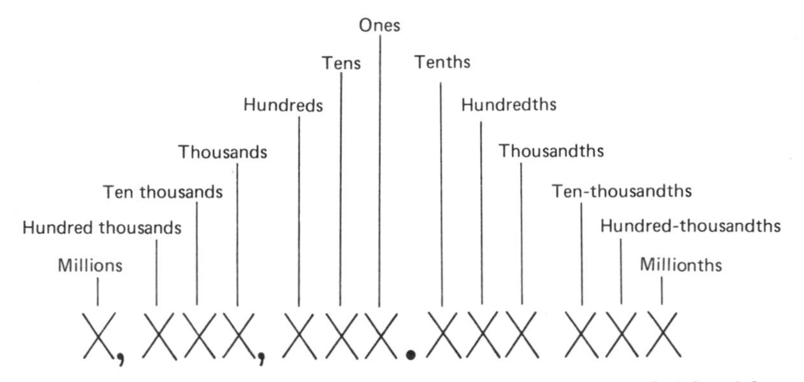

FIG. 2-1 Nomenclature of integer and digital positions, left and right of decimal point.

NOTE: X's represent any number from 0 to 9

d. Identification of integer positions Each number to the left of the decimal point has a name, as shown also in Figure 2-1.

NOTE: Here it is important that you not be the victim of carelessness or inattention. The problem is that the terms on both sides of the decimal point sound so much alike that it is easy to confuse them. Be careful. A hundred must not be confused with a hundredth, because a hundred is 10,000 times bigger than a hundredth. And a thousand is a million times bigger than a thousandth, and a million is a trillion times bigger than a millionth, and so on.

Here are some examples of decimal expressions in both words and numerals:

- Four and seven-tenths:

4.7

- Four and seven-hundredths:

4.07

- Four hundred and seven-hundredths:

400.07

- Four thousand six hundred fifty-five and seven hundred forty-three thousandths:

4655.743

NOTE: The word *and* is used to indicate the decimal point.

4. Commas and Spaces

It is obvious that these numerical expressions can get very long and cluttered. As always, in all written text where we want to clarify, commas can help. The rule is, for the whole-number side of the expression to the left of the decimal point, a comma should be inserted every three digits, going from right to left. This means that commas separate thousands, hundreds of thousands, millions, etc. But the rule is different for the decimal digits to the right of the decimal point. There, no commas are inserted:

16,984,000.3298792

The SI metric system (Section 2.1.6) recognizes that confusion can result from decimal numbers with many digits that are not broken up. It inserts spaces every three decimal digits, counting from left to right, beginning at the decimal point.

16,984,000.329 879 2

5. Percentages

A percentage is a special kind of decimal. It is a number always understood to be divided by 100. That is what *percent* means—per hundred.

So 17 percent means 17/100, and the fraction 17/100 is also the decimal 0.17. This means any decimal can be converted to the equivalent percentage by multiplying by 100: $0.17 \times 100 = 17$ percent.

NOTE: Have you ever thought that you use decimals every time you use money? What is a dime after all if it isn't 0.1 dollar? And a penny is certainly 0.01 dollar. They are fractions too: a dime is a tenth of a dollar, 1/10; and a penny is a hundredth of a dollar, 1/100; and a quarter is a quarter of a dollar, 1/4.

2.1.4 Exponents

It should be evident that much of what we have dealt with thus far has had to do with different shorthand ways to add up numbers (Σ), or multiply numbers (!), or divide numbers (fractions and decimals). The *exponent* is a way to tell that a number must be multiplied by itself. How many times? The exponent tells how many times.

Thus 6^5 means multiply 6 by itself five times.

$$6 \times 6 \times 6 \times 6 \times 6 \times 6 = 7776$$

1. Base

The number which is being multiplied—in this case, 6—is called the *base.*

2. Power

The expression that tells how many times to multiply the base by itself is known as the *power*—in this case, 5.

So "six to the fifth power" means 6 is to be multiplied by itself five times (6^5).

3. The Exponent

One uses the term *power* when one is describing this manipulation in words. But when writing it as a mathematical expression, the power is referred to as an exponent. It is always recognizable because it is written a half line up, as a superscript.

NOTE: The author of mathematical text never differentiates between numbers and letters. To such a person they are both the same thing—symbols. So the author experiences no strain in raising the base number to the *k*th power. Then the exponent will be k and the expression (for the same base of 6) will be

$$6^k$$

The point is that there are no restrictions. Anything goes. Anything can be a base or an exponent—numbers, letters, or any combination of these in any complexity.

4. Powers of 10

When we raise the base 10 to a power we are doing something special. It can be thought of as adding zeros to the base 10.

One hundred is 10 to the second power, which is the same as 2 tens multiplied together (10×10); and 1000 is 10 to the third power, which is the

TABLE 2-1
RAISING 10 TO A POWER

	Multiply Together	*As a Number*	*In Words*
10^1	1 ten	10	Ten
10^2	2 tens	100	One hundred
10^3	3 tens	1,000	One thousand
10^4	4 tens	10,000	Ten thousand
10^5	5 tens	100,000	One hundred thousand
10^6	6 tens	1,000,000	One million
10^9	9 tens	1,000,000,000	One billion
10^{12}	12 tens	1,000,000,000,000	One trillion
10^{15}	15 tens	1,000,000,000,000,000	One thousand trillion
10^{18}	18 tens	1,000,000,000,000,000,000	One million trillion

same as 10 multiplied by itself three times ($10 \times 10 \times 10$). Some powers of 10 are shown in Table 2-1.

a. Relation to the metric and the SI system Does this look familiar? It should. This raising of 10 to a power is the basis of the *metric system,* where all quantities are based on the number 10 multiplied by itself different numbers of times. In the metric system it becomes convenient to give a name to each power. (These are shown in Tables 2-2 and 2-3.) These names will be discussed in further detail when we get into prefixes and suffixes and when we explain the SI System, the International System of Units (Sections 2.1.6 and 1.1.3).

b. Any number expressed in powers of 10 Any number can be expressed in terms of powers of 10. Such a way to express numbers is known as *scientific notation* and is a basis of the SI system.

Take the number

$$421{,}987$$

It is the same as 4.21987 multiplied by 100,000:

$$421{,}987 = 4.21987 \times 100{,}000$$

Since 100,000 is the same as 10 multiplied by itself five times it can be written

$$100{,}000 = 10^5$$

Then the original number can be written in a way that is more convenient than writing the strung-out series of digits. That convenient way now will look like this:

$$4.21987 \times 10^5$$

This is a common way to show a number with many digits in technical writing.

5. Writing Exponents on the Natural Base

The number 2.1783 has a special mathematical significance (see "The Natural Base e" in Section 2.1.5). It is called the *natural base e,* and it is just as important as the base 10. While exponents may be applied to it, as in

$$e^{3k}$$

where e is the base and $3k$ is the exponent, there is a special rule when the exponent applied to the base e includes radical signs, integral or summation signs with limits, or special symbols in small type. The rule is to replace the e by exp.

CORRECT: $x = K_0 \exp \dfrac{\sqrt{m+n}}{5ab}$

INCORRECT: $x = K_0 e^{\sqrt{m+n}/5ab}$

CORRECT: $y = \theta \exp \left(- \int_0^t \phi \, dt \right)$

INCORRECT: $y = \theta e^{-\int_0^t \phi dt}$

6. Fractional Exponents

When an exponent is a fraction it has a special meaning. It means that the expression may be expressed as a root and still have the same value. (See "$\sqrt{}$ The Root or Radical Sign" in Section 2.1.1.) The rule is that the denominator of the exponential fraction tells the level of root to be taken.

Thus, the sixth root of 81 can be shown as

$$81^{1/6} \quad \text{or} \quad \sqrt[6]{81}$$

They mean the same thing. The technical author chooses the one that best fits the text and the context.

NOTE: It is desirable to print exponential fractions in a small typeface of a sort that will not be found on ordinary typewriter keyboards. To cope with this problem, one may either resort to normal-sized numerals or may draw them in neatly by hand. Note that an exponential fraction is always a shilling fraction.

7. Negative Exponents

All exponents shown to this point have been positive (the + is always understood when it is not written). Exponents may also be negative.

12^4 is 12 raised to the (plus) fourth power.

12^{-4} is 12 raised to the negative (or minus) fourth power.

a. Negative exponents convert expressions into fractions The rule is that the negative exponent becomes positive when the whole expression to which the exponent applies (the root) is made into the denominator of a fraction of which the numerator is 1.

So

$$12^{-4} \quad \text{is the same as} \quad \frac{1}{12^4}$$

b. Negative powers of 10 Negative exponents can be applied to any base, but there is special significance when the base is the number 10 (see "Powers of 10" on p. 2.14). Again we are dealing with the metric system (see "Relation to the Metric and SI System"), where everything is related to everything else by multiplying or dividing by different numbers of tens. This is illustrated in Table 2-2, which shows the use of both positive and negative exponents with the base 10.

TABLE 2-2
EXPONENTS—POSITIVE AND NEGATIVE

1,000,000,000,000,000,000	$= 10^{18}$	One million trillion
1,000,000,000,000,000	$= 10^{15}$	One thousand trillion
1,000,000,000,000	$= 10^{12}$	One trillion
1,000,000,000	$= 10^{9}$	One billion
1,000,000	$= 10^{6}$	One million
1,000	$= 10^{3}$	One thousand
100	$= 10^{2}$	One hundred
10	$= 10$	Ten
0.1	$= 10^{-1}$	One-tenth
0.01	$= 10^{-2}$	One-hundredth
0.001	$= 10^{-3}$	One-thousandth
0.000 001	$= 10^{-6}$	One-millionth
0.000 000 001	$= 10^{-9}$	One-billionth
0.000 000 000 001	$= 10^{-12}$	One-trillionth
0.000 000 000 000 001	$= 10^{-15}$	One thousand-trillionth
0.000 000 000 000 000 001	$= 10^{-18}$	One million-trillionth

While positive exponents make the base 10 ten times bigger for each integer increase of exponent, the negative exponent makes the base into a fraction that is smaller than the preceding one by multiples of 10.

NOTE: Refer to Table 2-2 for a tabulation of numbers in multiples of 10 expressed in terms of both positive and negative exponents.

8. Review of Exponents

A good way to review the subjects of exponents and roots is to make sure that you understand each of these examples. They are given first in words, as they appear in dictation or in written text, and then in mathematical notation:

- Three and six-tenths squared

$$3.6^2$$

- The square root of three and six-tenths

$$\sqrt{3.6}$$

- Two times the expression 3*x* minus *k*, times ten to the third

$$2(3x - k) \times 10^3$$

- The cube root of forty-three

$$\sqrt[3]{43}$$

- Which is the same as forty-three raised to the one-third power

$$43^{1/3}$$

- Two thousand times the expression 3*x* minus *k*

$$2000(3x - k)$$

- Thirty-eight to the minus fourth power (*or* to the minus four)

$$38^{-4}$$

- Which may also be expressed as one over thirty-eight to the fourth power

$$\frac{1}{38^4}$$

- Nine point two times ten to the seventh

$$9.2 \times 10^7$$

- Which may also be written as ninety-two million

$$92{,}000{,}000$$

- Nine point two times ten to the minus seven

$$9.2 \times 10^{-7}$$

- Which may also be written as oh point oh oh oh oh oh oh nine two

$$0.00000092 \qquad \text{or} \qquad 0.000\ 000\ 92$$

2.1.5 Logarithms

A *logarithm* is an exponent. In the conversion from a logarithmic expression to an exponential expression, the logarithm has the same value as the exponent.

So for the number 100

$$10 \times 10 = 10^2 = 100$$

where the exponent to the base 10 is 2. Then 2 is also the logarithm to the base 10 of the number 100, which is expressed as

$$\log 100 = 2$$

In the same way the logarithm of the number 1000 is 3 ($\log 1000 = 3$) and the logarithm of the number 1,000,000 is 6 ($\log 1{,}000{,}000 = 6$).

What about a number that does not end in zeros—can it have a logarithm too? Yes, any number can have a logarithm, which is a statement of how many times 10 is multiplied by itself to give that number. The lack of zeros merely means that the logarithm is not an integer but has a fractional, or decimal, part.

An example is the number 17. It turns out that if one multiplies 10 by itself 1.2304 times one gets the number 17, which may be expressed in mathematical notation as

$$10^{1.2304} = 17$$

and now we can say that the logarithm of 17 is 1.2304 ($\log 17 = 1.2304$).

1. Tables of Logarithms

Logarithms have been calculated for a great many numbers. These appear in reference books, available to every engineer or scientist, known, not surprisingly, as *tables of logarithms;* each entry is simply the exponent applied to the base 10 that will result in that number.

There are similar tables for antilogarithms (see below) and for logarithms to the natural base *e* as well (see "Writing Exponents on the Natural Base" in Section 2.1.4).

2. Antilogarithms

There are also tables of *antilogarithms* which allow the technical person to reverse the logarithmic process, to "get back."

So if the logarithm of 17 is 1.2304, the antilogarithm of 1.2304 is 17—you can look it up.

3. Other Bases

It is not necessary to be confined to the base 10, although that is the one most commonly used. There are two other bases that are employed frequently.

a. The base 2 This is the basis of the binary system of calculation used particularly in computer work (see "Binary Nature of Bits" in Section 3.2.1).

b. The natural base *e* The number 2.1783, the natural base, has special mathematical significance (see "Writing Exponents on the Natural Base" in Section 2.1.4 for the treatment of this number as an exponential base), and it is represented by the symbol *e*.

NOTE: A logarithm to the base *e* is not abbreviated *log* as with the bases 10 or 2; it is abbreviated *ln*.

2.1.6 Prefixes That Make Numbers Bigger or Smaller

1. The Metric System

So far, this chapter has shown two ways to make numbers larger or smaller by multiplying or dividing by 10. These are decimals and exponents in the form of powers of 10. Doing things by tens means using the *metric system*. That system really makes mathematical work much easier, since all you have to do is move the decimal point to the right or the left or add or take away zeros.

There are names for every level of 10, that is, for every level of size (see Tables 2-1 and 2-2). These names were worked out by an international committee that met in 1954. Their meeting was called the General Conference of Weights and Measures on the International System of Units. The name *International System* was adopted for their results (which deal with every kind of scientific notation, not just numbering by tens). The worldwide abbreviation for International System is SI (from the French Système Internationale). That is what is now almost universally used and what this book attempts to adhere to (see note below).

NOTE: Although some scientists and engineers received their training before the SI units were adopted, everyone now producing technical literature should use the SI units.

2. SI Prefixes and Symbols

Table 2-3 lists the SI expressions that designate prefixes to be used as parts of words in written text. It also shows the letter symbols that stand for these parts of words to be used in mathematical or scientific expressions. It can be seen by examining the table how all these terms are related to what we have just treated—exponents, decimals, and logarithms. (Also compare Table 2-3 with Tables 2-1 and 2-2.)

Notice that again you must be careful. Don't confuse M with m. That would be confusing a million with a thousandth, and since a million is a billion times bigger than a thousandth, that would be a billionfold mistake!

TABLE 2-3
PREFIXES FOR SI UNITS

Multiple and Submultiple	*Prefix*	*Symbol*
1,000,000,000,000,000,000 = 10^{18}	exa-	E
1,000,000,000,000,000 = 10^{15}	peta-	P
1,000,000,000,000 = 10^{12}	tera-	T
1,000,000,000 = 10^{9}	giga-	G
1,000,000 = 10^{6}	mega-	M
1,000 = 10^{3}	kilo-	k
100 = 10^{2}	hecto-	h
10 = 10	deka-	da
0.1 = 10^{-1}	deci-	d
0.01 = 10^{-2}	centi-	c
0.001 = 10^{-3}	milli-	m
0.000 001 = 10^{-6}	micro-	μ
0.000 000 001 = 10^{-9}	nano-	n
0.000 000 000 001 = 10^{-12}	pico-	p
0.000 000 000 000 001 = 10^{-15}	femto-	f
0.000 000 000 000 000 001 = 10^{-18}	atto-	a

Some examples of abbreviations of units of measure used in this system in text and in scientific writing follow. Abbreviations such as these are generally used with a number.

- Kilovolt, which means 1000 volts, when used with a number, will appear in scientific writing as kV (e.g., 100 kV).
- Millivolt, which means a thousandth of a volt, appears as mV in scientific writing.
- Megaton, which means a million tons, appears as Mt in scientific writing.
- Microfarad, which means a millionth of a farad, appears as μF in scientific writing.
- Gigahertz, which means a billion cycles per second, will appear as GHz in scientific writing.
- Centimeter, which means a hundredth of a meter, will appear as cm in scientific writing.
- Milliliter, which means a thousandth of a liter, will appear as ml or mL in scientific writing.
- Nanosecond, which means a billionth of a second, will appear as ns or nsec in scientific writing.

3. General Names Which Designate Quantities

In the case of decimals, there are also general names for each position to the left or right of the decimal point. These were described in "The Importance of the Location of the Decimal Point" in Section 2.1.3 and in Figure 2-1, which should now be compared with Table 2-3. Both sets of designations are equally valid and either or both may be used in the same text as required by what is being expressed.

2.1.7 Primes and Subscripts—Ways to Indicate Where Things Stand in a Sequence

The technical specialist frequently has occasion to describe a family of things that are similar to each other; yet they differ from each other in that this one is first and that one is second and that one is third, and so on.

1. Primes in a Sequence

A simple way to show where something is in a sequence is the use of what is called the *prime* system. It consists of the (typed) apostrophe symbol ′. (It should be labeled "prime" for the typesetter, who distinguishes between apostrophes and primes. One prime mark stands for the first item, two for the second, three for the third, and so on.

> Let us consider a sequence of terms relating to samples of the gas nitrogen. Say the engineer or scientist or chemist wants to discuss a set of four of these. Suppose it is necessary to consider them as being combined with each other, that is, added up. Then in words the following statement might be made: "We will add together samples nitrogen prime, nitrogen double prime, nitrogen triple prime, and nitrogen quadruple prime." That is certainly an awkward statement, but it becomes clear and lucid when written in mathematical notation as follows:

$$N' + N'' + N''' + N''''$$

2. Subscripts in a Sequence

It is not enough, in describing a sequence, just to say which is first, which second, and so on. It is useful to be able to give other, more descriptive designations.

a. Symbols as subscripts This is done with symbols used as *subscripts.*

b. Subscripts may have subscripts It is even possible to designate subclasses by placing subscripts onto subscripts.

In order for such procedures to have any meaning it is necessary for the text to have a section that defines each subscript symbol. Such sections are characteristically made up of equation statements of the form "Let $x =$."

These statements serve the function of identification. They are like a list of the cast of players in a play. Authors of technical or scientific material, like the author of a play, give us the cast list and go on to describe some events—events concerned with how members of the cast interact, events that proceed from one act to the other—until they "conclude." Let us make up some "playlike" text which will illustrate the rules of subscripts in a sequence.

> Suppose you were reproducing the writings of a nutritionist or food scientist and the subject is the cooking of fish, specifically, a comparison of the nutritional qualities of cod and haddock as a result of different cooking procedures. The text might have a pair of statements like this:

$$\text{Let } C = \text{cod}$$

$$\text{Let } H = \text{haddock}$$

> Now we begin to see the need for subscripts, because the nutritionist wishes to discuss the merits of fresh versus frozen fish. The next set of statements in the text could well be:

$$\text{Let } C_{uf} = \text{unfrozen cod}$$

$$\text{Let } C_{f} = \text{frozen cod}$$

$$\text{Let } H_{uf} = \text{unfrozen haddock}$$

$$\text{Let } H_{f} = \text{frozen haddock}$$

> NOTE: It is interesting that there is a little snare here: *f* is the first letter of both *fresh* and *frozen*. It was necessary to think up a symbol that would not be confusing; *u* would have been sufficient, but it is not as descriptive as *uf*, and there is no rule against two letters being used as a subscript symbol.

> But we're not done yet. Suppose it now becomes necessary to compare the quality of fish cooked in an ordinary oven with fish cooked in a microwave oven. That can be done by adding more subscripts. But now there is a complication. A string of drawn-out subscript letters can be confusing, something we want to avoid. Scientific text is complicated enough without adding further confusion by the way in which it is put onto the printed page. We could "index down," that is, provide another subscript level, designating C_{uf_o} as "fresh cod baked in an oven." But that becomes awkward, occupies too much vertical space, and presents a difficult task for typesetters. The preferred method is to set off higher-level subscripts with commas:

$$\text{Let } C_{uf,o} = \text{fresh cod cooked in an oven}$$

$$\text{Let } C_{uf,m} = \text{fresh cod cooked in a microwave oven}$$

$$\text{Let } H_{uf,o} = \text{fresh haddock cooked in an oven}$$

$$\text{Let } H_{uf,m} = \text{fresh haddock cooked in a microwave oven}$$

> We are not done yet. Suppose we wanted to indicate the first and the second and the third pieces of fish. We simply use primes, since both primes and subscripts may be used for the same term. Then

$$\text{Let } C'_{uf,o} = \text{the first piece of fresh cod cooked in an oven}$$

> And so on.

NOTE: It is usual to align subscripts and superscripts.

INCORRECT: $C'{}_{\mathrm{uf,o}}$

CORRECT: $C'_{uf,o}$

The cast of characters of this little drama has been identified. The time has come for some action. In this case let us say the nutritionist wishes to compare weight changes for the different fish samples for the two different modes of cooking. It then becomes necessary to introduce one more definition:

$$\text{Let } W = \text{weight}$$

and apply all the applicable subscript and prime modifiers to the different Ws.

$$\text{Let } W'_C = \text{weight of first piece of cod}$$
$$\text{Let } W''_C = \text{weight of second piece of cod}$$
$$\text{Let } W'_H = \text{weight of first piece of haddock}$$
$$\text{Let } W''_H = \text{weight of second piece of haddock}$$

and so on.

Now we are ready to provide some conclusions. Enough information has been given so that a mathematical statement may be made about weight loss for different kinds of fish cooked in different ways:

$$W'_{C,uf,o} + W''_{C,uf,o} + W'''_{C,uf,o} < W_{C,uf,m} + W''_{C,uf,m} + W'''_{C,uf,m}$$

or

$$\Sigma W_{C,uf,o} < \Sigma W_{C,uf,m}$$

(See "Σ (The Summation Sign)" in Section 2.1.1.)

or

$$\Delta W_{C,uf,o} > \Delta W_{C,uf,m}$$

which says that the change in weight is greater in oven cooked than microwave oven cooked. (See "Δ (The Increment Sign)" in Section 2.1.1.)

NOTE: All subscripts are shown in the preceding example in the same line. This is the rule. Multiple subscripts are written in this manner, separated by commas. Yet in a preceding example it was stated that $C_{uf,o}$ was the way to indicate fresh cod baked in an oven. There is no inconsistency here. That statement used *uf* and *o* as subscripts for *C*. In this case *C*, *uf*, and *o* are subscripts for *W*.

NOTE: Some new symbols are used here which will be discussed in the algebra section (see "The Family of Equals Signs" in Section 2.3.1). They are the symbols $>$ (which means "is greater than") and $<$ (which means "is less than").

Now we know the answer to one of the questions being investigated by our food scientist (all three equations say the same thing): If one cooks cod in an ordinary oven, it will come out dryer (it will lose more moisture) than if it is cooked in a microwave oven—a fact that has nothing to do with the subject matter of this book but which shows how the use of subscripts and primes allows such a statement to be made very concisely in a shorthand way, which, after all, is what technical notation is all about.

By using similar equations, the same kind of conclusion can be stated in comparing the two kinds of fish after, let us say, only oven cooking:

$$W'_{C,uf,o} + W''_{C,uf,o} + W'''_{C,uf,o} > W'_{H,uf,o} + W''_{H,uf,o} + W'''_{H,uf,o}$$

or

$$\Sigma W_{C,uf,o} > \Sigma W_{H,uf,o}$$

or

$$\Delta W_{C,uf,o} < \Delta W_{H,uf,o}$$

What is being said is that haddock loses more weight than cod loses in conventional oven cooking. The point is we can say this as we just did, in words, or we can say it, as in the example, using mathematical notation. Both say the same thing, but the mathematical expression says it more exactly—which is why people (mostly engineers and scientists) use mathematical expressions.

2.2 GEOMETRY AND TRIGONOMETRY

When we describe space we use the branch of mathematics known as *geometry*. It deals with position, distance, angles, lines, surfaces, and shapes. As in all scientific and technical matters, *deals with* means to treat concisely and quantitatively, as well as descriptively (qualitatively). And *treating* means, among other things, to commit to paper. There are some special procedures you must know in reproducing this kind of material, which are described here.

Geometry is the most "concrete" part of mathematics. It is used by artisans as well as by scientists and engineers. Who can think of the carpenter, the surveyor, the tailor without their geometrical tools—the ruler, the plumb line, the protractor, the tape measure? But there is a difference. Artisans generally don't have to write about what they do, as scientists and engineers must. Technical professionals have to know how to put geometry into text, and to do that correctly, they must know the meanings of certain important words and concepts that describe space, as well as know how to present and symbolize them on the printed page (see Section 2.2.1).

Trigonometry is the special part of geometry that deals with the study of triangles, in particular with the relationships between the angles and sides of a right triangle.

2.2.1 Words and Concepts That Describe Space

1. Point

Everything in geometry starts with the *point*. It is simply a position. It can be made with the period on your typewriter, but it really has no size—no length, width, or height. It just indicates a place.

NOTE: Do not confuse the position point with the decimal point described in "The Importance of the Location of the Decimal Point" in Section 2.1.3.

2. Line

When a point is moved, the path it takes is a *line*. The line can be straight or curved.

3. Plane

When a straight line moves, the path it describes is known as a *plane*.

Thus a plane is a surface such that a straight line joining any two points will be on that surface.

4. Area

When boundaries are put on a plane, the measure of the size of the surface within the boundaries is known as an *area*.

The size of an area is indicated by square units such as square inches, square feet, square yards, or acres (which consist of square feet) or square meters, square centimeters, square millimeters, square micrometers, etc.

NOTE: The engineer usually refers to these quantities with the words reversed: feet squared, centimeters squared, etc.

5. Angles

When two lines extend away from a single point (or two planes extend from a common line), an *angle* is formed.

a. Vertex and apex The point of meeting of the lines is the *vertex* (sometimes called the *apex*, especially when it is the angle opposite the base of a geometric figure).

b. The angle symbol Angles may be symbolized in written text by the symbol $\sphericalangle$ or $<$, which are drawn freehand. The author may use them instead of the word *angle*.

c. Naming angles Angles are usually referred to in text by merely writing the three capital letters that name the two lines making up the angle, with the vertex letter as the second letter.

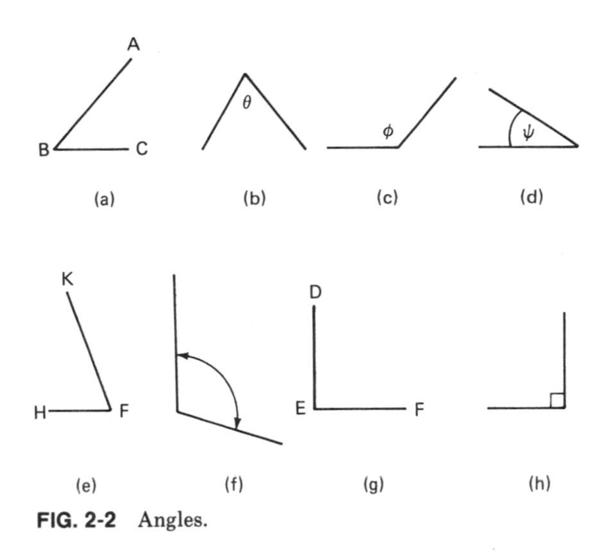

FIG. 2-2 Angles.

So the text can "name" the angle *ABC* (Figure 2-2*a*), angle *HFK* (Figure 2-2*e*), and angle *DEF* (Figure 2-2*g*).

d. Greek-letter angle names Angles may also be designated with Greek letters (Figure 2-2*b*, *c*, and *d*). In that case letters at the ends of angle lines are often (but not always) left out.

Then it is not necessary to use the three capital letters to designate an angle; instead we may say angle θ (theta) or angle ϕ (phi) or angle ψ (psi). These three are the Greek letters most commonly used to indicate angles. (See Section 1.1.1 on the use of Greek letters.)

e. The arc sign An arc may be used to "point out" any angle (Figure 2-2*d* and *f*). The arc may or may not terminate in arrows; their use is optional.

6. Right Angle

When a vertical line meets a horizontal line, a *right angle* is formed.

Figure 2-2*g* and *h* shows a right angle.

a. Perpendicular A line that forms a right angle when it touches or intersects another line is known as the *perpendicular* (noun form). The two lines are then *perpendicular* to each other (adjective form).

b. The perpendicular sign When the writer wishes to call attention to the fact that an angle is a right angle, it is customary to draw a small square inside that angle (Figure 2-2*h*).

NOTE: When just lines are concerned and not triangles, another way to show a perpendicular is with a sign that truly pictures it. It is the ⊥, drawn with the aid of a straightedge.

7. Triangles

When we draw a third line to close up an angle, we have made a triangle, which means "three angles" (count them). The area enclosed is the area of the triangle, and it is measured in square units, just like any other area.

NOTE: Here we can see the origin of the word *trigonometry.* It is made up of three separate Greek roots.

tri—three, for the three corners of a triangle

gono—angle

metry—measure

So trigonometry is that branch of mathematics which describes and measures the angles of triangles; although, in practice, it includes all angles, even those that are not parts of triangles.

a. Right triangle Triangles that contain a right angle are known as *right triangles* (Figure 2-3).

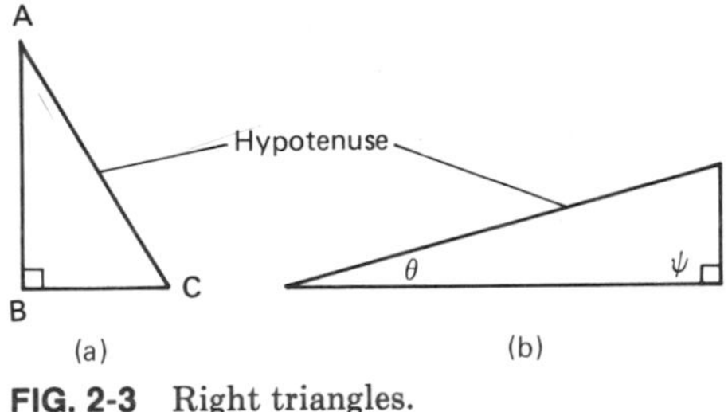

FIG. 2-3 Right triangles.

b. Hypotenuse There is a special name given to the side of a triangle which is opposite the right angle in a right triangle. It is the *hypotenuse.*

In Figure 2-3*a,* side *AC* is the hypotenuse, as is the side opposite angle ψ in Figure 2-3*b.*

8. Parallel and Nonparallel Lines

a. Parallel When two lines are so "lined up" with each other that they never meet, they are known as *parallel.* The symbol for parallel lines is ||.

b. Nonparallel When they are not so well lined up, they are known as *nonparallel,* and if they are in the same plane they will eventually meet, forming what we have just been discussing—an angle. The symbol for nonparallel lines is $\nparallel$.

9. Parallelogram

A *parallelogram* is a four-sided figure whose angles are not necessarily right angles and the opposite pairs of sides of which are parallel.

a. Rectangle A *rectangle* is a four-sided figure in which all four angles are right angles. It is a special case of a parallelogram.

b. Square A *square* is a four-sided figure in which all four sides are equal in length and all angles are right angles. It is a special kind of rectangular parallelogram.

10. Polygons

A *polygon* is any closed plane figure the boundaries of which are made up of straight lines. It may have any number of sides from three to infinity. Some polygons are listed in Table 2-4.

TABLE 2-4
POLYGONS

Three-sided	Triangle	Eight-sided	Octagon
Four-sided	Quadrilateral	Nine-sided	Nonagon
Five-sided	Pentagon	Ten-sided	Decagon
Six-sided	Hexagon	Twelve-sided	Dodecagon
Seven-sided	Heptagon		

11. Circle

A *circle* is a closed plane curve all points of which are at an equal distance from a given point called the *center.*

a. Radius The distance from the center of a circle to a point on the outside of a circle is the *radius* (see Figure 2-4*c*).

b. Diameter A line that passes through the center of the circle and whose two *endpoints* are on the circle is the *diameter.*

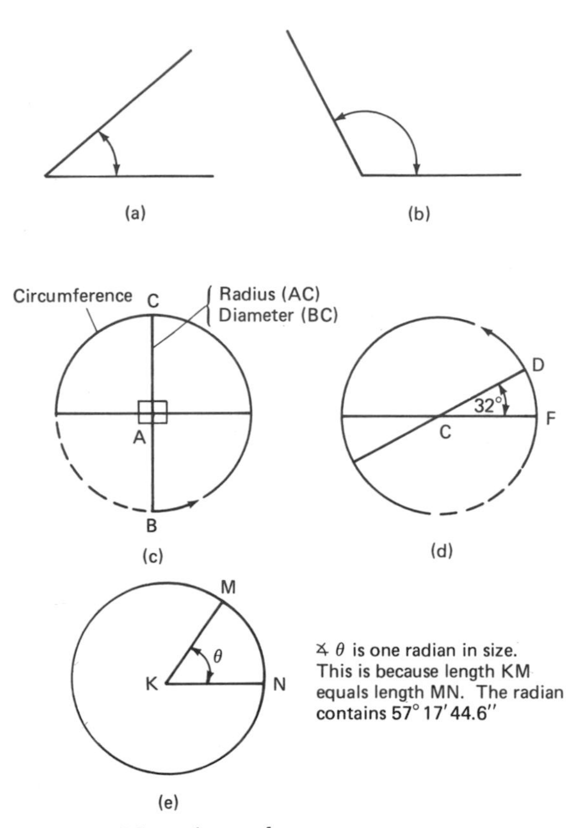

FIG. 2-4 Measuring angles.

c. Circumference The length of the outside perimeter of the circle is the circumference (see Figure 2-4*c*).

NOTE: The ratio of the circumference to the length of the diameter is 3.14159. . . . This is pi (see p. 2.3).

12. Degrees

Angles are measured in *degrees*. The wider (the more open) the angle, the more degrees it contains. (See Figure 2-4*a* and *b*.)

a. Ninety degrees in a right angle A right angle (see "Right Angle" in Section 2.2.1) is defined as containing 90 degrees (written 90°).

b. Three hundred sixty degrees in a circle Figure 2-4*c* shows four right angles in a circle. Thus all circles contain 360° (90° × 4).

c. Symbol The symbol for degree is a small circle printed as a superscript after the number (of degrees), as above.

Example:

In Figure 2-4*d*, angle *DCF* is a 32° angle.

13. Minutes and Seconds

Subdivisions of degrees are known as *minutes* and *seconds*.

The Greeks divided the circle into 360 degrees. The sixtieth part of the circle was called a *minute* and the sixtieth part of a minute was called a *second*. We still use clock faces in the form of a circle divided into minutes and seconds. In measuring angles, however, one *degree* (which is like one-sixth of a minute on the clock) is subdivided into sixty *minutes* of sixty *seconds* each.

a. Definition A minute is a sixtieth of a degree and a second is a sixtieth of a minute.

There are 60 minutes in a degree and 60 seconds in a minute, so there are 3600 seconds in a degree.

b. Symbols The minute appears as a superscript just as the degree does. It is a prime mark (see "Primes in a Square" in Section 2.1.7). The second is used similarly; it is a double prime.

An angle of 14 degrees, 32 minutes, 51 seconds is written

140°32′51″

with no space after the symbols for degree, second, and minute.

14. Radians

Another way to measure angles is by constructing a portion of a circle (Figure 2-4*e*). When the portion of the circumference intersected by two radii (plural of *radius*) is equal to the radius, the angle is then of a particular size. That size is given the name *radian*. So, in Figure 2-4*e*, the length of the arc *MN* equals the length of the radius (*KM* or *KN*) and the angle is exactly 1 radian in size.

It turns out that a radian is equal to 57 degrees, 17 minutes, and 44.6 seconds (57°17′44.6″).

The SI abbreviation for radian is *rad*.

15. Solid Geometry

Everything discussed thus far in geometry has been in two dimensions—flat shapes that can be drawn on a flat piece of paper. But the real world is in three dimensions. It is solid. Solid geometry is more complicated. All definitions described for two-dimensional space hold.

There are only a few new terms, such as *solid angle*, and the illustrations are more complex.

16. Summary of Words and Concepts That Describe Space

All words and concepts described in Section 2.2.1—and many more—are summarized, with illustrations, in Table 2-5.

TABLE 2-5
WORDS AND CONCEPTS OF GEOMETRY

TERM	FORM IN TEXT	SYMBOL
Position and Lines		
Point	•	•
Line	 A—B	
Parallel	//	‖
Nonparallel	 /\	∦
Angles		
Angle	 B, A, C; ϕ	∠ or ∡
Vertex or apex (A)	 A	
Right angle (90°)		⊥
Complementary (two angles that add up to a 90° angle)	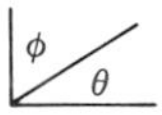 ϕ θ	
Supplementary (two angles that add up to an 180° angle)	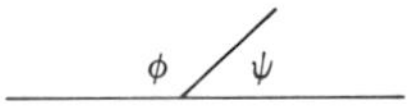 ϕ ψ	
Acute angle (smaller than 90°)		
Obtuse angle (larger than 90° and smaller than 180°)		
Degree (angular measurement: an angle 1/360 of a circle)		Superscript ° or deg
Minute (angular measurement: an angle 1/60 of a degree)		Superscript ′ or min
Second (angular measurement: an angle 1/60 of a minute, 1/3600 of a degree)		Superscript ″ or sec
Radian (angular measurement: an angle whose intercepted arc is the same length as the circle's radius—an angle of 57° 7′ 44.6″)	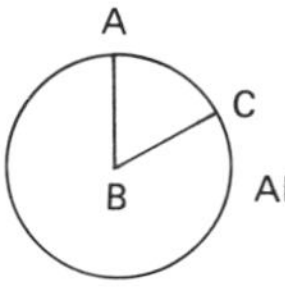 A, B, C; AB = AC	rad

TABLE 2-5
WORDS AND CONCEPTS OF GEOMETRY (continued)

TERM	FORM IN TEXT	SYMBOL
Plane Figures		
Triangle		
Circle		
Ellipse or oval		
Square		
Diamond		
Rectangle		
Parallelogram (opposite sides are parallel to each other)		
Diagonal (AB)		
Trapezoid (two sides are parallel)		
Pentagon (five sides)		
Hexagon (six sides)		
Octagon (eight sides)		
Polygon (n sides); see Table 2-4)		
Perimeter (the distance around any plane figure—AB + BC + CD + DA)		
Area (the space enclosed by any plane figure)		
Triangles		
Scalene (three sides of unequal length)		
Isosceles (two equal angles and two equal sides)		
Equilateral (three equal angles and three equal sides)		
Right (one angle is a right angle)		
Hypotenuse (AB, the longest side of a right triangle)		

TABLE 2-5
WORDS AND CONCEPTS OF GEOMETRY (continued)

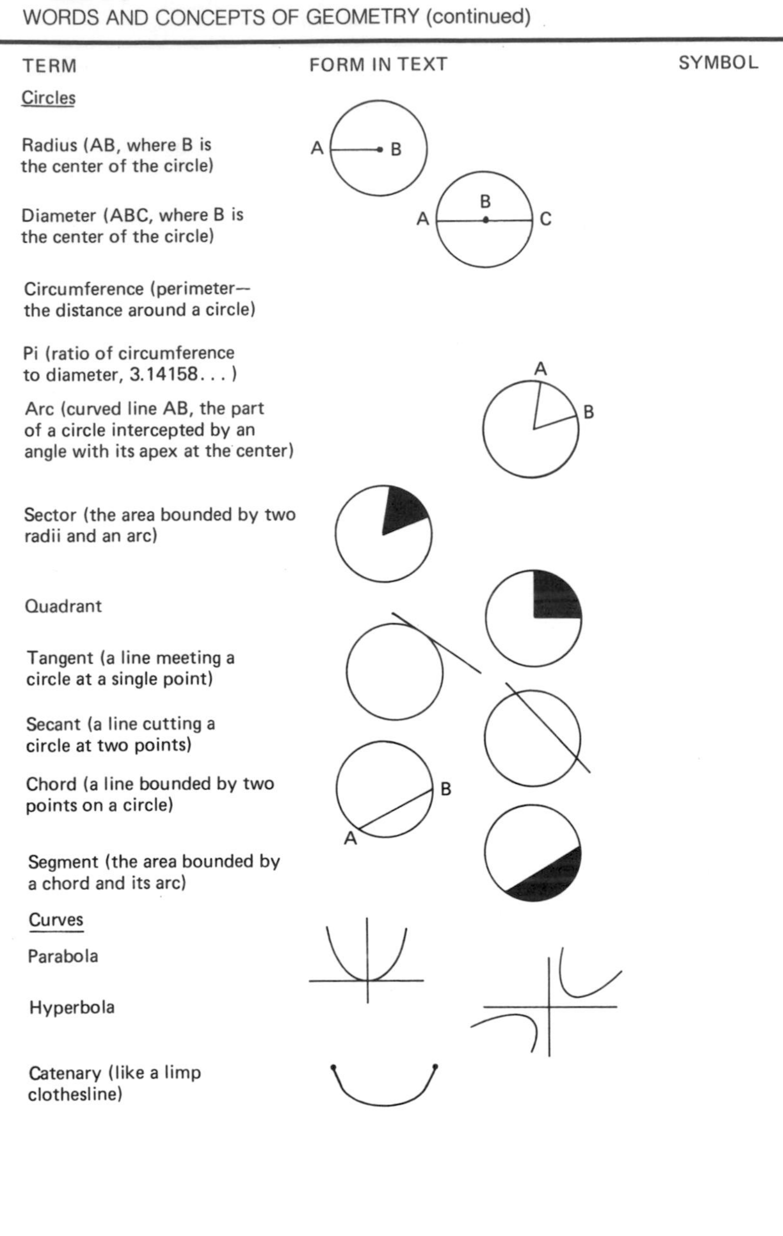

TERM	FORM IN TEXT	SYMBOL
Circles		
Radius (AB, where B is the center of the circle)		
Diameter (ABC, where B is the center of the circle)		
Circumference (perimeter—the distance around a circle)		
Pi (ratio of circumference to diameter, 3.14158. . .)		
Arc (curved line AB, the part of a circle intercepted by an angle with its apex at the center)		
Sector (the area bounded by two radii and an arc)		
Quadrant		
Tangent (a line meeting a circle at a single point)		
Secant (a line cutting a circle at two points)		
Chord (a line bounded by two points on a circle)		
Segment (the area bounded by a chord and its arc)		
Curves		
Parabola		
Hyperbola		
Catenary (like a limp clothesline)		

TABLE 2-5
WORDS AND CONCEPTS OF GEOMETRY (continued)

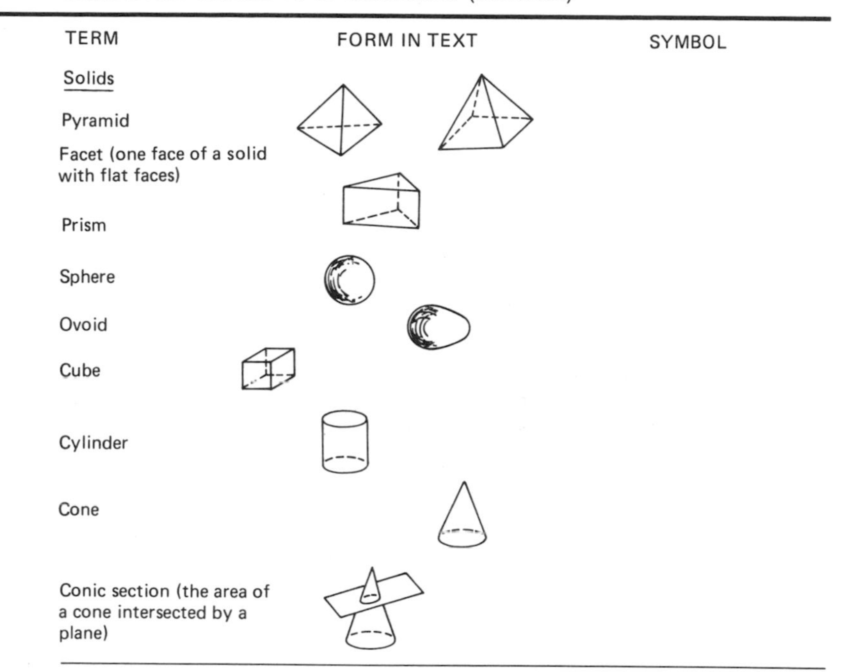

TERM	FORM IN TEXT	SYMBOL
Solids		
Pyramid		
Facet (one face of a solid with flat faces)		
Prism		
Sphere		
Ovoid		
Cube		
Cylinder		
Cone		
Conic section (the area of a cone intersected by a plane)		

2.2.2 Making Calculations in Space—Trigonometry

1. The Trigonometric Ratios

The most common way to deal with angles mathematically is by comparing the lengths of the sides of the triangles of which these angles make up the corners. Such comparisons of lengths are *ratios*.

There are six possible ratios for any angle in any right triangle.

NOTE: For any triangle that is not a right triangle, two right triangles can be constructed by drawing a perpendicular line. (See Figure 2-5.) So in effect the ratios can be calculated even if the triangle is not a right triangle.

These ratios are

Sine	Cosecant
Cosine	Secant
Tangent	Cotangent

Table 2-6 gives the meaning and abbreviation for each of these ratios, which are also known as the *trigonometric functions*.

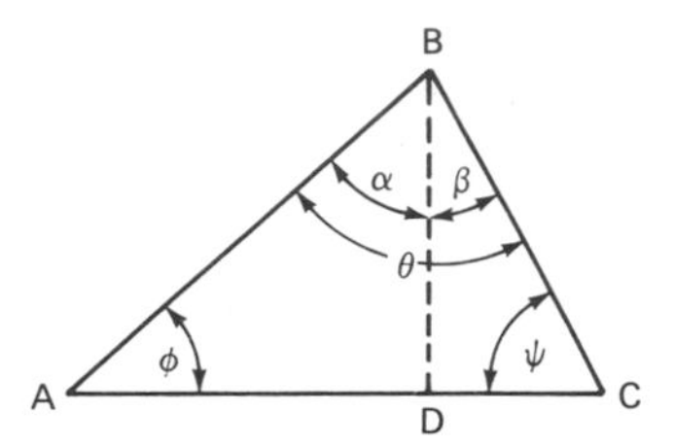

FIG. 2-5 Drawing the right angle line *(BD)* for purpose of trigonometric calculations. (This creates two new angles α and β, the sum of which is θ.)

TABLE 2-6
THE TRIGONOMETRIC RATIO

Sine θ	=	$\sin\theta$	=			$\frac{CB}{AC}$
Cosine θ	=	$\cos\theta$	=			$\frac{AB}{AC}$
Tangent θ	=	$\tan\theta$	=			$\frac{CB}{AB}$
Cosecant θ	=	$\csc\theta$	=	$\frac{1}{\sin\theta}$	=	$\frac{AC}{CB}$
Secant θ	=	$\sec\theta$	=	$\frac{1}{\cos\theta}$	=	$\frac{AC}{AB}$
Cotangent θ	=	$\cot\theta$	=	$\frac{1}{\tan\theta}$	=	$\frac{AB}{CB}$

C
θ
A B
Angle CAB is angle θ

NOTE: The abbreviations for the trigonometric functions all consist of three lowercase letters, always followed by the designation of the angle being described—$\sin\theta$, $\tan\psi$, etc. The angle being described can be a Greek letter (usually one of the five: θ, ϕ, ψ, φ, β). It can also consist of the three letters that mark the ends of the lines that make up the angle ($\sin\theta = \sin CAB$).

2. Exponents in Trigonometric Expressions

There is a rule for using exponents in trigonometry. (Sines and cosines and the others can be raised to a power just like anything else.) The exponent always comes just after the three-letter ratio designation and before the angle—never after.

CORRECT: $\tan^3 \theta$
INCORRECT: $\tan \theta^3$

3. Writing Trigonometric Expressions

a. Part of algebraic notation Trigonometric expressions may occur by themselves, but more often they are found as parts of equations. They become ingredients of algebraic text, in which they are merely another kind of algebraic symbol (see Section 2.3.1).

b. Anything goes The same rule holds in writing trigonometric expressions as for any other mathematical notation—anything goes.

Anything can be the angle; it can be a complex algebraic expression, it can be a complex fraction, a Greek letter—anything.

c. Be careful Make sure the reader is not confused.

Make sure that the expression which stands for the angle never gets mixed up with the rest of the mathematical statement. There should never be any doubt about where the angle is in the text: where it starts, where it ends. The use of parentheses and brackets can be helpful here. (This practice of employing parentheses and brackets will be treated in greater detail in Section 2.3.)

4. Translation and Rotation

These are two words that are used a great deal in mathematical text dealing with geometrical matters. They are worth mentioning here because to the engineer and scientist they do not mean exactly what they mean in conversational or written English.

a. Translation In mathematics this does not mean putting words from one language into another. Rather it refers to moving a geometric figure from one position to another, and in so doing every point on that figure describes a path through which the movement of that point takes place. We know from the definition of a line (see "Line" in Section 2.2.1) that the path taken during the translation of a point is a line.

So the path taken by the translation of a whole figure is an assemblage of lines (Figure 2-6*a*).

b. Rotation This differs from translation in that the figure remains in the same location—that is, translation does not occur—

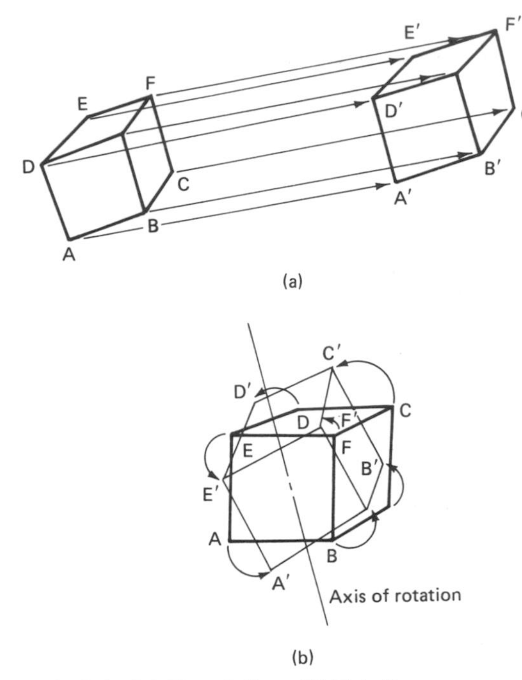

FIG. 2-6 (*a*) Translation. (*b*) Rotation.

rather, every point in the figure describes a path which is a circle around some previously selected line which is then the axis of all the circles. This is shown in Figure 2-6*b*.

2.3 ALGEBRA

Algebra comes from the Arabic *al-jabr*. It is the branch of mathematics that deals with numbers by means of *letters* and *symbols* and with *signs* that designate different mathematical operations. In the end the purpose is to *solve equations,* which means essentially to come up with the answer. The answer is usually some quantitative statement that wasn't previously known and that tells how big something is or how things relate to each other.

2.3.1 Letters and Symbols

Using a letter or a symbol to designate a quantity is a kind of shorthand. It allows the engineer or scientist to say a lot more on the printed page than would otherwise be possible. It is then necessary to define what each letter or symbol means.

1. Defining Terms

There are some simple conventions that are used in the defining of terms. They involve the use of the words *let* and *where* and the sign of operation—the equals sign =. Most of the time in designating an algebraic unknown the text will use two or three letters; x, y, and z are the most common, but any letters may be used.

- A kind of list is generated which says:

  Let x = the first quantity
  Let y = the second quantity
  Let z = the third quantity

- Next there will usually be a list of "where" statements that tell what the unknowns stand for and the units in which they are measured. Here is a sample:

Where x = wavelength, nm
y = amplitude, dB
z = frequency, Hz

- Next, the mathematical expression is given. It looks like gibberish until the let and where lists are examined. Then it becomes clear because all terms have been defined.

NOTE: As is clear from the preceding examples, it is correct practice to line up these statements so that all the equals signs are exactly under each other.

NOTE: See similar treatment for primes and subscripts (Section 2.1.7).

2. Signs of Operation

These are the symbols that tell what actions to take with the symbols that have just been defined. The most common ones are simply the arithmetic signs we all learned in grade school: +, −, ×, ÷. There are a few others which need a little explanation.

3. Multiplication

a. The times sign × The × is the most common way to show multiplication, but there are other ways.

b. The dot · The dot, raised half a line, ·, means multiplication just as the × does.

NOTE: This raises a question. Why use both × and the dot for multiplication? The answer is that they are used under different circumstances. It is customary to use the × in front of the 10 when a number is expressed in scientific notation. Yet in an algebraic expression, one wants to avoid using × as a sign of operation because it can be confused with x used in the same expression as a symbol for a quantity.

- $7x^2 \cdot \frac{y}{9a}$
- 6×10^3

c. Adjacency Merely placing two quantities next to each other also means that they are being multiplied by each other.

$$5x^2k$$

means that the three factors 5, x^2, and k are being multiplied times each other. (See Section 2.3.2.)

d. Parentheses, brackets, and braces A variation on the adjacency method of showing multiplication is by making use of *parentheses* ( ), *brackets* [ ], and *braces* { }. Merely placing such parenthetical statements next to each other means multiplication of the enclosed quantities. See "Signs of Aggregation" (next page) for treatment of how they are sequenced. This is a special case of adjacency. (See also Section 2.3.2.)

$$z = (x^2 - 3)(y + 3\sqrt{k} - x)$$

$$2Mz^2 = \left[(z + 3y)\left(x - \frac{3x^3}{4}\right)\right][x(3y^3 + k)]$$

4. Division

Division may be shown with the division sign $\div$, but there is another way to show division. It is by the fractional form of an expression. Here it is always understood that the number or expression on top, the numerator, is being divided by the number or expression on the bottom, the denominator. Then the horizontal line in the fraction is a symbol indicating division, just as $\div$ indicates division.

- $12 \div 4 = 3$
- $\frac{12}{4} = 3$
- $\frac{12x^2}{4} = 3x^2$

5. The Plus-or-Minus Sign

There is a sign that gives a choice. It is $\pm$, the *plus-or-minus sign.* It means literally "plus or minus," and that is how it will be referred to in dictation or in text. It is used to designate a spread between limits.

Thus if your weight varies between 130 and 140 pounds, you may say that it centers at 135 pounds plus or minus 5 pounds, and it will appear in symbolic notation as

$$135 \pm 5 \text{ lb}$$

6. Signs of Aggregation

Parentheses, brackets, and braces, known as *signs of aggregation,* have already been mentioned as ways to indicate multiplication (see "Parentheses" on p. 2.40). They are also used as a means to mark off material that "goes together." And [as in written text (if you want to place a parenthetical expression inside of another parenthetical expression)] you can use a second or a third sign of aggregation to show levels of groups. (Fortunately, a fourth level rarely is needed—either in English or in algebra.) The three pairs of symbols always appear in a particular order:

$$\{[(\quad)]\}$$

Thus a mathematical expression using these symbols might be

$$3\{x^2 + 2[3x - 4(5 \quad y)]\}$$

7. The Equals Sign

The most important symbol used in algebra is the equals sign =. It is not exactly a symbol of operation like the multiplication or subtraction sign. Rather it indicates a state of being. This is so because = means "is" or "the same as."

The equals sign is used in algebraic *equations,* or statements of equality, such as

$$y = 3x + 2 \qquad \text{or} \qquad x + 6y = 7y - 5x$$

What is indicated by the equals sign is that, after all the operations have been done to the factors and terms (see Section 2.3.2) on the left side of the equation, it will be the same (have the same value) as the right side of the equation, after all the indicated operations have been done to all the factors and terms on the right side.

So all equations are merely statements of identity. They are the engineers' and scientists' way of saying "This is that."

8. The Family of Equals Signs—The Many Ways to Say "Is"

Has it ever occurred to you how many ways you can vary the concept of "is-ness"? There are many, and algebra provides a family of symbols to describe those variations. These symbols are all variations on the equals sign. They are:

Sign or Symbol	*Definition*
$<$	less than
$>$	greater than
$\ll$	much less than
$\gg$	much greater than
$=$	equals
$\equiv$	identical with
$\sim$	similar to
$\approx$	approximately equal to
$\cong$	approximately equal to, congruent to
$\leq$	less than or equal to
$\geq$	greater than or equal to
$\neq$	not equal to
$\doteq$	approaches
$\propto$	proportional to

Let us take a simple statement: The Boston Red Sox is a baseball team. That can be expressed algebraically as

$$RS = BT$$

Now, how many variations can we give to that statement? The possibilities are shown in the accompanying table.

Statement	*Applicable Sign*
The Red Sox is indeed (identical with) a baseball team.	$\equiv$
The Red Sox is less a team than the Yankees (Y).	$<$
The Red Sox is more a team than the Yankees.	$>$
The Red Sox is much more a team than the Yankees.	$\gg$
The Red Sox is approximately equivalent to a baseball team.	$\cong$
The Red Sox is similar to a baseball team.	$\sim$
The Red Sox is not a baseball team at all.	$\neq$

This is not a particularly mathematical example, but it does illustrate the nuances of equality that are possible with this family of signs.

Let us now make the same statements mathematically by using the symbols—the shorthand of algebra.

$$RS \equiv BT$$
$$RS < Y$$
$$RS > Y$$
$$RS \gg Y$$
$$RS \cong BT$$
$$RS \sim BT$$
$$RS \neq BT$$

What has been given is a series of variations, or nuances, on a state of being. These are all deviations from exact identity, but some form of identity is involved. So all may be stated in mathematical notation as equations.

A list of the symbols that designate these subtle variations and short definitions of what they stand for are given in Glossary G.

2.3.2 Factors and Terms

These are two words that are used all the time by engineers and scientists to refer to different parts of an algebraic expression, and so you should know what they mean.

1. Factors

Factors are those parts of an expression that are multiplied or divided by each other.

2. Terms

Terms are those parts of an algebraic expression that are added to or subtracted from each other.

- As always, anything goes; so factors and terms can be of any length, and any factor may include terms or other factors, and any term may include factors or other terms.
- Factors may be recognized by the fact that they are separated from each other by the signs of multiplication—$\times$, $\cdot$, and the signs of aggregation (parentheses, braces, and brackets)—or by simple adjacency with no sign between.
- Factors may also be recognized by the fact that they are separated from each other by the signs of division. Thus the expressions that make up the top (numerator) and bottom (denominator) of a fraction are also factors since they are separated from each other by the horizontal line (the fraction bar), the sign of division. The sign $\div$ also separates factors.

- Those parts of an expression which are terms may be recognized by the fact that they are separated from each other by the signs of addition and subtraction, which include $+$, $-$, and $\pm$.
  In the expression

$$7k(4x^2 - 3y)(k + \sqrt{1 - B^3}) + \left[\frac{x}{y}(1 - B)\right][A(x^2 + xy) + 3]$$

the factors are

$$7$$
$$k$$
$$4x^2 - 3y$$
$$k + \sqrt{1 - B^3}$$
$$\frac{x}{y}(1 - B)$$
$$A(x^2 + xy) + 3$$

and the terms are

$$7k(4x^2 - 3y)(k + \sqrt{1 - B^3})$$
$$\left[\frac{x}{y}(1 - B)\right][\mathrm{A}(x^2 + xy) + 3]$$

And within these factors and terms are other factors and terms—so that within $4x^2 - 3y$, factors are

$$4$$
$$x^2$$
$$3$$
$$y$$

The terms are

$$4x^2$$
$$-3y$$

Within $\frac{x}{y}(1 - B)$, factors are

$$x$$
$$y$$
$$1 - B$$

and terms are

$$1$$
$$-B$$

And so on.

2.3.3 Solving Equations

What is the engineer or scientist trying to say with all these algebraic expressions—all these symbols and signs, all these terms and factors put together as equations? We have already explained that an equation is a kind of sentence stated in a kind of shorthand. It always says (more or less) the same thing: "This is that."

1. An Equation Describes a Situation

The person making such a statement is setting up a new situation by describing something that is happening or that might happen. It is a kind of story that is being told. The characters are the symbols; the actions to be taken are the signs; the situation is the statement of the equation itself.

2. The Characters Interact

Now, all the symbols interact (as the signs tell them to).

3. The Solution

The result is the *answer* (also known as the *solution*).

The solution amounts to knowing something about the symbol or symbols not known before. For example, to find a solution to the equation means to discover a quantity or quantities which, when substituted for the symbol or symbols, make the equation a true statement.

Thus, for the algebraic equations:

- $3x = 18$ the solution is $x = 6$.
- $2(y - 1) = 8$ the solution is $y = 5$

2.3.4 Variables

In algebra, when a quantity is represented by a symbol, it is implied that its value is not known—until the solution has been worked out. It is further implied that it can have any value—until that value is known. Thus it is known as the *variable*, or the *unknown*.

In the preceding examples, x and y are variables, or unknowns. After the equations have been solved and it has been determined that $x = 6$ and $y = 5$, they are known and thus are no longer variables.

1. Interacting Variables—The Function

When there are two or more variables, the important thing is that they are interacting. Each affects the others. A change in "this" will cause a corresponding change in "that."

Thus, in

$$y = 2x$$

- If x is 3, y has to be 6
- If x goes up to 4, y has to be 8
- If x is 9, y has to be 18

And so on. So this story of how y varies with x can be told in words, like any story, or by mathematical notation as in the equation $y = 2x$. It can also be told in a picture, with a graph (see Section 1.3). But however it is done, the same story is being told—how y varies with x.

We have a word for this kind of interactive, dependent variation. It is *function*. We say "y is a function of x."

This is stated in mathematical notation as

$$y = f(x)$$

which is read "y is equal to f of x." What function? Why (in this case), $y = 2x$, of course.

NOTE: We are not limited to x and y. The symbols are the ones the writer happens to be using. For example, they could be

$$m = f(n)$$

or

$$q = f(p)$$

In any functional relationship, each symbol is a variable.

As we have just seen, variables not only vary, they are constrained to vary *together*. A variation in one always causes a corresponding variation in the other.

a. Independent variable It is customary for the symbol after the f, the one in parentheses, to vary first and thus to "have an effect" on the other variable. The variable that varies first—and thus causes a change in the other variable—is known as the *independent variable*.

b. Dependent variable The symbol that is being affected is known as the *dependent variable*.

Thus in the preceding example

- x is the independent variable.
- y is the dependent variable.

c. Variables as functions of each other We can say that the variables are dependent on each other, which is another way of saying that one of them must be a function of the other.

In our economy, we know that supply varies with (is a function of) demand. Thus

$$S = f(D)$$

where S = supply
D = demand

Since demand usually determines supply, we can say that demand is the independent variable and supply the dependent variable. Supply is a function of demand.

d. Direct function When one increase causes another increase—or, to say it another way, when an increase in the independent variable causes an increase in the dependent variable—that is known as a *direct functional relationship,* or a *direct function.*

The preceding example of supply and demand is a direct functional relationship. Supply is a direct function of demand.

e. Inverse function When an increase in one variable causes a decrease in the other variable, that is an *inverse functional relationship,* or an *inverse function.*

Such a case is the relation between supply and price. If the available amount of a product on the market goes up—and thus becomes an oversupply—then the price of the product will go down.

Inverse functions are expressed

$$y = f\left(\frac{1}{x}\right)$$

So the supply-price functional relationship can be stated as

$$P = f\left(\frac{1}{S}\right)$$

where P = price
S = supply

In this case supply S is the independent variable since it determines the price P. At the same time this is an inverse functional relationship, and P is an inverse function of S. You will often encounter examples like this, and they will usually be stated as this example has been.

f. Order Functional relationships may be said to have different orders. The *order* of a functional relationship (of an equation) is the same as the largest exponent that occurs in that equation.

Thus

$$y = 16x - 13 + K$$

is a *first-order* equation.

$$y = 16x^2 - 13 + K$$

is a *second-order* equation.

$$x^3 - 3xy + 24 = 4(y + 3)$$

is a *third-order* equation.

$$x^2 = y^7$$

is a *seventh-order* equation.

2. Graphs

It is said that a picture is worth a thousand words. Perhaps a picture is worth a thousand symbols too. Just as a picture can tell a story—as words do—so a *graph* has the same capability. A graph is a mathematician's picture. It depicts the same functional relationship that an algebraic expression describes.

Just as a story may have illustrations to show something about the story, so mathematical notation may have graphs to show something about the functional relationships that are being described.

a. Coordinate system The graph is constructed on a framework of two lines, usually drawn with one vertical and the other horizontal. They meet, or intersect, in a right angle at a point called the origin. These lines are called *coordinate axes*. They may be marked off in equal intervals, each denoting a given quantity. Fine (nonobtrusive) lines may extend up and down and left and right from these interval points. The horizontal axis is usually called the x axis, and the vertical axis is the y axis.

b. Coordinates When we plot a point on the coordinate axes, it will have both an x value and a y value. The x value, or coordinate, is called the *abscissa* of the point. The y value, or coordinate, is called the *ordinate* of the point.

c. Origin The coordinate axes meet at a point called the *origin*, where both the x value and the y value are 0.

d. Negative quantities If the coordinates are drawn upward for the ordinate and to the right for the abscissa, they describe positive quantities. If they are drawn down and to the left, they describe negative quantities.

Figure 2-7 illustrates the framework for the construction of graphs.

NOTE: This coordinate system, where axes are at right angles to each other, is known as the *cartesian coordinate system.*

3. Plotting Graphs

a. Plotting a first-order direct function This is best explained with an example.

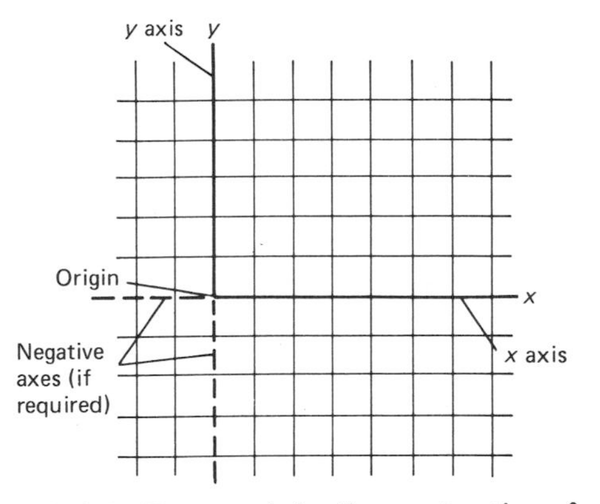

FIG. 2-7 Framework for the construction of a graph.

Let us say that a secretary types copy at some rate (so many words per minute). Let us next hypothesize that we have a word processor that prints copy eight times faster than it can be inputted by the operator. At the same time it will print nothing unless something indeed is inputted. Thus we know that there is a functional relationship that can be expressed

$$C_p = F(C_i)$$

where

C_p = the printed copy, in units of number of characters outputted per minute
C_i = the amount of copy inputted, in units of number of characters keyboarded per minute

Because of the nature of this relationship, it is evident that what is printed is dependent on how much is inputted; so C_i is the independent variable and C_p is the dependent variable. Now the exact equation (or function) can be written; it is

$$C_p = 8C_i$$

To draw a graph from this function proceed through the following steps while referring to Figure 2-8.

(1) Draw the two lines—the coordinate axes—that will form the frame of the graph.

(2) Label each axis by typing in the name of each variable with the pertinent units.

The independent variable is placed under the horizontal axis as it should be. The dependent variable is placed by the vertical axis as it should be. But this is not a hard and fast rule; the placement could be reversed.

(3) Lay down a grid of lines running horizontally and vertically.

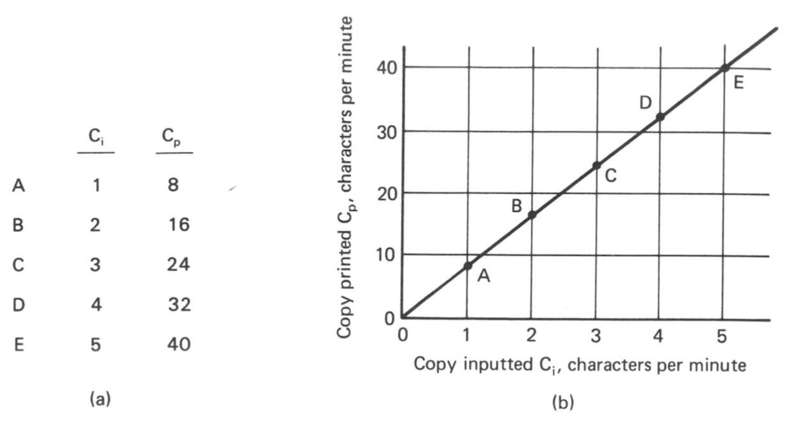

FIG. 2-8 (*a*) Pairs of values calculated for the function $C_p = 8C_i$. (*b*) Graph of function $C_p = 8C_i$

NOTE: This is the graph of a first-order, linear function.

NOTE: If graph paper is used in the text, the lines are already there. In either case the grid lines should be fine (press down lightly) as compared with the axes, which should be drawn thick and dark. Often it is preferred that the grid lines not show at all, in which case they should be drawn in light pencil and used just as an aid to plotting, and then erased.

(4) Number the grid lines.

Use values for each grid space of such a size that the final graph will approximately fill the available space within the axes.

(5) To draw the graph it is necessary first to plot the points that make up the graph. To do this construct a table of values by choosing representative values for the independent variable and substituting them into the equation to find the corresponding values for the dependent variable.

TABLE OF VALUES FOR THE FUNCTION $C_p = 8C_i$.

Point	C_i	C_p
A	1	8
B	2	16
C	3	24
D	4	32
E	5	40

Figure 2-8 shows the graph that can be constructed from the table of values.

The usual situation is that these pairs of numbers will already have been worked out by the writer of the text. But they are shown above so that you may understand what has been done.

Now each point of the graph can be drawn (Figure 2-8) by using each pair of data from the table. Mark off to the right, along the horizontal axis, the value of the independent variable (the abscissa), C_i, of the first pair. From the table of values it is 1. Now mark off upward, along the vertical axis, the value of the dependent variable (the ordinate), C_p, of the first pair (which is 8). This is the first point of the graph, here labeled A. (Sometimes points are labeled, sometimes they are not. If they are, an arrow is sometimes used.)

Now all the other points from the table are plotted as well. So point B is positioned by the data pair 2 and 16, point C from the data pair 3 and 24, and so on.

(6) Draw the line that connects the points.

Draw it firmly and boldly; it should be the same darkness and weight as the axes. This graph is a straight line.

All *first-order equations* (see "Order" on p. 2.47) will graph as straight lines. All such functions (relationships between variables) which give straight-line graphs are also known as *linear* functions.

NOTE: Even though it is a straight line in this case, the author may nevertheless refer to it as a "curve." All lines drawn as graphs are known as curves.

(7) The points should be drawn so that they still show, boldly.

They are an important part of the graph, they indicate where actual data had been taken.

NOTE: The steepness of the line indicates the degree to which the dependent variable depends on the independent variable. Here the graph tells us that the printer is eight times faster than the typist. If it were ten times faster the line would be steeper; if it were four times faster the line would be less steep. The amount of steepness is known as the *slope,* a term which will be encountered very often when graphs are being discussed.

b. Graphing higher-order equations When the equation being graphed has a term raised to a power, a curve will result.

If the relationship betwen the variables is a "square" one (the exponent 2 appears and, therefore, this is a second-order equation), then the table of values might appear as does the table for $x = 2y^2 - 1$ on the next page. A graph for this function is shown in Fig. 2.9.

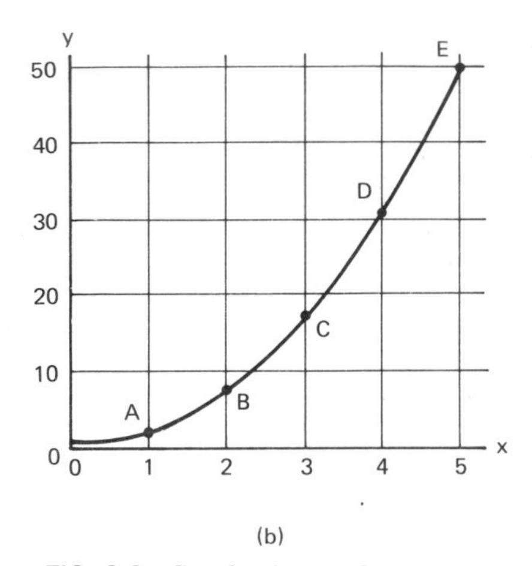

(b)

FIG. 2-9 Graph of second-order function, $x = 2y^2 - 1$. (*b*) Graph of second-order function, $x = 2y^2 - 1$, $x \geq 0$.

TABLE OF VALUES FOR $x = 2y^2 - 1$, $x \geq 0$

Point	x	y
A	1	1
B	2	7
C	3	17
D	4	31
E	5	49

NOTE: It should be evident from examining such a curve that the slope is different at every point. The curve starts shallow and gets steeper.

You can try to draw a curvy (higher-order) curve freehand, but you will be better off if you use the *french curve*. It is pictured in Figure 2-10.

What is it? It is a "curved ruler." The ordinary everyday ruler will give straight lines and is what you need for drawing coordinate axes and straight-line graphs. But to draw a curved line, the french curve is a good aid.

c. Asymptotes There is a special shape of curve that is frequently encountered when graphs are drawn. It is a curve that approaches a line known as an *asymptote*. Such a curve is one that approaches a certain value represented by the asymptote, but never gets there. This is shown in Figure 2-11. The upper right end of the curve approaches—but never reaches—a value of 1. The line $y = 1$ is the asymptote.

FIG. 2-10 French curves. *(Courtesy Charrette Corp. Woburn, Mass.; Photographer, L. Jones.)*

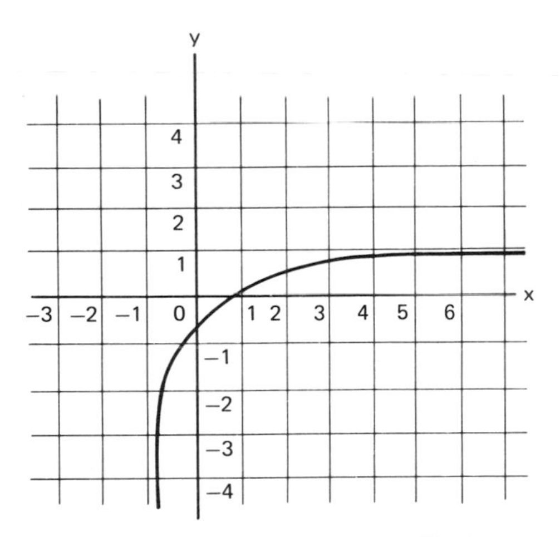

FIG. 2-11 An asymptotic curve. (The lower left portion of the curve approaches but never reaches the line $x = -1$. The right portion of the curve approaches but never reaches $y = 1$.)

2.3.5 Vectors

Numbers usually merely state a quantity. They rarely tell us anything about that quantity. But there is one sort of number or variable that tells something. It tells that the quantity has direction. Such a number is a *vector*.

Take the ordinary (nonvector) number 157. That is all it means—157. We can apply units to it like pounds or volts, as in 157 lb and 157 V. But what if we were talking about miles? Let us say the distance from Peoria to Chicago is 157 miles—*and we want to make a point of the fact that the distance is north.* We do that by converting the ordinary number 157 to the vector number 157.

Vectors are indicated by using boldface (**157**), an overbar ($\overline{157}$), or an arrow ($\overrightarrow{157}$). Which one do you use? The one your author prefers.

1. Scalars

Nonvector quantities are called *scalar* quantities. In fact, all quantities are scalar to start with. Vectors are the special case.

2. Vector Algebra

When we use vectors and related concepts in algebraic operations, we are using *vector algebra.* Its notation does not differ from other algebraic notation except for the different ways of expressing vectors and scalars and some new concepts and symbols used specifically with vectors.

a. Vector expressions Each one of these tells the reader that a complex manipulation of vector quantities has taken place. We will not describe these operations except to say that each is meaningful and distinct from the others. They are

gradient (abbreviated *grad*)
divergence (abbreviated *div*)
curl (not abbreviated, use *curl*)

Abbreviations for these terms are inserted directly into an equation much in the same way that the trigonometric abbreviations such as sin, cos, and tan are used. Like these, they are always followed by a number or a mathematical expression.

NOTE: In text or in speech the engineer or scientist might say either "the gradient of *y* squared plus 19" or "the grad of *y* squared plus 19." In either case, the statement will be written

$$\text{grad}(y^2 + 19)$$

Divergence or div and curl may be used similarly.

b. Vector symbols There are some alternate terms and symbols that also may be used to designate these three terms. They are the expression *del* and the symbol ∇, known as *nabla* (nabla is the name of an ancient stringed instrument, which we might suppose had a shape similar to this symbol). It also looks like a capital Greek delta Δ upside down. Incidentally, it is not a Greek letter.

- In speech or written text del may be used instead of grad. In mathematical notation the nabla may be substituted for grad.
- In speech or in written text the term *del dot* may be substituted for the term *div*. In mathematical notation it would appear as $\nabla\cdot$, with the dot centered like a multiplication sign—a half line up.
- In speech or in written text the expression *del cross* may be used instead of *curl*. In mathematical notation, it is written $\nabla\ \times$, with a space between the del and the cross.

2.3.6 Matrix Algebra

Like vector algebra, *matrix algebra* is a subdivision of algebra which entails a particular way of presenting numbers and symbols on a printed page. Specifically, it presents them in regular arrangements of rows and columns.

Such an arrangement may be any size. Of the examples given below, the first has three rows and three columns and the others have four rows and four columns, but there could be any number of rows and columns—in which case the problem comes down to presenting them neatly on the printed page.

Determinants and Matrices

Such arrays of numbers and/or symbols are known as *determinants* and *matrices.*

Determinants are enclosed in vertical bars and matrices in brackets. Matrices may also be enclosed in parentheses.

$$\begin{vmatrix} k_1 & k_2 & k_3 \\ l_1 & l_2 & l_3 \\ m_1 & m_2 & m_3 \end{vmatrix} \qquad \begin{bmatrix} 2 & 4 & 6 & 8 \\ a_1 & a_2 & a_3 & a_4 \\ b_1 & b_2 & b_3 & b_4 \\ Q & R & S & T \end{bmatrix} \qquad \begin{pmatrix} 2 & 4 & 6 & 8 \\ a_1 & a_2 & a_3 & a_4 \\ b_1 & b_2 & b_3 & b_4 \\ Q & R & S & T \end{pmatrix}$$

Determinant Matrix Matrix

2.4 CALCULUS

With calculus, algebra plunges into the modern world. Calculus was invented by two of history's greatest minds, independently, around the same time, with neither knowing that the other was going through the same creative act hundreds of miles away. These were Sir Isaac Newton in England and Baron Gottfried Wilhelm von Leibniz in Germany. The year was approximately 1675. Not exactly modern, you say. But everything is relative. Algebra had been around for more than five centuries by then. Viewed from that perspective, the invention of calculus does bring mathematics into the beginnings of modernity.

Calculus introduces a new, more refined, more delicate way to look at functional relationships than the way described by an ordinary algebraic equation. An equation merely states a relationship. Calculus takes that statement and puts it through a number of operational steps and as a result extracts more information from it than at first was evident. So calculus is a way to uncover "hidden meanings" in an equation.

It does this by examining the variables in the equation bit by bit—all the bits being the smallest possible parts of each quantity in the equation.

Since any equation can be restated as a graph, calculus is also a new way to look at any graphed curve and to find out the deeper meanings in that curve.

Calculus is based on the concept of the *differential.* It is thus often called *differential calculus.*

2.4.1 The Differential and Differentiation

1. The Differential

The smallest part of something is so small we can't mentally conceive a picture of it. But we can say it exists and we can give it a name. It is called the *differential,* or the *differential part* of something—which means the smallest possible part of that thing.

2. The Differential Symbol *d*

The letter d is the symbol for differential or differential part.

It comes more or less from the idea of the increment (see "Δ (The Increment Sign)" in Section 2.1.1), which is a small (additional) part of something and which is symbolized by the Greek capital letter Δ. Of course, differential also starts with d, and maybe that is where the symbol came from.

3. The Whole Is the Sum of Its Differentials

If you add up all these smallest parts of a number, you come with just that, the number, since everything is equal to the sum of its parts. That adding-up process is known as *integration.* The expression that results from the process of integration is the *integral.* These concepts will be treated in detail in Section 2.4.2.

4. Differentiation

Then if the integration process is reversed and we operate on an integral to convert it back to the smallest part again, the process is called *differentiation.*

So we have two sets of words, both in noun and verb form, that are the most commonly used ones in calculus; both sets have meanings that are the reverse of each other:

Noun: *differential,* which is the opposite of *integral.*

Verb: *differentiate,* which is the opposite of *integrate.*

5. To Show Differentiation

a. The fraction of the form *dy/dx* To show that differentiation is taking place, the symbol d is placed in front of a pair of variables. They are then shown as a fraction which has a new value, obtained by a mathematical calculation which need not be explained for our purposes but for which a typical example will be useful.

Thus for the equation

$$y = 3x^2$$

differentiation may be shown by creating the ratio or fraction

$$\frac{dy}{dx}$$

which replaces the y on the left side of the equation. The right side turns out to have a value of $6x$ (by the aforementioned calculation). Thus

$$\frac{dy}{dx} = 6x$$

This process of differentiation is also called finding the *derivative of y with respect to x.*

NOTE: Use the built-up fraction in displayed mathematical notation. The slant, used with the shilling fraction, is acceptable in written text, as $dy/dx = 6x$.

b. *dy* = · · · *dx* Multiplying both sides of the equation by dx allows it to be expressed as

$$dy = 6x\ dx$$

This form of $dy = \cdots dx$ is another familiar form of expressing the differentiation operation, and you will see it frequently.

NOTE: When the differential (such as *dy* or *dx*) follows an expression, it is always separated from that expression by a space.

c. The dot There is another way to show differentiation. It is by the use of a dot over the dependent variable.

Thus

$$\dot{y} = 6x\ dx$$

means the same as the preceding differential equations

$$dy = 6x\ dx \text{ and } dy/dx = 6x.$$

d. The prime There is yet another way to show differentiation. It is by use of the prime.

Thus

$$y' = 6x \qquad \text{or} \qquad f'(x) = 6x$$

are also ways of expressing the derivative of y or $f(x)$ with respect to x.

6. Higher Differentials

It is possible to differentiate a derivative. A derivative is often used to express a rate of change, and the mathematician may want to know the rate of change of a rate of change. If differentiation is done twice to the same algebraic expression, the process is known as finding the second derivative. It is possible to go on—to the third derivative, and the fourth, and so on to higher and higher derivatives. There is no limit.

To show higher differentials the practice is to place a superscript (like an exponent) after the *dy* in the numerator and the same superscript after the *d* in the denominator. That number tells how many times the differentiation process is to take place.

Thus

$$\frac{dy}{dx} \quad \text{is the first derivative}$$

$$\frac{dy^2}{d^2x} \quad \text{is the second derivative}$$

$$\frac{dy^3}{d^3x} \quad \text{is the third derivative}$$

$$\frac{dy^4}{d^4x} \quad \text{is the fourth derivative}$$

and so on.

7. Partial Derivatives

So far we have discussed differentiation in expressions with only two variables. If there are more than two variables and differentiation is to be performed, it is necessary to designate which variables are to be differentiated (compared) with one another. This is done by employing the concept of the *partial derivative.*

When partial derivatives are to be taken, the signal is given to the reader by using the Greek lowercase delta ∂ instead of the d.

Thus we might see the equation

$$\frac{\partial y}{\partial x} = 7x - 4y^2(1 - z)$$

which expresses the partial derivative of y with respect to x.

NOTE: The character ∂ will probably not be on your keyboard. In that case you must draw it neatly with a fiber-tipped pen.

8. Higher Partial Derivatives

Just as higher derivatives may be taken (see item 6), so it is possible in equations with more than two variables to take higher partial derivatives. The same rules hold.

Thus

$\frac{\partial y^3}{\partial^3 x}$ is the third partial derivative

$\frac{\partial y^6}{\partial^6 x}$ is the sixth partial derivative

and so on.

2.4.2 Integrals and Integration

Integration occurs when the differentation process is reversed.

We differentiated $y = 3x^2$ to get the differential equation $dy = 6x\ dx$. We can integrate $6x\ dx$ to get back to $y = 3x^2$, our original equation.

1. To Show Integration

The symbol for integration—which means adding up all the differential parts—should have something to do with the concept of adding or summing. So a symbol resembling an S is used.

The symbol appears on few keyboards; it is a strange, long-waisted one:

$$\int$$

It is placed in front of an expression and it says that the process of integration will be performed. That process is the inverse of the mathematical process of differentiation.

Then

$$\int dy = \int 6x\ dx$$

when integrated gives

$$y = 3x^2$$

NOTE: A space should be left between the integral sign and any expression that follows it.

2. Multiple Integrations

Just as it is possible to repeatedly differentiate an equation, so it is possible to repeatedly integrate.

Thus we have a second, or double, integral:

$$\int \int (\tan x + 3y)\, dx\, dy$$

or a third, or triple, integral:

$$\int \int \int \left(zx^4 - 3 + \frac{1}{3y} \right) dx\, dy\, dz$$

NOTE: The integral symbol and the summation symbol Σ (see "Σ (The Summation Sign)" in Section 2.1.1) are related to each other in that both are signs of summation; also both make use of a form of the letter S. The summation sign is the Greek uppercase Σ, or sigma, and the integral sign is a modified script S. It must be remembered that they are different from each other in that the former sums a set of sequential numbers by simple addition while the latter sums by a complex algebraic operation.

3. Limits of Integration

Just as the summation sign adds up a series of known numbers—and no more than just the numbers in the given sequence—so integration will take place over a certain spread of numbers, and no more. That spread is defined by showing the extremes, or *limits,* of the integration.

The limits are written at the top and bottom of and following the integral sign.

In the following example the statement is being made that integration will take place for the variable n between the limits of 4 and 11.

$$\int_4^{11} (M - n^2)\, dn$$

2.5 STATISTICS

Statistics is the branch of mathematics that organizes data in ways that make the meaning of that data more easily understood. On the basis of those meanings, statistical mathematics may make predictions about the future.

Every scientist and engineer collects data. In order to derive the greatest value from that data, most scientists and engineers will employ statistical methods. Thus the technical secretary and technical writer must be able to deal with these methods as they may appear in text which must be reproduced, and this requires acquiring familiarity with some new terms, new concepts, new symbols, and a few new ways to draw graphs and set up tables. But for the most part the material will be merely an extension of what has already been discussed in the sections on algebra and calculus.

2.5.1 Organizing Data

1. The Table

The best way to organize data is to group them logically in a table.

Consider the example of buying a refrigerator. The buyer, whether he or she is a technical or a lay person, should ask (among other questions) what is the most economical (energy-efficient) model among the dozens that could be purchased.

The U.S. Department of Energy has determined ratings of cost of electricity for a year of normal operation for each make and model. Those ratings are the data we require for statistical analysis. They are tabulated in Table 2-7 by imposing a grouping in terms of $10 intervals of running costs. (Any interval could have been chosen, but it is evident that a $10 interval makes more sense, has more practical meaning, than $1 or $50 intervals.)

TABLE 2-7
STATISTICAL TABULATION OF REFRIGERATOR YEARLY RUNNING COSTS

Running Costs per Year with Electricity at $0.05 per kilowatt-Hour, $	*Frequency, Number of Refrigerators in Each Category*	*Percentage of Refrigerators in Each Category*
40–50	1	2
50–60	5	10
60–70	18	36
70–80	14	28
80–90	9	18
90–100	3	6
	50	100%

The techniques of descriptive statistics now require that we count up the number of refrigerators that fall into each interval (or category). They also require that we calculate the percentage of a group in each category. Table 2-7 has indeed organized the data in this manner.

One might say that the table's organization is obvious, but it allows a better presentation of the data than that offered by simply a general picture in the buyer's mind. The prospective buyer is now in a better position to do comparison shopping than before the table was put together. The technical secretary will be required to prepare similar tables in technical text.

2. Graphs

Tables are only the beginning of the statistical organization of data. Using the groupings of such tables it now becomes possible to draw graphs in a variety of forms which present the same data pictorially—and thus more clearly.

NOTE: The *procedures* for drawing graphs for statistical purposes are no different from those used generally and as described in "Graphs" and "Plotting Graphs" in Section 2.3.4.

a. Histograms *Histograms* are graphs that plot the number in each grouping against the different groupings. Histograms may be drawn as a set of vertical contiguous bars or as a smooth curve.

Figure 2-12 is a histogram of the data of Table 2-7. Both the bar and smooth curve form are shown here.

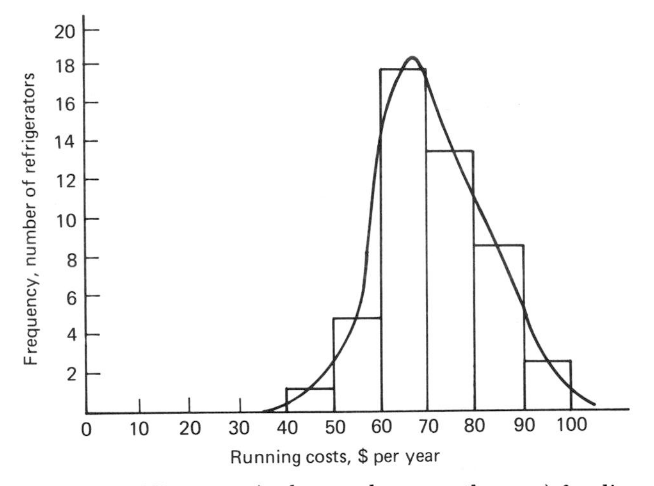

FIG. 2-12 Histogram (as bars and as smooth curve) for distribution of yearly running costs of refrigerators.

b. Bar charts These are a form of histogram. They differ only in that bars are drawn with spaces between and the bars are drawn horizontally. (See Figure 2-13.)

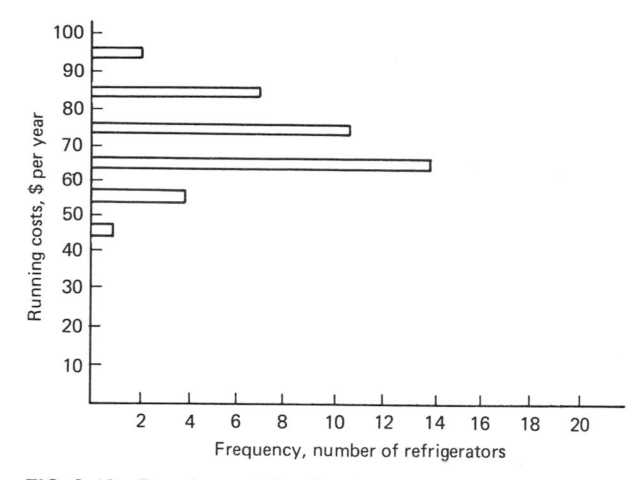

FIG. 2-13 Bar chart of distribution of yearly running costs of refrigerators.

c. Pictograms Another variation of the histogram is the *pictogram.* (See Figure 2-14.)

d. Pie charts The same information can be presented as a pie chart (Figure 2-15).

2.5.2 Analyzing Graphs of Data Distributions

The histogram is one of the most important tools for further statistical analysis. It describes how data is *distributed.* Thus it is known as a *distribution.*

1. Shape of Distributions

Data may be distributed in a *normal* or a *skewed* manner, thus affecting histogram shape.

a. The normal histogram A normal histogram is symmetrical (see Figure 2-16*a*).

This is also known as a *gaussian distribution,* after the eighteenth-century mathematician Karl Gauss.

b. The skewed histogram A *skewed* histogram is nonsymmetrical (see Figure 2-16*b*).

Whenever data takes on this form it means that it contains "oddballs." For example, a group of men with a normal distribution of height would

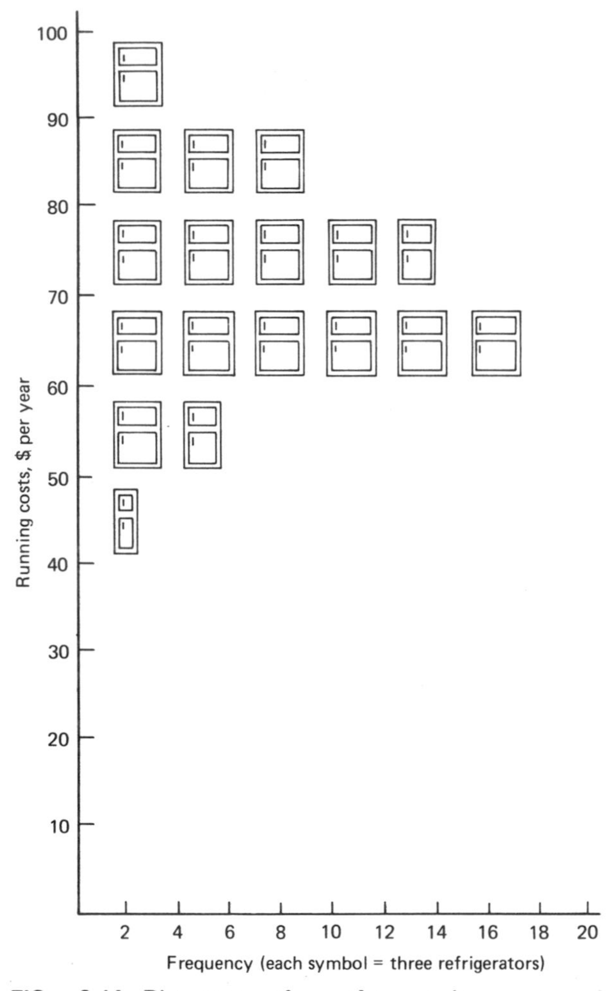

FIG. 2-14 Pictogram of yearly running costs of refrigerators.

probably range between 5 feet 3 inches and 6 feet 3 inches tall. But the presence of a few tall men at 7 and 8 feet would cause the histogram to have an extended "tail" at the high end—and to be skewed.

2. Mode, Median, and Mean

There are three mathematical quantities that can be calculated and ascribed to any distribution of data that has been graphed as a histogram. These are mode, median, and mean.

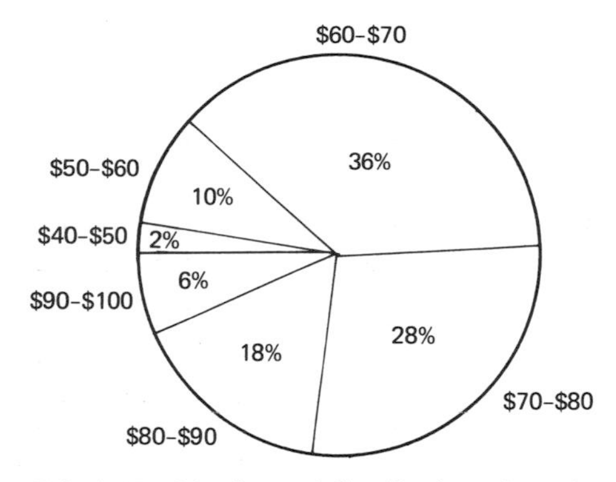

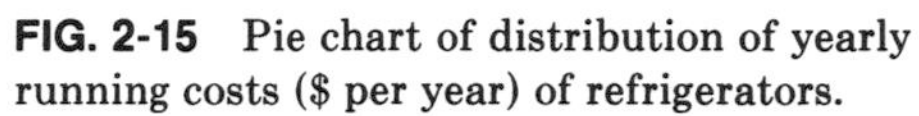

FIG. 2-15 Pie chart of distribution of yearly running costs ($ per year) of refrigerators.

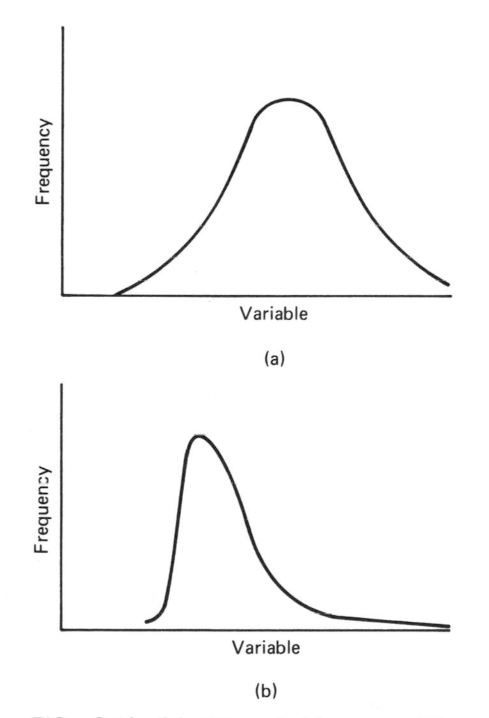

FIG. 2-16 (*a*) Normal histogram. (*b*) Skewed histogram.

a. Mode This is the value occurring most frequently in a given set of data.

In the refrigerator case, as shown in Table 2-7 and Figure 2-12, the mode is the $60 to $70 category.

There may be more than one mode in a single histogram. Such an occurrence usually means that more than one population is being examined. When there are two modes the histogram is called *bimodal.*

In the refrigerator case, suppose the survey included half 14-cubic-foot refrigerators and half 22-cubic-foot refrigerators. Each size would have its own mode, where the largest number of costs of yearly electricity would occur. Graphed on a single histogram, two modal peaks would appear (Figure 2-17).

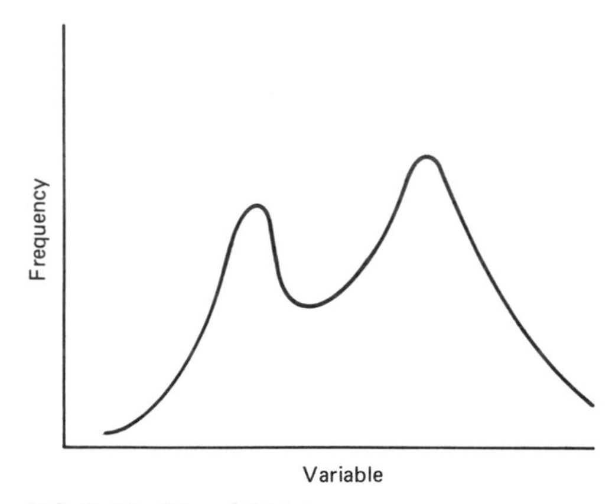

FIG. 2-17 Bimodal histogram.

b. Median This is the exact center of the distribution. It is the fiftieth *percentile.* Half the units are on each side of the distribution. On a histogram drawn as a curve, the median is the line which divides the distribution into two parts of equal area.

In the refrigerator case (Figure 2-12), the median is at about the $70 point.

c. Mean This is the average. Averages are designated by a symbol with an overbar. An average is computed by adding the numerical values of a given data set and dividing by the number of items in the set.

For example, consider buying six boxes of strawberries of different size fruit (but the same size box). If the boxes have counts of 45, 34, 55, 61, 21, and 33 strawberries, respectively, the average number of strawberries per

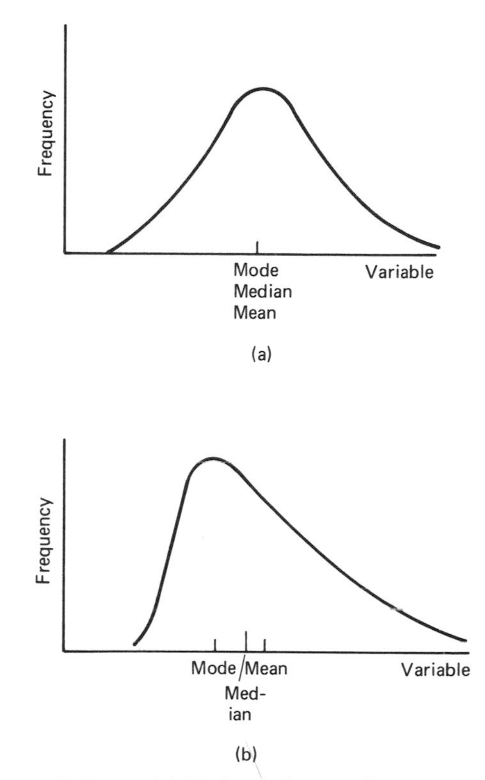

FIG. 2-18 (*a*) Mode, median, and mean for a normal distribution. (*b*) Mode, median, and mean for a skewed distribution.

box will be 41.5 (the sum divided by 6). And it will be written in mathematical text as $\overline{41.5}$.

Figure 2-18*a* shows that for a symmetrical (or gaussian) distribution of data, the mode, median, and mean are at the same point.

Figure 2-18*b* shows that for a skewed distribution of data, mode, median, and mean occur at three different locations and values.

3. Standard Deviation

For any one item located anyplace in a distribution, it is likely that its value will not be the same as the mean value. The difference is the *deviation from the mean.*

In Figure 2-19 the deviation from the mean is shown for a refrigerator that has a running cost of $56. That is $18.60 less than the mean value of $74.60. The deviation from the mean is thus $18.60.

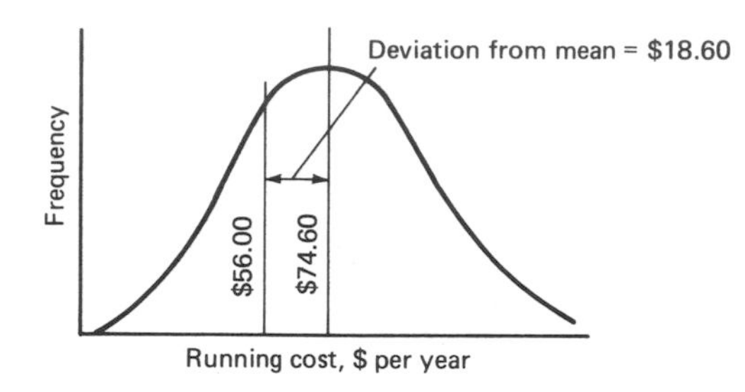

FIG. 2-19 Deviation from mean.

There is a particular defined value of deviation from the mean that characterizes the narrowness or the spread of the distribution of data. The formula that produces the value of this defined deviation will not be given here. The quantity is known as the *standard deviation.* A standard deviation can be calculated for every histogram. It is symbolized by the lowercase Greek sigma, σ.

In Figure 2-20, it is evident that for the broad histogram, curve *b,* which has widely spreading values, the standard deviation is a large number, and that for the narrow histogram, curve *a,* which has a narrow spread of values, the standard deviation is a small number.

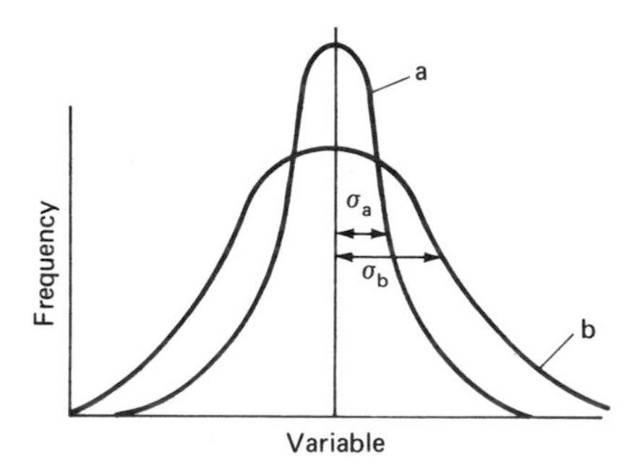

FIG. 2-20 Standard deviations for two normal distributions.

4. Probabilities

How "sure" is something to happen which hasn't happened yet? Some things are more likely to happen than others. The statistician assigns the number 1 to something that "will" happen. But he or she can't be sure. Unless we know something *must* happen, the best that can be said is that something is so likely that its *probability* of occurring is close to 1. Maybe .99. When something is sure not to happen, it has a probability of zero. But we can't be sure—so perhaps a more realistic probability would be .01. Probabilities always have a value between 0 and 1. If the probability is close to 1, the event is likely to happen; if it is close to zero, the event is unlikely to happen; if it is in between, the event is proportionately more or less likely to happen. The value of this kind of designation of probability is that it is quantitative.

When we toss a coin hundreds of times, the probability that it will come up heads is about .5. The probability that it will come up tails is also about .5. The probability that it will come up either heads or tails is 1 (which is the sum of the two probabilities—.5 + .5). The probability that it will come up tails a hundred times in a row is close to 0.

a. Probability of combined events The statistician often wants to know the likelihood of certain events occurring together. That probability can be calculated from the data about each event considered separately and together.

Suppose you are going to buy a shirt. Let us assume that the data says that 34 percent of the time you buy red shirts and 18 percent of the time you buy green shirts. Then the likelihood is that you will buy either a red or green shirt the next time you shop for one, with a probability of .34 + .18, or .52. (And that means that .48 of the time—48 percent of the time—you will buy a shirt that is neither red nor green.)

NOTE: In the field of probabilities, it is customary to use these decimal expressions, like .52, without introductory ciphers, which are usually used for all other decimals (see "Ciphers" in Section 2.1.3).

Probabilities are expressed by symbols for each event, with P to designate each probability and with the symbol $\cup$, which is a large, hand-drawn U to indicate *union* (or combination).

Thus, if A_r is the event of buying a red shirt and A_g is the event of buying a green shirt,

$$P(A_r \cup A_g) = P(A_r) + P(A_g) = .34 + .18 = .52$$

There is also a way to show events that do not add up using the $\cup$ upside down, $\cap$.

Consider the case of 100 sports fans buying season tickets for baseball and hockey in a particular city with major league teams for both sports. Previous statistics tell us that 72 percent of these people will buy baseball tickets and 55 percent will buy hockey tickets. Those are the probabilities for this year too. But they add to a probability of more than 1. It must be that some people buy tickets to both.

Thus, if A_b is the event of buying baseball season tickets and A_h is the event of buying hockey season tickets,

$$\begin{aligned} P(A_b \cup A_h) &= P(A_b) + P(A_h) - P(A_b \cap A_h) \\ &= .72 + .55 - .27 \\ &= 1.0 \end{aligned}$$

The fact is that 27 percent of the buyers will buy season tickets to both sporting events, and so we subtract that number in order not to count some buyer twice. Since everyone buys something, the probability of any sale is 1.

b. Conditional probabilities Sometimes the probability of an event is desired, but only under certain conditions. These conditions restrict the available "space" in which the event may occur. Such conditional probabilities are designated by a notation in the form of

$$P(A \mid S)$$

NOTE: While the vertical line which must be drawn, is more commonly seen, the slant is also acceptable:

$$P(A/S)$$

Suppose there are 60 job applicants for a single job, but there is a requirement that persons applying have accounting experience. If 12 have such experience, the chance to get that job is, for those 12, 1 in 12 ($P = 0.083$) rather than 1 in 60 ($P = 0.017$). But suppose there be a new restriction—that the person have a college degree in business. Then the probability of getting the job, $P(J)$, has a qualifier: it is D, the requirement for a degree. Thus if only 4 job candidates (of the original 60) have both accounting experience and a degree, the odds for any of these 4, and only these 4, is now 1 in 4 or 0.25.

Thus

$$P(J \mid D) = 0.25$$

It now becomes evident that job selection has become restricted. Originally there were 60 people who would be considered, then 12; now there are only 4.

NOTE: The numbers of a group for whom an event is to be considered in terms of its probability is known as its *universe.* The original universe in the example was 60. The universe was later restricted to 12 and finally to only 4.

Basic Practice—Techniques and References

2.6 MATHEMATICAL NOTATION

This section describes how to write (type, draw, arrange, space) the many symbols and abbreviations described in the first part of this chapter. It also shows how to present mathematical operations and constructions such as equations, fractions, exponents, primes, and subscripts. Accepted practices for depicting algebraic, trigonometric, and statistical material are presented, as are procedures employed in calculus. Procedures for drawing simple figures and graphs are also shown.

The reader is advised to refer back to the first part of this chapter for further explanations and for additional examples and illustrations. The reader is also advised to make reference to the glossaries, particularly those that deal with other alphabets, and with mathematical signs and symbols.

2.6.1 Numbers

When typing numbers, always be sure to use the number key (1) and not the letter key (l), the lowercase "ell." They are (except with a few typefaces) different symbols and that difference should be respected.

NOTE: Some typewriters do not have separate keys for the number key and the letter key. In that case, there is no choice but to use the letter key—the "ell" key. Differentiate them by labels for the typesetter.

1. Additional Symbols

Since many symbols (including ∞, π, Δ, Σ, and $\surd$) are not on most keyboards you will have to find insertion keys or special ball fonts or daisy wheels. Some are also available in templates which will aid in drawing them. If none of these are available, you will have to learn to draw them freehand with a fine-point, black fiber-tipped pen. Instructions on how to do this are given.

a. Drawing the infinity symbol ∞ The infinity symbol looks like the number 8 lying on its side. If you do not have it on your keyboard (or insertion key or ball font or daisy wheel), you can construct it by placing two o's close together—like this:

oo

Or you may draw it with a fiber-tipped pen.

∞

But be careful. Draw it neatly, to look the same every time.

b. Drawing pi π If you do not have a key for π, draw it neatly with a fine-point fiber-tipped pen.

π

c. Drawing the increment symbol Δ The symbol is the same for both the increment or the decrement. It is the Greek capital delta Δ (see Glossary A).

As for all symbols described in this section, if you do not have a key for this symbol, you must draw it freehand—neatly—or construct it neatly any way you think makes sense. Some people like to use the underline key for the base of this symbol:

—

Then they draw in the two sloping lines to make:

Δ

Should you use a ruler? Not necessarily—only if it looks neater that way.

NOTE: In dictation, the author will say "delta S," and the secretary will write ΔS.

This should be written with no space between the Δ and the mathematical expression which follows it. Thus

INCORRECT: $\Delta\ S$

CORRECT: ΔS

d. Drawing the sigma Σ If there is a sigma on your keyboard, use it. Ordinarily you won't have one, so as with most Greek letters, you must learn to draw it with a fine-point, black felt-tipped pen. Usually the symbol will occur several times on the same page. Try to make them all look alike.

NOTE: There is an issue of size. When the summation sign applies to an expression that is large, it should be drawn large too—so that it "fits." How do you decide how large to make it? By using good judgment. Here are two examples.

$$\sum_{x=1}^{x=20} \frac{x^2 + 4xy + 32}{7(x - 5)} \qquad \sqrt{\frac{\Sigma f(M - u)}{N}}$$

There are templates which are commercially available. These provide sigmas of different sizes: If you use this method, just select the right one and, using the fiber-tipped pen, let the hole pattern in the template be your guide.

Care should be taken to place the limits of summation neatly on top and bottom of the summation sign. They should be centered and they should be close to the sign (within the next line space up and down). Also, it is advisable to type the small equations which express these limits of summation without spaces on either side of the equals sign, because space is limited and because this is the way these equations will be set into type. On the other hand, for ordinary equations, spaces are always included on either side of the equals sign.

NOTE: How are summations "said"—as, for example, in dictation?

The scientist might say, "sigma of *M* between the limits of 4 and 7," and the secretary will write

$$\sum_{4}^{7} M \quad \text{or} \quad \sum_{M=4}^{M=7} M \quad \text{or} \quad \sum_{M=4}^{7} M$$

e. Drawing the factorial symbol !

NOTE: When you hear this symbol verbally, as in dictation, you may hear it with the word *factorial* given either before or after the number to which it applies. Thus you may hear "eight factorial" or "factorial eight." But in either case you will write the symbol after the number, like this:

$$8!$$

f. Drawing the Root or Radical Symbol $\sqrt{}$ There are two ways to write the root sign.

- You can construct this symbol by using the slash /. Then with a fiber-tipped pen add the remaining two lines.
- But it is probably neater to draw the whole symbol freehand with the fiber-tipped pen. Templates are available to aid in this drawing process.

Sometimes the expression under the root sign is quite long. In that case you just draw the horizontal line long enough to accommodate it. Here, one has no choice; the only good way to draw a long line is with a ruler.

Draw tall expressions to fit. Here again, you are well advised to use a ruler.

$$\sqrt{\frac{KA + 3X^2 - 56}{(1 - 2y)[(K - 2)/3]}}$$

g. Alphabetical symbols You must become familiar with all foreign alphabet symbols which authors use. When they are not on your keyboard, you will have to learn to write them—neatly. These are listed in Glossary A.

2. Fractions

a. Simple fractions Fractions are written in two ways. They may be built up, with numerator on the top and denominator on the bottom and a neatly spaced line (fraction bar) between.

$$\frac{6}{1 + 3y}$$

Or they may be written on a single line in the shilling style, with the fraction bar replaced by a slant bar, as $6/(1 + 3y)$.

NOTE: Fractions written on a single line are known as *shilling* fractions.

In shilling fractions, when more than one term appears in the numerator and denominator, either it must be set off by use of grouping symbols, or signs of aggregation. Parentheses are one such sign.

The choice of which form to use lies with the author. The secretary is advised not to improvise here. But the built-up format is more frequently employed in sections with displayed mathematical material, and the shilling format is more frequently employed within lines of narrative text.

Example:

Pressure in the fuel chamber varies as $T/(V - AB)$.

or

Pressure in the fuel chamber varies as

$$\frac{T}{V - AB}$$

b. Complex fractions You must write complex fractions clearly and with no danger of ambiguity. There should be no mixing up of what symbols are on top and what symbols are on the bottom.

In typing complex fractions, allow one line for the fraction bar and space a full line up for the top expression (the numerator) and a full line down for the bottom expression (the denominator). But use common sense. Small fractions in the numerator or denominator are best represented as shilling fractions. If fractions in the numerator or denominator are many and complicated, then use the built-up style (see sample page, Figure 2.21).

Example

$$\frac{zt^2 - 34Y + (6x/T)L - z}{y - 14/x}$$

c. Ratios There are three ways to show ratios.

These are:

- 3 to 5
- 3/5
- 3:5 (with no spaces around the colon)

Any is correct as long as it is used consistently.

3. Decimals

a. Ciphers Whenever a decimal stands alone (without a whole number to the left of the decimal point), insert a cipher, unless the decimal represents a probability, in which case the cipher is optional. The cipher is a way to keep the reader from overlooking the decimal point.

NOTE: Zero is a number, not a letter. Most typewriters have a separate character for zero. If your typewriter does not, always use an uppercase (capital) "oh." O is correct; o is incorrect. Label it "zero" for the typesetter.

b. Point In dictation authors usually say the word *point* when they mean *decimal point*.

You hear, "Microwave radiation is emitted at a frequency of two point four five gigahertz," and you type

2.45 GHz

NOTE: Do not confuse the decimal point with the geometric position point (see "Point" in Section 2.2.1).

c. Location of the point For decimal digits to the right side of the decimal point, commas are never inserted:

16,984,000.3756256

However, the SI system (Section 2.1.6) recognizes that confusion can result from decimal numbers with many digits that are not broken up. Thus spaces may be inserted every three decimal digits, counting from left to right beginning at the decimal point.

16,984,000.375 625 6

4. The Exponent

The exponent is always written a half line up, as a superscript.

It helps to write exponents in small type if your keyboard has such numbers; otherwise you have no choice but to use ordinary numbers and letters.

Some exponents are very complex, and the exponential expressions may be more complicated than the base expressions to which they refer. In such a case there is no substitute for care and neatness. And sometimes handwriting the expression is the only way to ensure that there is no confusion.

Example:

$$z = K\theta^{2(x-3y)} \qquad \text{or} \qquad z = K\theta^{2(x-3y)}$$

5. Logarithmic Expressions

a. Log and designation of base The rule for mathematical notation is that the base is written as a subscript after log with no space between. But the number of which the logarithm is being taken is written at the same level and after a single space.

The logarithm of 14 to the base 3 is written

$$\log_3 14$$

NOTE: When the base is not written, it is always understood to be 10.

Thus

$$\log 17 = 1.2304$$

is the same as

$$\log_{10} 17 = 1.2304$$

Either usage is acceptable, but the common practice is to leave out the base when it is 10.

b. Ln, the natural base When the natural base e is called for, accepted practice is to use ln rather than $\log_e$.

Thus the natural logarithm of 17 is written

$$\ln 17$$

and is pronounced "lynn 17" or "ell-en 17."

c. Abbreviations

- For logarithm: log
- For antilogarithm: antilog
- For natural logarithm: ln
- For anti-natural logarithm: antiln

d. Be careful Your task is to present logarithmic expressions so clearly that the reader will have no doubt as to which symbol is the base and which is the number for which the logarithm is being taken. There will be no ambiguity if the base is written as a subscript and if spaces are used—or not used—as outlined above.

6. The SI Prefixes and Symbols

This book attempts to adhere to the International System (SI) of weights and measures. Table 2-3 gives the SI prefixes and symbols to be used in written text to represent numbers.

7. Primes

Primes are written with apostrophes.

CAUTION: Primes, being superscripts, could be confused with exponents, which are also superscripts. The technical person will not confuse N''' with N^3. But N'^3 could be confusing; it could look like N to the thirteenth power. When there is danger of miscommunicating like that, the best thing is to use parentheses:

$$(N')^3$$

NOTE: Primes, being superscripts, should be lined up with subscripts, should any be present. (See example, Section 2.1.7.)

$$M'_{a,b}$$

8. Subscripts

With multiple subscripts, the preferred method is to set off higher-level subscripts with commas:

$$M_{a,b} \quad \text{is better than} \quad M_{a_b}$$

(See example, Section 2.1.7.)

2.6.2 Geometry and Trigonometry

1. Drawing Angles

You will often be called upon to draw angles within the text and to designate them with symbols, usually Greek letters or sets of Latin letters. Figures 2-2 to 2-5 show some angles. They should be drawn neatly, with a fine fiber-tipped pen. If the illustrator in the publications department does not draw them, the typist should. Notice that it is often common practice to letter the ends of the lines. Then the reader and the writer can conveniently specify which angles are being dealt with.

NOTE: If it is important to show exact angle sizes, a protractor may be used as an aid to drawing angles.

2. Degrees

Some keyboards have the degree sign (a small, raised circle). If it is not available, use the lowercase o, typed as a superscript.

3. Minutes and Seconds

For the symbol for the minute, use an apostrophe. For the second, use a quotation mark.

4. Geometric Figures

Table 2-5 provides an extensive listing of most of the shapes you will encounter.

5. The Trigonometric Ratios

The abbreviations sin, cos, tan, csc, sec, and cot occur frequently in mathematical text. They are generally followed by a symbol that designates an angle, usually a Greek letter, but a set of three uppercase letters may be used as well. The rule is to leave a space between these abbreviations and the angle. (See Table 2-6 and sample page, Figure 2-21).

When a three-letter set is used for an angle it is common to place a hand-drawn arc over them.

$$\tan \widehat{ABC}$$

But ABC is equally acceptable:

$$\tan ABC$$

There is even a fourth way. The letters can be underlined:

$$\tan \underline{\mathrm{ABC}}$$

When these trigonometric ratios are raised to a power, the rule is that the exponent comes just after the three-letter trigonometric abbreviation and before the angle designation—never after.

INCORRECT: $\cos \alpha^2$

CORRECT: $\cos^2 \alpha$

6. The Perpendicular Sign ⊥

This sign is best written by hand-drawing it with a straightedge and a fine-point felt-tipped pen.

$$\perp$$

NOTE: The base of this symbol is the length of an underline, and such a line may be used.

2.6.3 Algebra

1. The Plus-or-Minus ±

The best way to type the ± is in two steps. First input the +, then backspace once and adjust the vertical setting of the hyphen so that when

Expressed in terms of the ratios introduced in Eq. (3.6.7), this is

$$\alpha' = \frac{a'}{d'} = \frac{\alpha - \delta\left(\frac{1}{f_1} + \frac{1}{f_2}\right)}{1 - 2\alpha\left(\frac{1}{f_1} + \frac{1}{f_2}\right) + 2\beta\left(\frac{1}{f_1} - \frac{1}{f_2}\right) + \frac{4}{f_1 f_2}\delta}$$

$$\beta' = \frac{\beta + \delta\left(\frac{1}{f_2} - \frac{1}{f_1}\right)}{1 - 2\alpha\left(\frac{1}{f_1} + \frac{1}{f_2}\right) + 2\beta\left(\frac{1}{f_1} - \frac{1}{f_2}\right) + \frac{4}{f_1 f_2}\delta}$$

(3.6.20)

$$\gamma' = \frac{\gamma}{1 - 2\alpha\left(\frac{1}{f_1} + \frac{1}{f_2}\right) + 2\beta\left(\frac{1}{f_1} - \frac{1}{f_2}\right) + \frac{4}{f_1 f_2}\delta}$$

$$\delta' = \frac{1}{d'} = \frac{\delta}{1 - 2\alpha\left(\frac{1}{f_1} + \frac{1}{f_2}\right) + 2\beta\left(\frac{1}{f_1} - \frac{1}{f_2}\right) + \frac{4}{f_1 f_2}\delta}$$

Microwave energy transmitted can now be expressed as a waveform obtained by an image rotation about the y axis. This is done in the direction of decreasing θ by noting that the primed wave at $\theta + \delta$ is equal to unprimed, i.e., $b' \cos(2\theta + 2\delta) = b \cos 2\theta + c \sin 2\theta$, where

$$a' = a$$

$$b' = b \cos 2\rho - c \sin 2\rho$$

$$c' = b \sin 2\rho + c \cos 2\rho \qquad (3.6.21)$$

$$d' = d$$

FIG. 2-21 Sample page, mathematical text. *(Courtesy Raytheon Co., Lexington, Mass.)*

printed it just touches the bottom tip of the +. It amounts to setting the hyphen key a quarter space down—if there were such a thing as a quarter space. Practice first so that you are sure you can get it right every time before you commit to final copy.

2. The Dot and the X as Multiplication Symbols

When the dot is used to indicate multiplication, type it one-half space up and separate it from the factors which are being multiplied by a space on each side. The same spacing rule holds for the X as multiplying symbol.

$$9.3Y \cdot (1 - A)x^2$$
$$9.3Y \times (1 - A)x^2$$

3. Parentheses and Other Signs of Aggregation

These include parentheses, ( ), brackets [ ], and braces { }.

Draw these signs as large as they have to be in order to include what they have to include. Templates are available and can be helpful, and some keyboards have them in various sizes. (See example, Section 2.11.6, and Figure 2-21.)

4. The Equals Sign and Equations

Always leave one space on each side of the equals sign and other signs of operation. Otherwise the equation is difficult to read.

INCORRECT: $X^2{=}3(M{-}9)$

CORRECT: $X^2 = 3(M - 9)$

Always center an equation neatly on a page. Always line up the equals sign next to the fraction bar.

$$\sin \alpha = \frac{A^2 + 2AB(1 - Y)}{\sqrt{R_1 R_2}}$$

Always line up equals signs in a series of related equations. This is illustrated in the sample page, Figure 2-21.

5. The Family of Equals Signs

A list of the symbols that designate various subtle variations of the equals sign and short definitions is given in Glossary G.

Some of these symbols are on your keyboard, although it is more likely that only the equals sign is there.

- One of them can be constructed: $\neq$. It can be made by overprinting a slash on an equals sign.

NOTE: While you can overprint with a typewriter, you usually cannot with a word processor. This is one of the few instances where word processors are less flexible than typewriters.

- We do not recommend constructing the $\equiv$ sign, however, since the weight of the hyphen placed over the equals sign is never the same as the weight of the equals sign. The best way to handle this symbol is to draw all three lines of it with a fiber-tipped pen.
- The same goes for all the other symbols in this glossary. If they are not on your keyboard, draw them. And as always, do it neatly and with good

spacing and in a way that leaves no doubt in the mind of the reader what symbol is being presented.

6. Plotting Graphs

Simple graphs are entirely within the capability of the technical secretary or the technical writer since they demand only some common sense. Use the ruler and a few simple drafting tools (see "Drafting Tools" in Section 1.2.1 and "Plotting Graphs" in Section 2.3.4). Graphs can become very complex and critical, however, in which case they must be given to an experienced drafter or graphics specialist.

French curves are good aids to drawing curved graphs. They are used to connect the points that have been plotted from a table of values (see Figure 2-9). It takes some practice, but after a while you will find it easy. The trick is to slide the French curve along from point to point, all the time drawing a smooth curve.

The examples of Section 2.3.4 should be helpful in guiding you in the drawing of typical graphs.

7. Writing Vectors

There are several ways to indicate that a quantity is a vector. The preferred practice is to use boldface, and that is what will be employed in published books and articles and for manuscripts (Section 1.3). Since most keyboards do not have boldface type, you can indicate that boldface is required by hand-drawing a wavy line under the vector quantity.

If we designate the distance from Peoria to Chicago as M, then it would be correct to show the vector as

$$\underset{\sim}{\mathrm{M}} = 157$$

From this a typesetter would know to set $\mathbf{M} = 157$. However, the secretary or technical writer can also make the boldface indication by drawing it freehand, tracing the printed symbol with a fiber-tipped pen.

$$\mathbf{M} = 157$$

Drawing the boldface symbol is the preferred method for internal reports and memos, but there are two other ways to indicate that a quantity is a vector. They are the overbar and the arrow.

$$\overline{M} = 157$$
$$\vec{M} = 157$$

Usage depends on the author's preference or that of a journal for which the secretary or editor is preparing a manuscript. (See Section 1.3.)

NOTE: Your keyboard will probably not have an arrow. This means that you must go over the text after you have written it and draw in the arrows. You are now presented with a danger. Messy, inconsistent arrows will destroy the appearance of an otherwise neatly typed page—try to have them look like each other. And draw them straight, being sure that the arrow extends completely over the vector quantity.

Some examples:

BAD: $\vec{M} + 4Y^2 + \vec{K} = 3(1 - \vec{x})^4$

GOOD: $\vec{M} + 4Y^2 + \vec{K} = 3(\overrightarrow{1 - x})^4$

8. Vector Expressions and Symbols

The vector designations grad, div, and curl resemble the trigonometric designations (sin, cos, tan, etc.), in the manner in which they are used; that is, they are always followed by a mathematical expression and separated from that expression by a space. For example,

$$\operatorname{div}\ (A - 2z^3)$$

Vector symbols are hand-drawn.
The nabla resembles an upside-down delta.

$$\nabla$$

The del dot is a nabla followed by a dot one half line up

$$\nabla\cdot$$

The del cross is a nabla followed by a $\times$ and separated by a space.

$$\nabla \times$$

9. Drawing Matrix Arrays

If your keyboard does not have outside brackets or parentheses, draw them with a fiber-tipped pen. Use templates if they offer correct sizes.

CAUTION: As always, there is no substitute for neatness and logic or common sense. Lay out the matrix so that it is easy to read, with adequate spaces between columns and rows. Here special care must be taken during the actual typing not to let any of the columns or rows slip into positions where the reader might think they belong to the wrong column or row. You must pay attention to each row and column, making sure that they all fit together—horizontally and vertically—in the correct sequence and orientation.

Matrix arrays are just that—arrays of numbers and symbols. These arrays may be enclosed in vertical bars, brackets, or parentheses, according to how the author uses them.

Some examples are

$$\begin{vmatrix} k_1 & k_2 & k_3 \\ l_1 & l_2 & l_3 \\ m_1 & m_2 & m_3 \end{vmatrix} \qquad \begin{bmatrix} 2 & 4 & 6 & 8 \\ a_1 & a_2 & a_3 & a_4 \\ b_1 & b_2 & b_3 & b_4 \\ Q & R & S & T \end{bmatrix} \qquad \begin{pmatrix} 2 & 4 & 6 & 8 \\ a_1 & a_2 & a_3 & a_4 \\ b_1 & b_2 & b_3 & b_4 \\ Q & R & S & T \end{pmatrix}$$

2.6.4 Calculus

1. Differentials

Differentials are always made up of a *d* followed by an algebraic symbol, such as dx, dy, and dz. When they appear in mathematical expressions, they are always set off from the portion of the expression that precedes them by a space.

$$ds = \frac{2}{\sqrt{\pi a}} \, dt$$

Differentials are sometimes designated by a dot positioned exactly over the algebraic symbol. Thus

$\dot{s}$ is the same as ds

Partial differentials are designated by a Greek lowercase delta ∂, which is missing from most keyboards. Learn to draw it neatly, freehand. The preceding example, expressed as a partial differential equation, is

$$\partial s = \frac{2}{\sqrt{\pi a}} \, \partial t$$

2. The Integration Sign

This sign is in the shape of an elongated S $\int$.

You may draw it freehand with a fiber-tipped pen or with the aid of a template. It may be of different lengths, depending on the size of the mathematical expression it introduces.

Some keyboards have this symbol in several different lengths. And such flexibility is required, since some expressions are quite tall.

$$\int \frac{dx}{\sqrt{1 - x^2}} = \sin^{-1} x$$

$$\textstyle\int a \, dx = ax$$

CAUTION: Do not use outsize integral symbols unless you have to. However, a page of text will always look better when all the integral signs are the same size.

Multiple integral expressions will occur where a number of integral signs will appear together in a single expression. They are to be separated by single spaces. (See examples in "Multiple Integration" in Section 2.4.2.)

3. Limits of Integration

NOTE: the limits are written at the same level as the top and bottom of the integral sign—not over and under the top and bottom.

INCORRECT: $$\int\limits_{4}^{\sin(x2-3)} \int\limits_{a}^{9} (A - yx)\, dy\, dx$$

CORRECT: $$\int_{4}^{\sin(x2-3)} \int_{a}^{9} (A - yx)\, dy\, dx$$

Also, limits are always written to the right of the integral sign.

INCORRECT: $${}^{A}_{B}\int$$

CORRECT: $$\int_{B}^{A}$$

CAUTION: Since any algebraic expression may be a limit (it doesn't have to be a simple number), you must be careful to type the limits exactly in the right position at the top and bottom of the integral sign (which you have drawn by hand) so that there is no doubt in the mind of the reader that these are indeed limits and not part of the expression that follows the integral sign. Also, spacing on the page must be generous enough so that these limits do not look as though they belong to the preceding or the following mathematical expression.

2.6.5 Statistics

1. Data is Plural

Data is the plural form of datum, although data may be either singular or plural in construction.

Thus

This datum—CORRECT
These data—CORRECT
This data—CORRECT

NOTE: Although data is a plural noun, it is becoming more and more common to see it used as a singular noun in construction, especially in the computer field. Thus you might hear "Data is inputted," and this is perfectly acceptable and correct.

2. Graphs

Bar charts, pictograms, histograms, and pie charts are normally produced by a graphics department, but the secretary is often asked to prepare them if they are not too complex. Success will result from commonsense use of a straightedge, a felt-tipped pen, and a french curve. (See "Graphs" and "Plotting Graphs" in Section 2.3.4 and Figures 2-7 to 2-20.)

3. Probability Symbols

The union symbol is a large, hand-drawn U. It may appear in an equation right side up or upside down, in which case it represents an *intersection* of elements or sets:

$$\cup \qquad \cap$$

See the example in "Probabilities" in Section 2.5.2.

The *conditional probability* symbol is a vertical line, hand-drawn with a straightedge slightly longer than the height of an uppercase letter. For example:

$$P(K \mid L)$$

Computers

Basic Principles

A *computer* is a brain tool in the same way that a hammer is a hand tool and a microscope is an eye tool.

A tool is a device that extends and expands human capability while at the same time remaining under human control.

The particular brain function which computers perform is the *handling of information.*

We feed information into a computer. It processes that information, and when we require it, the computer gives us back that information—in processed form.

This chapter describes the most important information-handling operations computers perform—such functions as arithmetic and sorting and organizing data into different groupings and arrangements.

Engineers and scientists in every field of technology use this tool extensively for handling information (because, when you come right down to it, they are all in the information business). In fact, one wonders how engineers and scientists ever made any technological advances before we had computers—and that wasn't so long ago; computers were used very little in the first half of this century.

This chapter also describes the three major functioning parts of a computer: the input devices, the central processing unit, and the output devices.

As with other subjects in this book it is necessary that certain crucial concepts be understood in order that the special ways of presenting computer-related text make sense. Understanding of computer concepts is the basis of what is called *computer literacy.*

Today's technical secretary, technical writer, and editor must be well versed in computer concepts in order to do a professional job in reproducing text by means of or related to a computer.

It should not be surprising that text dealing with information handled by and derived from computers has its own jargon, its own notation, and its own appearance. These will be described here.

3.1 COMPUTER QUALITIES—AND HUMAN QUALITIES

We, as humans, can do everything the computer can do. If this is so, why should we require its services? The answer is simply because the computer is *not* human and thus it has few of our human deficiencies.

Science fiction writers like to play with the idea that computers, having virtually no human failings, will someday "take over." But that is unlikely, also because computers are not human. They have failings of another sort—computer failings.

3.1.1 Computers Can, Humans Can't

Computers can do certain things so much better than humans that their performance is "inhuman."

- The computer can perform boring tasks endlessly. Thus it is *inhumanly efficient.*
- The computer cannot change (unless it wears out). Thus it is *inhumanly repetitious,* since it performs every particular action in exactly the same particular way. Thus
- The computer is inhumanly *accurate.*
- The computer is *inhumanly neat* when it organizes data and prints copy.

NOTE: The electrical parts of a computer can't wear out. They are made of semiconductors and metals that never move and never overheat. Only the electrons move. And when materials sit still they cannot rub against each other (which is how things wear out). In a sense, solid-state devices will last "forever," and that is inhuman too.

On the other hand, the mechanical parts of a computer—the keyboard, the tape mechanism, and the printer—will some day wear out because they are made of parts that do rub against each other.

- The computer can act with the speed of electrons or with the speed of light (when fiber optics are used). Thus it is *inhumanly fast.*

3.1.2 Humans Can, Computers Can't

On the other hand, computers are incapable of doing certain thought functions which the human brain does all the time.

- The computer *cannot exercise judgment.* To the computer everything has the same significance.

It *can make choices;* for example, it can answer the question "Is 2 greater than 1?" It will choose 2, but that is not the same as making a judgment about relative size. In fact, we can program a computer to choose 1 as greater than 2.

- The computer has *no intuition.*
- The computer has *no imagination.*
- The computer has *no morality,* nor has it any immorality. It is amoral.

NOTE: However, lurking in the future is *artificial intelligence.* That means that certain judgment qualities will be imparted to computers. And some computers can "learn," profiting from the experience of previous errors, and learning improves judgment. Already, computers can judge which moves to make in the game of chess that will beat most human chess players. We suggest that even though some judgment qualities can be programmed into computers of the future, we should not panic. The overlap with humanness will be slight and vague, as well as "artificial."

3.1.3 Memory—The Storage of Information

On the other hand, there is one human quality that computers do have, and, as with speed, accuracy, efficiency, and neatness, they can be better at it than people. That is *memory.* Computers have a very good memory. Computer scientists refer to computer memory as *storage capacity.*

The idea of an inanimate device having a memory should not be upsetting. Consider that a book is a storage device for information. The book may have been printed 10 years ago, but all you have to do is open it to find the exact information on page 92 you would have found on page 92 when the volume was first published. Is not page 92 a memory in some way of the information on it? Similarly, we can argue that a phonograph record is a memory of music played long ago.

In the same way the information stored in a computer—on punched cards or on reels of tape or on a disk—is "remembered." And it can be recalled, or *called up*, when it is needed—out of the storage area of the memory of the computer.

NOTE: The memory function provided by our brains is similar. Somehow information is imprinted into certain brain cells and stored there. And when wanted, it can be remembered. But here is the difference: we are "only human," and as a result we sometimes forget. The computer is inhuman and it doesn't forget. Nor does it care how boring is the information it is asked to remember.

But it does have a limit. Each computer can remember (store) only as much data as it has storage capacity. Each computer has a different storage capacity. Where did it get that capacity? It was designed and built into it—by us. Humans are still in charge.

3.1.4 Calculators—Computers with Limited Capability

While some calculators are astonishingly broad in their capability, even approaching the performance capabilities of small computers in that they may be programmable or produce printouts, most are valuable just because they provide arithmetic processing in a small, convenient hand-held device. Such calculators do not arrange or organize data because they have little storage capacity.

Calculators have comparatively little memory. Computers do have memory, and in the case of large computers, the amount of memory, or storage capacity, is phenomenal.

3.1.5 How Computers Think—Bit by Bit

What goes through our minds when we think? Words? Pictures? Or something called *ideas?* The answer is all of these, but especially words.

1. Words

Words are not merely written symbols we see on a printed page, nor are they simply symbolic sounds we hear with our ears. They are mental symbols too. We think with words.

What goes through your mind when you decide to go down to the corner store to buy a carton of milk? Is it not the words, arranged more or less in a sentence that say, inside your mind, "I will go down to the corner store for a carton of milk"? Not only that, the words are in the English language (if that happens to be your native language).

Computers "think" with something similar to words. They do not think with pictures, nor do they think with ideas. (This is another instance in which computers are less capable than humans.)

2. The Two-Character Alphabet of Computer Words

As with ordinary words, computer words are constructed from an alphabet. But the computer alphabet is not made up of letters; rather, it consists of two digits, not 26 letters as is our alphabet. Each digit is known as a *bit* (see "Bits" in Section 3.2.1).

Even when a computer generates a picture, it is a "digital" picture. It is made up of an array of digits, or bits, constructed out of the two-character alphabet.

NOTE: The computer-created pictures we get from space of the surface of Venus and the ones a radiologist examines after a patient has been x-rayed (with digital apparatus) are called *digital pictures.*

3.2 COMPUTER LANGUAGES

Words mean nothing unless they are part of a language. The language computers understand is called *machine language.* The name makes sense since the computer is a machine.

The people who run computers are not machines. However, they are smart enough to learn and to use machine language. But that language is merely a *code* of seemingly meaningless numbers. That makes machine language awkward for most people. Thus the people who run computers have invented *bridging languages*—languages that are closer to their own (English, for example).

Bridging languages are known as *program languages.* Those who use them are *programmers.* A programmer gives a set of instructions to a computer on how to process data. Those instructions are known as a *program.*

Bridging languages—the languages with which we address computers—have names like BASIC, FORTRAN, PASCAL, and ADA. They will be described in Section 3.2.2.

So the computer operator talks to the computer in a program language which the machine can translate into its own machine language. Having done that, the computer acts (thinks) with the machine language, and in so doing performs the processing of information we desire.

After the information has been processed, it can be stored in the computer's memory or it can be given back to the person running the machine.

The form in which it is given back is one that is useful for humans. That is, the words are recognizable words—in English—and the numbers are recognizable numbers which are now arranged in some useful manner, as in tables or lists or plotted as graphs.

How is the information given back? In many ways, but usually from a printer in the form of continuous paper, folded like pleats.

It should be evident that the giving back of processed information from the machine to the human operator involves another step of translating—from machine language back to our language. The computer knows how to do that.

3.2.1 Machine Language

As with most languages, machine language is made up of an alphabet of characters which can be arranged into groupings called words.

1. Bits

Machine language has a very short alphabet. It has only two characters. Each character is known as a *bit*. The two characters are actually on-off states represented by numbers—the zero, 0, and the one, 1.

Here is a four-byte, 32-bit word constructed with this alphabet:

00100011000000111100011110010101

It is a nuisance for humans to read—but the computer understands it.

As humans, we have evolved a language based on a much richer alphabet.

With our mouths and tongues and palates we can make vowels: a, e, i, o, u. We can also make a variety of consonants, including m (used in humming) and s (used in hissing).

The electronic components in computer circuits can make only two things happen: They can let electrons flow or they can keep them from flowing. So the two characters in machine language are made to stand for *on* and *off*.

And that is all the computer can do. It can't make vowels and it can't hum and it can't hiss. Thus the character 1 means *on* and the character 0 means *off*, and that's all they mean.

a. The switch The electric component that exists for just this purpose is the switch (see "Switches" in Section 5.3.2). The switch is either open or it is closed. So it is either on or off.

NOTE: Well before we had computers we used another two-letter alphabet to communicate with on the basis of opening and closing a switch. That was the *telegraph*. The *telegraph key* was in fact the switch. There was a minor difference in that the two letters both stood for *on*. One letter was on for a short time (the *dot*) and the other letter was on for a longer time (the *dash*). With a language called *Morse code* we were able to combine dots and dashes to make ordinary English words.

b. Electronic switches—semiconductor diodes and transistors A more usable—in the sense of swifter and more flexible—kind of switch is the vacuum tube (no longer used in computers) or the semiconductor diode or transistor (see "The Vacuum Tube" and "Semiconductor Diodes and Transistors" in Section 5.3.3). These can let electrons flow (on) or stop them from flowing (off) by means of an electronic rather than a mechanical switch. In the case of diodes the process is known as *rectification* (see "Rectification" in Section 5.3.3).

Computers thus use a two-character alphabet because the electronic components in the computer circuits are admirably able to handle the signals on and off, but any more characters in the alphabet cause excessive complication.

NOTE: Since much of what the computer does is arithmetic calculation, you may be wondering about numbers. For the computer, numbers are words too. Using the two-character alphabet, the computer constructs numbers as it does words. In English we write "four" for the number 4. The number 4 in machine language is 0100.

2. Binary Nature of Bits

The two-character alphabet is known as a *binary alphabet.* The numbers are constructed based on the number 2 rather than on the customary multiples of 10. Thus the computer uses a *binary numbering* system, just as it uses a binary lettering system.

You will come across the abbreviation BCD. It means *binary-coded decimal,* and it refers to converting each digit of a decimal number into a binary number.

3. Bytes

Certain groupings of bits have been given a name by computer engineers. It is one they have invented; it did not exist before computers. It is the *byte.*

Bytes may comprise any number of bits; thus we have 2-bit and 4-bit and 6-bit bytes. But 8-bit bytes are most common.

Byte-size groupings of bits are easier for computers to manipulate and organize than groupings of random length. The more sophisticated and complex the computer, the longer are the bytes that it can handle.

Thus to call up an area of storage which might be, for example, data information listing number 2, the user of the computer looks up the program code for that area, which might be the alphanumeric DIL2. Then the user types DIL2 on the keyboard (inputs it), and the computer knows the corresponding byte, which might be 011010 (let's say), which is a word in *its* language. And the computer then gives, or *reads out,* the required storage area information on a video screen or printout, or into some other useful device for dealing with that listing, such as an external tape or disk recorder (which since it is external is known as a peripheral—see Section 3.4).

a. Alphanumerics One of the most important uses of bytes is to represent *alphanumerics.*

Alphanumerics are the characters we use every day to represent letters, numbers, and certain signs of punctuation. They include all the letters of the alphabet, the 10 numbers from 0 to 9, the comma, the period, and a few others.

TABLE 3-1
A SAMPLE OF CODING USED TO REPRESENT CHARACTERS IN A STANDARD 6-BIT CODE

Character	*Standard BCD Interchange Code*
0	00 1010
1	00 0001
2	00 0010
3	00 0011
4	00 0100
5	00 0101
6	00 0110
7	00 0111
A	11 0001
B	11 0010
C	11 0011
D	11 0100
E	11 0101

For example, the number 2 may be represented in a 6-bit byte as 000010, and the letter E may be represented as the 6-bit byte 110101 (see Table 3-1).

b. Addresses and mnemonics The user of the computer, by employing a program language, can instruct the computer to call up, or retrieve, a certain area of storage. This is done with a code in program language which is interpreted by the computer in the language to which it is accustomed, its machine language. The human just has to know the alphanumeric that represents the byte. Then the computer is said to be *byte-addressable.*

NOTE: What has just been explained is the practice of *translation* between different languages. Just as in the famous French phrase *la plume de ma tante,* which means "my aunt's pen," *plume* means "pen" and *tante* means "aunt," so DIL2 means "data information listing number 2" and, to the computer, DIL2 means 011010; the computer then proceeds to address data information listing number 2 and calls it up.

The program code which can call up any byte is known as an *address.* An address is called a *mnemonic* if it serves at the same time to remind the programmer of the meaning of the concept which the program code will call up. Mnemonic means "aid to memory."

In the preceding example, DIL2 is an address which is also a mnemonic, since DIL is an acronym for data information listing. And what better way to remind us of the number 2 than the number 2?

NOTE: There are no set rules on how mnemonics may be formed. For example:

MUL reminds us of *multiply*.

STA reminds us of *store action*.

MOV reminds us of *move*.

NOTE: Pronounce mnemonic "nemonic." The first m is silent.

c. Words As in ordinary languages, groupings of alphanumerics will form words. Thus groupings of bytes, which are representations of alphanumerics, will form computer *words*.

The more sophisticated the computer, the more words it can handle and the longer the words can be.

4. Reproducing Computer Text

The definitions of bit, byte, address, alphanumeric, etc., which are given in this section should be useful since the technical secretary and technical editor will encounter them with great frequency. But what are their implications for the person reproducing a technical report? Or preparing a technical paper from copy provided by an engineer or scientist who is using computer technology? Or who may be developing new computers? The answer is that, for the most part, the text reproducer is in luck. Most of the tedium of keyboarding address codes and other machine-language formats is given to the programmer.

With the advent of computers, a new body of people and a new discipline have been given identities. And these people, if they wish to command their computers, can do so only by typing in their commands. And they do the typing themselves. They don't ask a secretary to do it. They don't like the term typing though; they prefer *inputting* or *keyboarding*.

This means that most computer inputs, if they find their way into a report at all, are photocopied from a computer printout and used as illustrations (see Figure 3-1).

NOTE: This is not to say that the secretary is always saved from this sort of tedium. Sometimes the text will demand a typed assortment of "meaningless" bytes. The sample page of Figure 3-2 illustrates this.

3.2.2 Program Languages

Machine language (Section 3.2.1) is what the computer uses to talk to itself. *Program language* is what the programmer uses to talk to the computer (so that it may translate the program language into its native machine language).

If the computer is going to perform correctly, it must be very care-

```
ACR = $$200B LENGTH 1
PCR = $$200C LENGTH 1
IFR = $$200D LENGTH 1
IER = $$200E LENGTH 1
PADATA = $$ 200 F LENGTH 1

(* U17-6522 V.I.A. CONTROL REGISTERS *)

QBDATA = $$4000 LENGTH 1
KEYPORT1 = $$4001 LENGTH 1
QBDDR = $$4002 LENGTH 1
KEYPORTDDR = $$4003 LENGTH 1
ACR17 = $$400B LENGTH 1
PCR17 = $$400C LENGTH 1
IFR17 = $$400D LENGTH 1
IER17 = $$400E LENGTH 1
KEYPORT = $$400F LENGTH 1

ON = 0 ACTIVE LOW
OFF = 1

(* CONSTANTS *)

ZEROZ = #0 LENGTH 256
ONEZ = #1 LENGTH 256
TWOZ = #2 LENGTH 256
THREEZ = #3 LENGTH 256
.PAGE ZERO PAGE DATA AND MEMORY SPACE ALLOCATION
(* LOCATION COUNTER STARTS AT $00467 WHEN USING XMONG PROM *)

@ = $0047

DISPLAYINDEX .VAR 1 ZPAGE DATA BEING OBSERVED

TOP.PTR .VAR 1 ZPAGE DATA BEING OBSERVED

TOP.PTR .VAR1 POINTER TO TUBES [1,2,4,8]
BOT.PTR .VAR1 POINTER TO TUBES [1,2,4,8]

TOP.DELAY .VAR 1 180 LINECOUNTS AFTER FILAMENT, PLATE ON
BOT.DELAY .VAR 1 3 SECONDS TO THE T'

LINECOUNT .VAR 1 120 VAC LINE CROSSING COUNTER

TOP.LEVEL .VAR 1 TOP MAGS POWER LEVEL [0-100, . 100 = OFF]
BOT.LEVEL .VAR 1 DATA READ FROM INPUT PORTS
TOP.ACCUM .VAR 1 POWER CONTROL BUCKET
BOT.ACCUM .VAR 1
TOP.PULSE .VAR 1 ON OR OFF AS RESULT OF POWER CONTROL MODULE
BOT.PULSE .VAR 1
```

FIG. 3-1 Computer printout.

fully and painstakingly instructed. So the program language must be utterly unambiguous.

NOTE: Computers, for all their inhuman mental skills, are limited in that they believe everything they are told. Although a human, on hearing someone "misspeak," might say, "Oh, I know what you mean even though you didn't say exactly what you meant," the computer is incapable of that kind of understanding. Computers take everything they are told literally.

2.2.3.2 Privileged Instructions

These instructions may be used only in programs which run as privileged.

2.2.3.2.1 Set Virtual Address - OPCODE 2

This instruction sets the virtual address, or task number, of the CE.

Label	SETVAD		Comment

11111	000010	00000
15	10	4 0

Usage: The virtual address is specified as an integer on register A. Only the 6 LSBs of the integer are used as the virtual address.

2.2.3.2.2 Drop to Register Set Zero - OPCODE 6

This instruction is used by DOS-0 to abort an errant application program.

This instruction causes the function in progress to be aborted and the CE returns normally to operation at the level zero register set. The level zero PSW is placed in the PSW register; the PSW pointer is reset to zero; and the trace indictor (bit 2) in the PSW is cleared. All subroutine references on the higher register set levels are popped off the link stack.

Label	DROP		Comment

11111	000110	00000
15	10	4 0

2.2.3.2.3 Trace - OPCODE 10_8

This instruction enables or disables trace mode for the next lower level of program operation by modifying its PSW.

Label	TRACE	N	Comment

11111	001000	00	N	00
15	10	4 3	2	1 0

The operand, N, must be E or 1 to enable, or D or 0 to disable trace mode. The object contains bit 2 set to 1 or 0. The trace indicator (bit 2) in the next lower level PSW is modified as specified by N.

2.2.3.2.4 Set Real Time Clock - OPCODE 16_8

This instruction resets the real time clock as specified.

FIG. 3-2 Sample page—computers.

1. Some Program Languages

There are numerous computer languages, but the most commonly used are BASIC, COBOL, FORTRAN, PASCAL, and ADA (also written Basic, Cobol, Fortran, Pascal, and Ada).

a. BASIC Most programmers learn how to program with BASIC. The letters stand for beginner's all-purpose symbolic instruction code.

Since *to program* means to give commands to a computer, BASIC is a particular format used to give these commands. Other languages use other formats.

Commands in BASIC are usually given in three-part statements consisting of the following items:

- A line number
  This is a line on the printout. It designates a single statement and differentiates it from any other statement (or line).
- Type of command
  This is often a verb, such as MOVE, WRITE, LET. It designates a kind of action. It is always written in capital letters.
- Value
  This ascribes numerical values to a particular variable or describes the action the computer is to take. It is then the object of the verb.

Thus a simple BASIC program, keyboarded into a computer by the programmer, may appear as follows:

```
10 LET A = 6
20 LET B = 9
30 LET C = A + B
40 LET D = (3 * A)/4
50 END
```

where the numbers 10, 20, etc., are the line numbers, the LET and END statements designate types of commands, and the numerals to the right of the equals signs are the value statements.

NOTE: BASIC makes use of some unconventional operational symbols which are shown in Table 3-2.

b. FORTRAN and some others Each programming language listed here has features in common with BASIC insofar as fundamental approaches to simple command statements are concerned. For our pur-

TABLE 3-2
SOME SYMBOLS OF BASIC

Operation	*BASIC Symbol*
Addition	+
Subtraction	−
Multiplication	* (asterisk)
Division	/ (slash)
Exponentiation	∧ or ↑
Equality	=
Less than	<
Less than or equal to	<=
Greater than	>
Greater than or equal to	>=

poses it is necessary only to be aware of their existence and to know that each has it own special area of applicability.

- FORTRAN (*for*mula *tran*slation)
- PL/1 (*p*rogramming *l*anguage 1)
- ALGOL (*algo*rithmic *L*anguage)

c. High-level languages These have greater complexity and flexibility than BASIC, and they may be especially oriented toward the specific kind of problem to be solved or to the procedures being used. They are notable for using symbols that are closer to human understanding than the 1s and 0s which the computer finds understandable and which most humans find meaningless.

Thus high-level program languages will employ convenient letters or symbols or even English-like text. While high-level languages are inherently more complex than simple program languages, they may be easier to learn and easier to write—for the programmer. FORTRAN is a high-level language. COBOL (*co*mmon *b*usiness-*o*riented *l*anguage) is perhaps most similar to English. Two other important high-level languages are *PASCAL* and *ADA*.

TABLE 3-3
KEYWORDS OF COMPUTER LANGUAGES*

ADA	input/output
address	instruction
ALGOL	instruction set
ANSI standard	interpreter
APL	language level
assembly language	logic
BASIC	machine language
bit	mnemonic
byte	object program
character	op code
COBOL	operand
CODASYL	operation code
code	PASCAL
compiler	PL/1
D/A	RPG
FORTRAN	source program
hexadecimal	symbolic address
high-level language	

*Refer to "Dictionary of Technical Terms" in Part 2 for definitions.

NOTE:

- PASCAL was named for Blaise Pascal, seveneenth-century French mathematician.
- ADA was named for Ada Augusta, daughter of the great British poet Lord Byron. She later became Countess of Lovelace. She lived in the early nineteenth century and was a developer of the concept of binary numbers. She has been called the first programmer.

d. Compilers A program written in any program language must be translated into machine-usable code before the computer can follow its commands.

A program that performs that translating function is connected to the computer for that purpose. It is called a *compiler.*

2. Key Words of Computer Languages

Some of the most commonly encountered words and terms associated with computers and program languages are listed in Table 3-3.

These are listed in this table for reference as to their identification and format. For definitions, refer to the dictionary in Part II.

3.3 PROGRAMMING

The programmer must perform the act of programming—the act of commanding the computer as to how information is to be processed. Since, in most cases, the commands can be very complicated, the best procedure for the programmer is first to plan the flow of the program before plunging into the line-by-line steps of the program itself.

3.3.1 The Algorithm

First, the programmer must define the problem which the computer will confront. Second, the programmer must decide how to solve that problem in a step-by-step procedure.

Sometimes the problem is a simple one of organizing data: in a given sequence, in order by size, by date, or by any other kind of criterion set up by the programmer. For engineers and scientists the problem may be mathematical and thus involve a calculation. In all cases the programmer must determine the steps which, when done in order one by one, will solve that problem.

That set of steps which will solve the problem is known as the *algorithm.*

For example, the algorithm for the area of a rectangular surface in square yards (given the length l and width w in feet) is the product of l times w

divided by the number of square feet in a square yard (9). So the algorithm is

Step 1: Multiply the length in feet by the width in feet.

Step 2: Divide the product obtained in step 1 by 9.

3.3.2 The Flowchart

Knowing the algorithm to use in generating the desired program, the programmer now incorporates it into a larger sequence of events that will detail the flow of the entire program from start to finish. Most programmers like to use the *flowchart* as a planning aid for this purpose.

A flowchart makes use of a number of symbols which can be used in any program, no matter how complicated. Each symbol represents a particular program function and thus serves as a shorthand way to chart the flow of the program's action. Seven symbols are most commonly used, although there are others for particular purposes.

All the symbols are in the form of geometric shapes, or boxes, in which certain kinds of messages are inscribed (by typing them in). Each different shape is used for only one sort of message.

The seven major flowchart symbols are those for input/output, process, terminal, decision, connector, predefined process, and annotation flag.

Templates may be used to draw these symbols (see Figure 3-3).

A more comprehensive list of computer flowchart symbols may be found in Glossary L.

- The input/output symbol designates the information being read into the central processing unit (see Section 3.4.2) of the computer or the information being printed out.

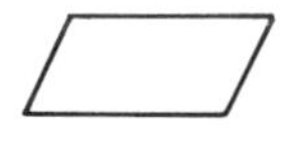

- The *process* symbol designates any operation performed by the computer.

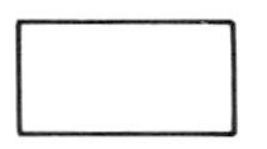

- The *terminal* symbol designates the start or stop of a program.

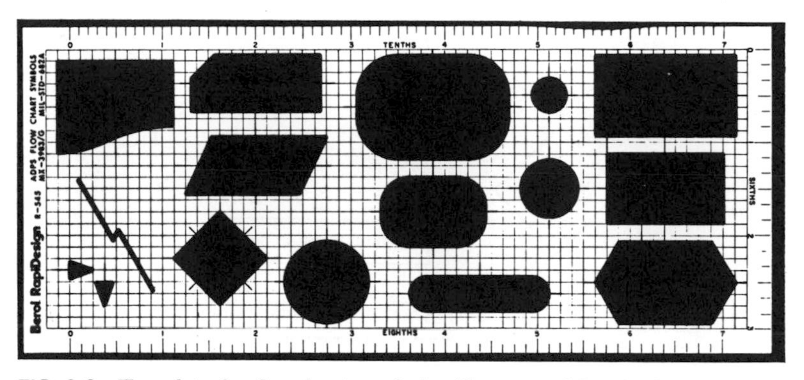

FIG. 3-3 Template for flowchart symbols. *(Courtesy of Berol USA, Danbury, Conn.)*

- The *decision* symbol indicates that there is a choice to be made and the flow of information has more than one way to go.

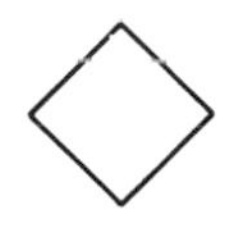

- The *connector* is a symbol inserted to ensure that there is no ambiguity in the flow of information; it indicates where to proceed to or from another part of the flowchart.

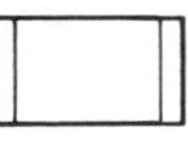

- The *predefined process* symbol, like the process box, indicates an operation performed by the computer. However, it represents a number of operations that frequently repeat (known as *subroutines*). Use of this symbol is a way to avoid repeating numerous steps that operate in a single, predefined manner.

- The *annotation* symbol is a clarifying device which indicates that information is to be added as required.

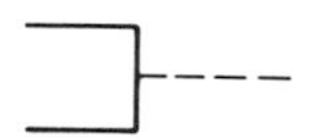

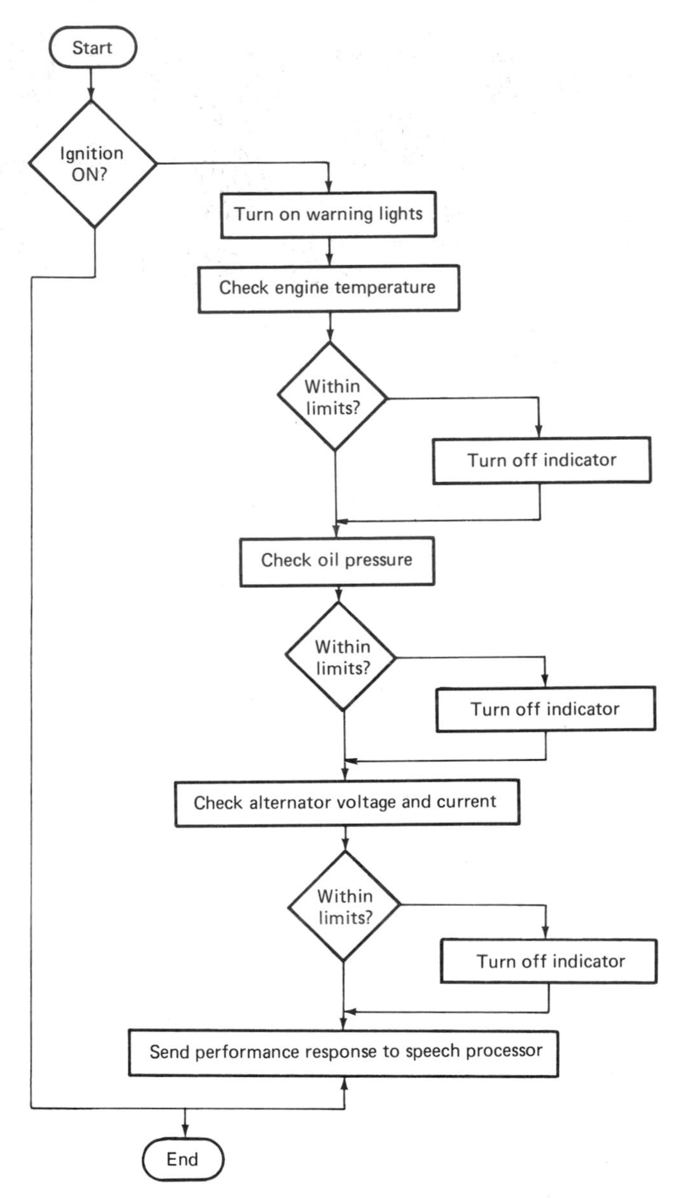

FIG. 3-4 A typical flowchart.

A typical flowchart using some of these symbols is shown in Figure 3-4.

3.3.3 The Program—Software

Once the programmer has decided which computer, which programming language, and what algorithms will be used, and how the flow should be constructed and sequenced, he or she is ready to embark on writing the program.

1. Writing the Program

As has been noted (Section 3.2) the program is a series of commands written in simple terms which the computer can convert into machine-language instructions. The machine then "acts," which means it proceeds to process the information which has been fed to it and then produces the result desired by the programmers.

Figure 3-5 illustrates part of a typical program written in BASIC. It is only a tiny fragment, since this one goes on for hundreds of lines.

NOTE: This program was not typed by the secretary but was keyboarded, or inputted, by the programmer. It was imprinted on the paper, or printed out, not by a typewriter but by the printer attached to the computer after the programmer decided to issue the command for print. Usually a secretary is not asked to reproduce material such as that shown in Figure 3-5, but there are always exceptions. The secretary will on occasion be required to reproduce sections of program text (see sample page, Figure 3-2).

2. Debugging

As has been pointed out (Section 3.1.1), the programmer, no matter how skilled in the art of programming and in the art of instructing the computer (*interfacing* with it or *addressing* it), is still human. Mistakes will be made. Consequently either the program will not run (the machine will know enough to shut itself down—a phenomenon known as *crashing*) or the program will not compile properly.

This means that when the programmer attempts to compile the program, the compiler will reject the program by printing out the numbers of lines where there are errors along with some clue as to what the error is in each indicated line.

The programmer now has the task of painstakingly reviewing the program, step by step, line by line, until all errors are detected and corrected. This review process is known as *debugging*.

NOTE: The expression has an interesting history. In 1945, during the earliest days of computers, before transistors, one of the pioneer computer operators, Grace Murray Hopper, found that her computer had stopped working. She opened the computer and examined its insides and was drawn by her sense

```
450 ' the previous figure is erased here and the new one is displayed
460 LOCATE Y-2,X : PRINT "          "
470 LOCATE Y-1,X : PRINT "          "
480 LOCATE Y+0,X : PRINT "          "
490 LOCATE Y+1,X : PRINT "          "
500 LOCATE Y+2,X : PRINT "          "

510 LET Y = TY
520 LET X = TX
530 LOCATE Y-2,X : PRINT "    *     "
540 LOCATE Y-1,X : PRINT "   ***    "
550 LOCATE Y+0,X : PRINT "  *****   "
560 LOCATE Y+1,X : PRINT "   ***    "
570 LOCATE Y+2,X : PRINT "    *     "
580 RETURN
590 ' these are the increment and decrement cursor routines
600 LET TY = Y - 1
610 GOSUB 460
620 RETURN
630 LET TY = Y + 1
640 GOSUB 460
650 RETURN
660 LET TX = X - 1
670 GOSUB 460
680 RETURN
690 LET TX = X + 1
700 GOSUB 460
710 RETURN
720 ' the parallel line border with corners is printed here
730 LOCATE 1,1 : PRINT CHR$(201)
740 LOCATE 1,80 " PRINT CHR$(187)
750 FOR I = 2 to 79
760 LOCATE 1,1 : PRINT CHR$(205)
770 NEXT I
780 LOCATE 22,1 : PRINT CHR$(200)
790 LOCATE 22,80 : PRINT CHR$(188)
800 FOR I = 2 to 79
810 LOCATE 22,1 : PRINT CHR$(205)
820 NEXT I
830 FOR O = 2 to 21
840 LOCATE I,80 : PRINT CHR$(186)
850 NEXT I
860 LOCATE 2,80
870 FOR I = 2 to 21
```

FIG. 3-5 Fragment of a BASIC program.

of smell to a location that was "smoking." Across two high-voltage terminals a moth was frying. That moth was extracted with tweezers and pinned to the computer logbook. That was the first "bug." Once the bug was removed, the computer operated effectively again. It had been "debugged."

The bugs found by the debugging process are not actual bugs: they are lines of programming instruction which are incorrect and which the programmer must correct.

3. Software

Finally the program is debugged and complete. When it is printed out, it is usually in the form of a sheaf of continuous pleated paper that may be 20 feet long or so when unfolded, and the typical one may weigh a pound or two. The program, including the stack of paper on which it is printed as well as the form in which it is stored in the computer, is known as *software*.

TABLE 3-4
SOME KEYWORDS OF COMPUTER PROGRAMMING*

address	instruction
algorithm	loop
alphanumeric	master file
annotation flag	menu
application software	object code
architecture	op code
assembler	operation code
bug	program
computer-aided design (CAD)	pseudocode
CAD/CAM	random-access memory (RAM)
computer-aided manufacture (CAM)	read-only memory (ROM)
code	sign on
command	simulate
conditional operation	software
crash	source code
data	source program
data base	storage
debug	subroutine
debugger	system software
digitize	terminal symbol
file	time-sharing
flowchart	virtual storage

*Refer to "Dictionary of Technical Terms" in Part 2 for definitions.

The program is the major element of software, but the term includes all other documents, procedures, and routines associated with the computer system.

NOTE: The term *software* should be considered in contrast with the term *hardware*. Hardware refers to all the tangible elements of the computer, that is, those which are capable of being held in the hand or touched, and which indeed are hard (Section 5.3.4). These include all the electronic, magnetic, and mechanical components of the computer.

It is the *combination* of the software and the hardware which provides a functioning computer system that will do work for us. One without the other is useless.

4. Key Words and Terms of Programming

Some key words and terms of programming are given in Table 3-4.

3.4 THE PARTS OF A COMPUTER

Every computer, no matter how small or how large, no matter how elementary or how sophisticated, consists of three major hardware elements. These are input devices, the central processing unit, and output devices.

The central processing unit is the heart of the system. It provides the means to perform processing of the data. It contains the circuitry to control the interpretation and execution of the instructions given by the program. It can organize data; it can calculate mathematically. It is indeed, as its name indicates, central. This means that all other parts of the computer system, while essential, are peripheral. Thus the input and output devices described here (Sections 3.4.1 and 3.4.3) are known as *peripherals.*

The term *peripheral* also includes auxiliary storage (memory) systems.

The central processing unit is frequently referred to in text by its initials as the CPU. Input/output devices are referred to as I/O devices.

Figure 3-6 indicates how input, central processing unit, and output are arranged relative to each other. While there may be more peripherals which may increase the capability of the input and output devices, all arrangements in all computer systems are some variation of this simple figure.

3.4.1 Input Devices

Since computers are machines which efficiently process data, they require means for acquiring the data. Those means are known as *input devices.* There are many kinds of input devices, including keyboards, disks, tapes, cards, etc.

NOTE: The information that is fed into the computer via the input devices is an assemblage of data that is then organized by the computer. It is known as the *data base* and in turn consists of smaller groupings of data known as *files, records,* and *fields.* The data base is considered to be software.

All input devices are peripherals. All input devices are also hardware.

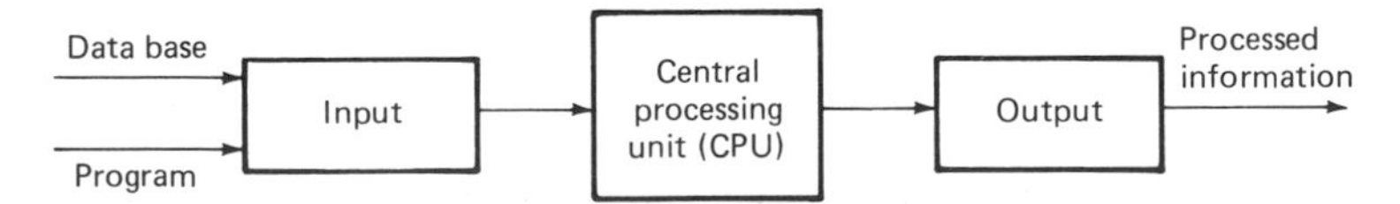

FIG. 3-6 The basic construction of a computer.

1. Keyboard

The *keyboard* is the most direct way for a human to communicate with (talk to, interface with, address) a computer. What is keyboarded goes first to a video terminal and then to storage devices such as disks, tapes, or cards.

NOTE: The *video terminal* (also known as a *video display terminal*) consists of a *cathode-ray tube* (CRT) which looks like a television picture tube. It displays input information and also receives processed information from the computer.

A computer keyboard resembles a conventional typewriter keyboard, and anyone who knows how to type can operate it, although the operator must be instructed in the use of various function keys.

It differs from both electric and mechanical typewriters in that striking a key does not activate any mechanical hammer to impact on paper through a typewriter ribbon so as to leave an ink imprint of the letter that has been struck. Rather it activates a sequence of electric signals that will be interpreted by the machine as machine language—in bits and bytes—and the letter that has been struck is then "inside the computer" and ready to be processed.

NOTE: The keyboarder can keep track of what has been typed in by observing a video screen situated just above the keyboard. That screen is, in that instance, merely a monitor of what has been done. The combination of keyboard and screen is known as a *terminal.*

a. On-line When the keyboard feeds information directly into the computer so that the information can be called up at any time, the system is said to be *on-line.*

b. Off-line When the keyboard feeds information to a device, such as a disk or a tape or cards, which must be hooked up to the computer in order to call up the information, the system is said to be *off-line.*

2. Magnetic Systems

The electric signals from a keyboard can be transmitted and imprinted (as memory) onto a disk or tape which has particles bonded to its surface that are capable of being magnetized.

NOTE: This condition of the particles of being magnetized or not magnetized is what is required to provide bits—that is, a way to have the on or off or yes or no signals which make up the binary basis of the two-character computer alphabet (see "The Two-Character Alphabet of Computer Words" in Section 3.1.5). These magnetic on-off signals activate the semiconductor diode switches in the CPU (see "Bits" in Section 3-2-1).

a. Disks A *disk* is a circular sheet of plastic which is usually 5¼ inches in diameter. It is coated with magnetizable particles.

NOTE: The data in the form of magnetic impulses is inputted to the disk either from a keyboard operated by a keyboarder (human) or from other data sources which supply electronic information.

Here are some qualities of disks:

- Once a disk has been imprinted with information (magnetically), it is a form of storage or memory.
- The information on a disk can be erased by subjecting it to a demagnetizing field.
- The information on a disk can be retrieved, or read, from it with a piece of hardware known as a *disk reader*. The disk reader may be a separate device located adjacent to the computer or it may be incorporated into the computer.
- The disk is small, easy to store, and inexpensive, and it lasts "forever."
- The disk is made of soft, pliable plastic; thus it is called a *floppy disk*. But it is not really "floppy." It is held in a paper envelope which makes it rigid enough to be handled easily. It is also called a *diskette*.
- There are also nonflexible disks, in the sense that they are more rigid. They are larger than diskettes and are known as *hard disks*. They are also known as *Winchesters*.

NOTE: Disk may also be spelled disc. The former spelling is preferred.

b. Tape *Tape* is a long narrow strip of thin flexible plastic which is easily spooled up and easily unspooled in the input or output mechanism of the computer. The surface of the strip is coated with a thin layer of magnetizable-demagnetizable particles just as the disk is. Indeed the tape is exactly the same as the disk except for its geometric form.

Here are some qualities of tapes:

- The combination of tape and its spool is known as a *reel*.
- Tapes vary in size, from only a few feet of tape to hundreds of feet on a reel. One convenient type is the *cassette*, which is like the cassette used in home hi-fi systems or cassette tape players.
- Tapes are imprinted or read in a separate mechanism called a *magnetic tape reader* which may be positioned adjacent to the central processing unit. A miniature tape reader, also used in the same way, is the *cassette reader*.
- There is a kind of tape reel in which the tape moves in a single direction endlessly since both ends of the tape are joined together. It is known as a *streaming tape*.

c. Magnetic ink It is possible to combine ordinary human-readable typed text with magnetic systems by making use of an ink which contains magnetizable-demagnetizable particles. Many personal checks are so imprinted and can be read by magnetic readers at banks. The information on the check is inputted to the bank's central processing unit of its computer and processed by the computer calculating and recording the new account balance.

Such a system includes a magnetic ink printer and a magnetic ink reader.

3. Punched Systems

The oldest way to record data used in computing is the punched-hole system.

It dates back to the time of the American Revolution, when a French weaver used such a system to control his looms.

Once holes are punched in a particular pattern they are a form of permanent memory.

They are entirely suitable as storage devices for computers because a hole is either present in a particular location or not present. Thus, as it is scanned, with either fine metal feelers or narrow beams of light, it gives either a yes or a no signal, which, of course, is all that is required for the two-character alphabet used by computers to make bits and bytes (Section 3.2.1).

Here are some qualities of punched systems:

- Holes are mechanically made by punches.
- Punches are activated, usually by a human operator who manually punches a keyboard, each key of which is directed to a particular region where the hole is to be made.
- The process of punching in data is known as *keypunching*. The person who performs this task is a *keypunch operator*.

a. Punched cards We are all familiar with punched cards in our daily lives and have all encountered them from our childhood days.

Punched cards have the following special qualities:

- Cards are arranged to receive punches in a pattern of rows and columns, usually 80 or 96 rows and 12 columns.
- Cards are *human-readable* in that they may be imprinted with ordinary characters as well as with the array of punched holes.
- Cards are used not singly but in stacks in which each card maintains a particular position in a sequence.
- A number of mechanical devices exist for the handling of cards; these include:

  Card punch machine (at which keypunching takes place)

  Card verifier

Card sorter

Card reader

- Punched cards are nonerasable. A punched hole can't be unpunched.

b. Punched tape A continuous strip of paper tape can be punched with holes and used with suitable spools and spooling mechanisms to function just as a stack of punched cards functions.

4. Optical Systems

The use of light to provide data input for a computer has already been mentioned as an alternative way to read the presence or absence of a hole in a punched card. Two other devices in which light serves the same function are laser disks and optical scanners.

a. Laser disks A *laser disk* is about the same size as a magnetic disk (diskette). It differs from a magnetic disk in that it receives its signals from fine beams of light rather than from magnetic signals. Consequently, very much more information (more bits) can be imprinted on the same size disk.

b. Optical readers All systems described for inputting to this point are compatible in some way with the central computing system. But what is not compatible is the ordinary printed page—like the one you are now reading.

There is no automatic way to enter the data from the page into the computer. To do this requires an operator to keyboard the data into one of the systems that has been described.

The *optical reader* (or *scanner*) eliminates that problem. It scans each letter of each word of each line on the page. It has its own built-in computer which "recognizes" each letter and converts it into a set of electric signals that eventually magnetize the particles of a diskette. In a few seconds the data from the printed page has been transferred to a floppy disk which in turn becomes the input to the computer.

NOTE: There is a limitation in the applicability of optical scanners. They are set up to accept only one size and form of typeface. Copy intended for scanning must be produced with a particular type font, usually a rotating ball used with a particular kind of typewriter. The correct font is inserted as required.

The value of the optical scanner is that the normal operating procedures of an office need not be disrupted, and sophisticated equipment such as word processors need not be installed. The typist types pages of text—as usual—on a typewriter rather than on some special keyboard. (See Section 1.5 on office automation.)

5. Words and Terms of Input Devices

Words and terms of input devices are described in Table 3-5.

TABLE 3-5
SOME COMMONLY ENCOUNTERED WORDS AND TERMS OF INPUT DEVICES*

card reader	light pen
cassette	machine-readable
cathode-ray tube (CRT)	magnetic tape
cursor	mouse
digitize	network
direct-access storage device (DASD)	optical character recognition (OCR)
disk drive	optical reader
diskette	optical scanner
disk pack	punched card
dumb terminal	punched tape system
floppy disk	reel
hard disk	stylus
input	tape drive
intelligent terminal	video terminal
keyboard	voice coder
keypunch	visual display terminal
key-to-diskette entry station	Winchester disk
laser disk	word processor

*Refer to "Dictionary of Technical Terms" in Part 2 for definitions.

3.4.2 The Central Processing Unit

Once data has been inputted by means of the (peripheral) input devices described in Section 3.4.1, the central (nonperipheral) portion of the computer will have its way with them. It will process them, which means it will act on them in accordance with the directives of the program. Processing may be divided into two major categories: organizing (and reorganizing) and mathematical calculating. This section describes the *central processing unit* of the computer, the *CPU.*

The CPU has five major functions:

- ***Input:*** Receives data and program instructions from input devices (see Section 3.4.1).
- ***Primary storage:*** Stores data.
- ***Arithmetic logic:*** Manipulates data according to instructions (see "Carrying Out Program Instruction from Storage" in Section 3.4.2).
- ***Processing:*** Moves data to primary storage; controls arithmetic-logic; controls output media.
- ***Output:*** Accumulates and/or presents results of processed data.

1. Organizing and Reorganizing—Primary Storage

The first thing we do—as humans—when we want to organize and reorganize data is to assign each datum to some category. Once assigned, the data remains until we need it. In effect, we "store" it. The CPU does the same thing. The portion of the CPU which accomplishes this storage of designated groups of data is known as *primary storage.*

It has been pointed out that a quality shared by humans and computers is memory (Section 3.1.3). Storage is a form of memory and the primary storage portion of the CPU is a group of memories.

a. Addresses When we, as humans, want to remember something, we ask our brains to "remember." We get specific. We make a particular demand on our human memories, and if we have a "good memory," the brain gives us back the information we desire. (In the process, it may even supply organized information for us.) When we want the computer to remember something, we also specify what it is by using a description or label of that something. That description or label is known as an *address.*

Suppose we ask our brain, "What is the name of that blonde-haired woman I worked with back in 1977 who always wore turtleneck sweaters?" Just by asking that question we have given our brain an "address."

The memory portion of our brain has in fact once stored that information. It did so by using a process similar to the one used by the primary storage of the CPU. It has stored the information because in 1977 it was inputted—at the moment we met her. Then the brain "programmed" the information to be "stored" (memorized). We haven't thought about her for all these years; yet the information was there, tucked away someplace. And all we have to do is to call in the question to our brain (address our brain), and the answer comes out. Her name was Elvira.

Addresses are *codes,* usually two to four digits—numbers such as 04, 27, 0036, and so on. These in turn are constructed from bits and bytes. (See Section 3.2.1.). Addresses are constructed so that the user can get to the proper storage location.

NOTE: Storage locations may be *byte-addressable, character-addressable,* or *word-addressable.*

Addresses work both ways—that is, for the purpose of putting data in or taking data out.

The act of *entering* data is called *read in, read into,* or *write.*

The act of *retrieving* data is called *read out, read from,* or *read.*

NOTE: It is useful to compare the address code with a telephone number. It too is a number code used to address (call up) or call from.

b. Codes Other entering or retrieving is done with *codes* based on mathematical principles of the binary system.

Some of these are:

- *BCD,* or binary-coded decimal ("Binary Nature of Bits" in Section 3.2.1)
- A *6-bit code* (a 6-bit byte).
- An *8-bit code* (an 8-bit byte).

c. Specialized storage elements In addition to the active memory of the primary storage section of the CPU, there are memories that are special in the sense that they are "put away"; they are "on the shelf," ready to be used as required.

In effect they are little parallel programs called *microprograms.*

They function as though they were "scratch pads" of information to be used when needed in the main active programs.

These specialized programs are kept in what is called *buffer storage,* also known as *cache memory.* Information in buffer storage is available at a much faster rate than that in primary storage, thus increasing the speed of processing. Buffer storage usually contains frequently used instructions or data needed by the arithmetic-logic unit of the CPU (see "Carrying Out Program Instructions from Storage" in Section 3.4.2).

d. Chips We know that all the categories of memory and storage that have been described are contained in the whorls and convolutions of our living brains.

We do not understand a large part of the physics and chemistry of how these living cells succeed in carrying out these functions—let alone the nonphysical psychological principles. But we do know that different brain regions carry memory storage.

In the same way, we know very well which pieces of hardware in the CPU are responsible for memory. We know because those parts are installed by humans.

They are electronic components which contain millions of electronic on-off switches which can decipher a two-character alphabet, one character of which is *on* and the other *off*—the bit. The components constitute an array of semiconductor diodes and transistors and other integrated electronic components such as capacitors. (See Sections 5.3.2 and 5.3.3.)

Those parts are *chips.*

"Chip" in Section 5.3.3 points out that when a great number of components are constructed at one time on one slice of semiconductor material and the resulting assemblage packaged in a suitable little plastic container, that structure is known as a *chip*. A chip is a small rectangular slab about 1 by ½ by ⅛ inch. (See Figure 5-26.)

These are not too different in size from a wood chip one might make while chopping wood. Perhaps that is where the name came from.

NOTE: A typical office or commercial computer may have only a few chips or several hundred chips, depending on its size. Section 3.5 describes some differently sized computers, such as microcomputers, minicomputers, and mainframes.

There are different classes of computer chips which perform different kinds of functions and have different capabilities.

Some of these are:

- *Microprocessor chip*—A single chip which functions as a complete CPU for a small (personal or micro-) computer.
- Chips used for different storage functions. The ones listed here are known by their acronyms, which describe the manner in which they function.

    RAM (random-access memory): A chip that can be addressed at any time.

    ROM (read-only memory): A chip with a set memory, fixed at the time of manufacture, that cannot be reprogrammed.

    PROM (programmable read-only memory): A chip for which memory can be programmed after manufacture but which then cannot be reprogrammed.

    EPROM (erasable programmable read-only memory): A chip for which memory can be programmed after manufacture and which can be erased and reprogrammed.

2. Carrying Out Program Instructions from Storage—Arithmetic-Logic and Control Functions

On the basis of the stored information in the various portions of the primary storage section (just described), both our brains and the processing portion of the computer's CPU stand ready to make use of the information (*retrieve* it) and to carry out the essential functions of processing. This is done by the *arithmetic-logic unit*, the *ALU*.

Like the primary storage unit, it is made up (in the sense of hardware) of chips. When a single chip performs the entire ALU function, in a small computer, it is known as a *microprocessor*.

a. Calculations The different chips of the ALU perform different functions, mostly of a mathematical nature. Thus they have different names, usually related to a mathematical function.

These are:

- *Adder*
- *Accumulator*
- *Comparer*
- *Storage Register*

b. Control There are a number of similar functions which seem mathematical in nature but which are actually control functions. Some of the more common control functions take the form of *instructions* or *commands.* These lead to specific operations such as *multiply, subtract,* or *store,* for which the programmer issues commands such as MUL, SUB, or STO. Instructions may also take the form of addresses to different locations in the memory. The data retrieved from the memory then is used in the mathematical operations of multiplying, subtracting, storing, and so on.

3. Words and Terms of Central Processing Units

Some of the more commonly encountered words and terms of central processing units are listed in Table 3-6.

3.4.3 Output Devices

The data has been inputted (into one or more of the input devices described in Section 3.4.1), as has the program (by the programmer).

TABLE 3-6
SOME COMMONLY ENCOUNTERED WORDS AND TERMS OF CENTRAL PROCESSSING UNITS (CPUs)*

accumulator	nondestructive storage
address register	nonvolatile storage
arithmetic-logic unit (ALU)	off-line
buffer	on-line
byte-addressable	output
cache	primary storage
chip	programmable read-only memory (PROM)
core storage	random-access memory (RAM)
decode	read-only memory (ROM)
erasable programmable read-only memory (EPROM)	register
instruction	semiconductor storage
microprocessor	sequential access
microprogram	working storage

*Refer to "Dictionary of Technical Terms" in Part 2 for definitions.

And the CPU has done its work of calculating or organizing. But nothing is of any value unless the newly organized data or newly calculated solutions can be communicated back to the humans who run the machines. The humans say, "Speak to me," and the computer speaks—through its *output devices.*

Like the input devices, output devices are peripherals.

NOTE: Not all computer output is of the sort that can be understood directly by humans, although most is—like printed words, numbers, and words and numbers, or graphs or even pictures, on a video screen. Some computer output is communicated to disks and tapes, devices much like input devices. Such disks and tapes are known as auxiliary, or secondary, storage. (see "Auxiliary Storage" in Section 5.4.3).

1. Printed Output

Many computers have a *printer* the function of which is to give a *printout.* The printout can be in many forms, depending on the type of computer. For example, a word processor can produce a printout much like the typescript produced by an ordinary typewriter. Or a printout may be in the form of a sales slip like the one that accompanies your supermarket order (Figure 3-7). Or it may be a continuous sheet of paper stacked by means of accordion folds.

We are always impatient while waiting for the printout. This is only natural. Usually the CPU has done its work in a fraction of a second. That makes the printer a slowpoke—even if it can print 10 times faster than any human can type. Yet the technology of computer printing is a marvel in itself. Movable type is never used because it can't be handled fast enough. Instead there are many different ways of making marks on paper that are as new as this minute. For example, many are based on marvelous tricks played with flying droplets of ink.

NOTE: Gutenberg, the inventor of printing with movable type, did very well. His invention has lasted over 500 years, and we still use it when we turn to our typewriters and some of the older printing presses. But his day is about over. We predict that the printing devices descended from his invention won't make it into the next century.

2. Video Display

Almost every computer has a cathode-ray tube video display terminal (see "Keyboard" in Section 5.4.1). Not only can the operator see the data on a screen after it has been processed, he or she can also see it as it is being inputted, as anything keyboarded is instantly displayed.

There is then the opportunity to make corrections—one of the most valuable qualities of a word processor, for example.

The video screen can display data in any form—as words, numbers, graphs, or pictorial representations.

```
KELMAN'S MARKET

PRODUCE        1.59 H
PRODUCE         .24 H
PRODUCE        1.49 H
PRODUCE         .63 H
FRZ FOOD       1.15 H
FRZ FOOD        .69 H
FRZ FOOD       1.69 H
FRZ FOOD        .24 H
FRZ FOOD        .25 H
MEAT           1.49 H
PRODUCE        2.09 H
H.B.A.         1.09 H
GROCERY        1.22 H
GROCERY         .53 H
DAIRY           .40 H
DAIRY           .55 H
DAIRY          1.09 H
GROCERY        1.99 H
GROCERY        1.99 H
DEPOSITS        .10
DEPOSITS        .10
DEPOSITS        .10
SUBTOTAL      20.61
ST TAX         1.40
TOTAL         22.01

CASH          30.00

CHANGE         7.99

# ITEMS    22

      THANK-YOU
8037 C108 R01 T14:55
```

FIG. 3-7 A supermarket sales slip—a computer printout.

3. Auxiliary Storage

When data is stored on cards, tapes, or disks, such media are known as *auxiliary,* or *secondary, storage devices.* They function in mechanisms (e.g., for a disk, a *disk drive*) in which they may be used either for input, as when information is to be retrieved from them, or for output, as when information is to be printed out.

Such devices are auxiliary (or secondary) in the sense that they are separate from the computer, but they function together with the computer in that they can be "plugged in." Thus they belong to the class of devices known as peripherals.

NOTE: These auxiliary media can be used for three purposes: data entry (input), secondary storage, or information output. They are thus known as *triple-purpose media.*

Auxiliary media are

- Punched cards
- Punched paper tape
- Magnetic tape
- Magnetic disks
- Optical systems such as laser disks or laser tape

4. Microfilm

Output can be recorded directly from a computer on film; thus *microfilm* and *microfiche* and associated techniques can serve as output media.

5. Computer Graphics

Computer graphics is really a subclass of video output and, to some degree, of printed output. A sketch can be made to appear on a video terminal, and it can be manipulated by the computer operator to produce design drawings for machine parts or other structures, mathematical curves and charts, maps, and a number of other graphic representations.

NOTE: There is a kind of inputting that is used primarily in graphic processing. It is "writing" directly on the video screen with a penlike instrument known as a *light pen;* or *stylus.* It is a touch system. Any point on the screen that is contacted by the tip of the light pen will "respond." The combination of touch and response is as much input to the computer as the contact between the fingertip of the keyboarder and the key of the keyboard.

a. Computer-aided design (CAD) When the system is used to produce design drawings it is called *computer-aided design,* or *CAD.*

CAD in one sense is making the design drafter obsolete. In another sense it is making the designer—the person who sits in front of the terminal—more skilled and more productive and more sophisticated.

NOTE: CAD is to the drafter as the word processor is to the secretary.

b. Computer-aided manufacture (CAM) When design drawings produced by CAD are then upgraded and modified into instructions for the machines that will produce the objects that have been designed, the process is known as *computer-aided manufacture,* or *CAM.*

For this purpose the video information is outputted onto a tape, either punched or magnetic. That tape then serves as the input for a fabricating machine, instructing it how to cut and form metal, for example.

NOTE: Computer-controlled fabrication machines are known as *numerically controlled*, or *NC*, machines. They machine metal as a machinist might, but there is no machinist. The machinist is becoming obsolete, as is the drafter.

CAD and CAM are often used together. They are then known as *CAD/CAM*.

6. Voice Response

It is possible to construct sounds electronically which simulate the human voice, and as a result a computer output can be constructed to deliver its information in "spoken" words which are easily understood. To get output from some computers one merely has to listen.

7. Words and Terms of Output Devices

The most commonly encountered words and terms of output devices are listed in Table 3-7.

3.5 THE KINDS OF COMPUTERS

Computers can be classified by the range of their capabilities, usually indicated by the size and complexity of their CPUs. That in turn implies a corresponding range of sophistication of their input systems and their output systems. Types of computers include microcomputers, minicomputers, and mainframes, among others.

3.5.1 Microcomputers

Microcomputers are the smallest computers in that they have only a single logic chip (a microprocessor chip) in their CPU, although this may be supplemented by several random-access memory chips (RAMs) and some read-only memory chips (ROMs).

TABLE 3-7
SOME COMMONLY ENCOUNTERED WORDS AND TERMS OF OUTPUT DEVICES*

daisy wheel	microfilm
disk drive	optical reader
display	plotter
dot-matrix printer	printout
line printer	punched tape system
magnetic tape	tape drive
microfiche	triple-purpose media

*Refer to "Dictionary of Technical Terms" in Part 2 for definitions.

Personal computers are the most popular form of microcomputer. It is possible that one day they will be as common in our homes as the television set.

Although microcomputers are most popular in the home, they are also used in the following commercial and industrial applications:

- Payroll computations
- Machine tool control
- Desktop graphic displays for engineers, technicians, physicians, travel agents, etc.
- Business calculations
- Production of text (in word processors)
- Inventory control
- Cash registers in stores

Microcomputers are also known as *micros.*

3.5.2 Minicomputers

Minicomputers are one step up in complexity from microcomputers. They contain many logic chips in their CPUs. One minicomputer can control many stations, known as *satellites.* But the basic functions are the same as for microcomputers. They differ from micros in their greater storage capacity, speed, and ability to support a greater variety of rapidly operating peripheral devices.

Minicomputers are also known as *minis.*

3.5.3 Mainframes

Mainframe is the term used for computers which share all the qualities and functions of other computers such as micros and minis but on an enormously larger scale.

Mainframes are significantly larger than other computers when considered just in terms of the space they occupy and the tons they weigh. The typical mainframe requires many hundreds of square feet of air-conditioned (reinforced) floor space. It consists of dozens of 6-foot-high racks of electronic equipment and costs millions of dollars.

Mainframes are usually employed in *time-sharing,* a system in which *networks* of computer inputs and outputs, located many miles apart, are connected to the mainframe CPU by telephone lines and other communication devices. These *remote terminals* can all make use of the CPU at the same time.

NOTE: As with all technology, there is no "largest." There are computers that are much larger than mainframes and are more capable of performing computer functions at much higher speed. These are computers that dwarf mainframes both in physical size and in first cost. Not having come up with a better name, the scientific community calls them *supercomputers*.

3.6 MORE ABOUT COMPUTERS

What has been described in this chapter to this point only scratches the surface of this new field of technology (hardly 3 decades old). But if you have absorbed it you should have acquired enough computer literacy to function within this field—a must if you work with the production of scientific and engineering text in any field.

As for what has necessarily been left out, Table 3-8 lists some of the more commonly encountered words and terms.

TABLE 3-8
MORE WORDS AND TERMS OF COMPUTERS*

access	hard copy
analog computer	heuristic
artificial intelligence	hybrid
back-end processor	interface
baud	K
branch	large-scale integration (LSI)
bus	logic diagram
call	master file
canned program	modem
channel	multiplex
clock	nanosecond
collate	node
console	octal
counter	patch
cybernetics	picosecond
DASD	real time
diagnostics	routine
digital computer	run time
downtime	serial access
edit	streamer
editor	word length
electronic data processing (EDP)	zone bits

*Refer to "Dictionary of Technical Terms" in Part 2 for definitions.

Basic Practice—Techniques and References

It helps to possess at least some measure of computer literacy, and this is dealt with in the first part of the chapter, "Basic Principles." But even if you do not have this understanding, you can process computer-based text if you master the following special symbols and procedures.

3.7 SPECIAL SYMBOLS AND PROCEDURES

There are special ways to present text dealing with computer material, most of which are based on formats to depict code statements. Thus there may be unconventional usages of capital letters and spacings. These will be encountered primarily in the reproduction of program fragments.

As with any technical field, there are special symbols. These often resemble and are used like mathematical symbols already encountered in the mathematics chapter. But also employed are ordinary typewriter symbols normally used in nontechnical material, such as the asterisk * or the slash /. Be prepared for these to have a mathematical function such as multiplication or division.

Special charting procedures are used in the presenting of flowcharts. Generally these involves drawing a box of a particular shape—preferably with a template—and typing copy inside them. Boxes are interconnected with arrowed lines.

Most of all, computer text uses words—new words not encountered in ordinary language, or ordinary words in new combinations with new meanings. These are so numerous and so special that lists of them, without definitions, are provided here for quick reference (Tables 3-3 to 3-8). It is considered useful to the text processor just to know that expressions such as *byte* and *op code* are key terms relating to computer languages, that CAM is a commonly used computer expression, and that terms such as *bus, mouse,* and *floppy* have distinct computer-related meanings with no connection to transportation, rodents, or limpness.

Where particular definitions are desired, either to enhance one's computer literacy or just to explain a certain usage, they are available in the dictionary in Part II of this book.

Additionally, these tables should serve as guides to how expressions are to be shown, indicating where spaces may be required, where hyphens or slash may be used, and so on.

3.7.1 Program Languages

Program language names may be written in all capital letters or with an initial capital letter. For example, BASIC, COBOL, FORTRAN, PASCAL, and ADA or Basic, Cobol, Fortran, Pascal, and Ada.

The rule, if there is one, is that any single text should be consistent.

Do not mix BASIC and Basic in the same text.

3.7.2 The Flowchart

Practice varies on who should draw the flowchart's boxes and interconnecting lines for technical and scientific text. (See Section 3.3.2 for a presentation of flowchart box shapes.) Most organizations give the job to the graphics department. But some give the task to the secretary. Then the secretary or technical typist either types the message and then draws the box or draws the box first and then types in the message. Of course, for neat text the box must be just the right size to accommodate the message. Thus if the person who draws the box and the one who types the message are not the same, they must communicate with each other.

NOTE: Templates are available for drawing these flowchart symbols, and the person responsible for committing the symbol to the page should have a good supply of these in different sizes. A flowchart symbol template is illustrated in Figure 3-3.

Should a box larger than what is available in premade templates be required, draw it neatly, with exactly the right shape, with a fine-felt-tipped pen.

Figure 3-4 illustrates a typical flowchart.

3.7.3 Programs

Usually the secretary is not required to reproduce a program or program fragment like Figure 3-5, which was typed by the programmer and printed by the printer of the computer. However, there are always exceptions. There is no set rule on how to do this other than to reproduce exactly what the author has put down in exactly the way the author did it. Every letter or symbol must be exactly as shown in exactly the position shown.

Even ordinary punctuation signs may have new meanings. For example, in Figure 3-1, the sequence [0-100, · 100 = OFF] appears. The comma, the period, the space between them—all have meaning—and all must be written exactly as shown. (See Table 3-2.)

NOTE: Actually, there is one rule. It is that there are no lowercase letters in computer programs. All letters are capital letters.

When program instructions are to be reproduced in text, the secretary must not be surprised by encountering some rather strange words and phrases which seem like English but not altogether.

Words such as:

DO UNTIL

DO WHILE

ENDIF

ENDO

ELSE

IF-THEN-ELSE

THEN

These code words are constructed in ways that thwart our attempt to set rules. Thus, some require internal spacings and some do not. Some have internal hyphens; some do not. We must accept that each expression has its own reason for existing in the form in which it exists. The rule is: Do them exactly as they are presented by the author.

Thus, in the previous example:

CORRECT	*INCORRECT*
DO WHILE	DOWHILE
ENDIF	END IF
IF-THEN-ELSE	IF THEN ELSE

The secretary could expect to be presented with program text in which program symbols, lined-off codes, program instructions, and miscellaneous formats may all appear together. Figure 3-2 is a sample page of this sort of text.

CHAPTER FOUR

Physics

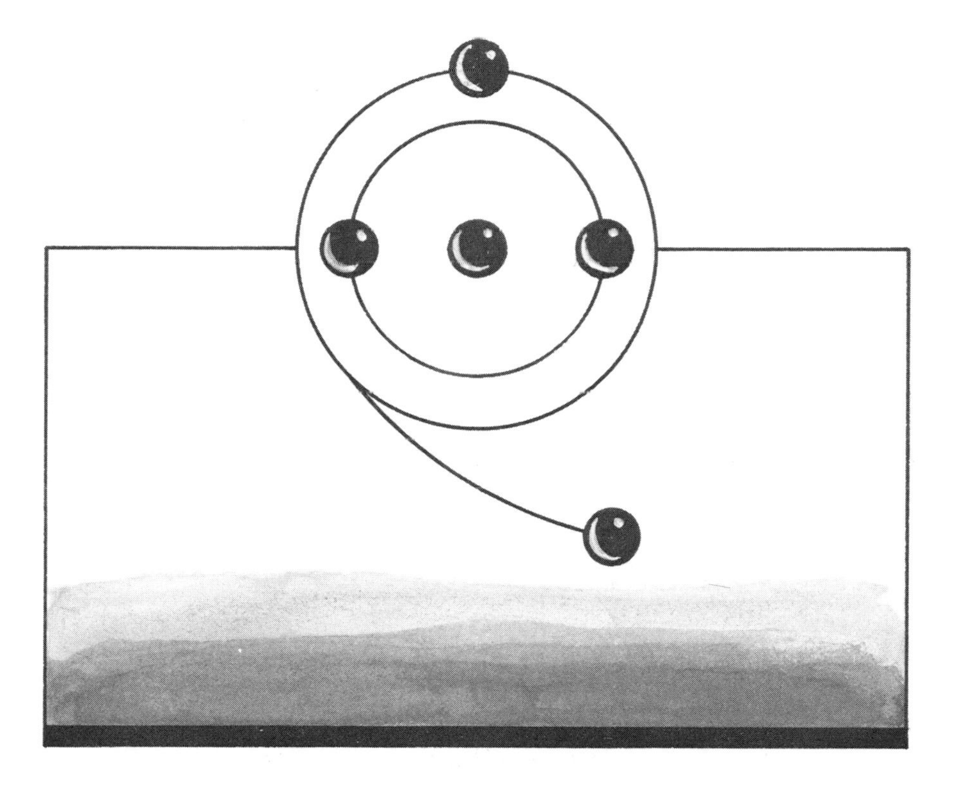

Basic Principles

Physics is one of the three basic sciences; the others are mathematics and chemistry. They are considered basic because virtually every other field of science and engineering is seen to be some variation or combination of these three.

(A case can be made that biology is a fourth basic science, but even biology can be interpreted as a kind of combination of chemistry and physics, and, like these, it can be treated mathematically.) Physicists—an arrogant lot who say that *everything* is physics—will even claim that chemistry is physics too. Let us go along so far as to say that physicists are partially right, that there is some overlap between each of the sciences and physics.

Chemistry is the scientific study of the properties, composition, and structure of matter, and how matter interacts with other matter; physics goes on to describe not just matter, but also energy, in terms of principles and laws that may be expressed exactly—in mathematical terms. To the physicist, *nature* means the interactions between matter and energy. So the physicist describes all nature.

4.1 PHYSICS IN GENERAL

There are dozens of different specialties and subspecialties in the field of physics, which is as broad as nature itself. But they all have some important connecting threads. All areas of physics have an important historical division—classic and modern. And all areas of physics may be considered from two points of view—theoretical and applied. In addition, the subject has its own way of using the English language, with its own vocabulary.

4.1.1 The Special Vocabulary of Physics

Although many words and concepts are used primarily by the physicist and have no meaning other than their scientific one, many terms have both everyday and scientific meanings. To the physicist the everyday meanings are inexact, so the physicist has invested them with new delicate nuances of meaning, transforming them into *scientific words.*

For example, did you know that to the physicist, *speed* does not mean the same thing as *velocity?* Both describe movement in terms of distance per time, but velocity is a *vector*—a quantity with a direction (see Section

2.3.5)—and speed is a *scalar* (see "Scalars" in Section 2.3.5)—a quantity that makes no reference to direction. For example, when we say that an airplane is flying from St. Louis to Seattle at a velocity of $\overrightarrow{630}$ miles per hour, we place an arrow over the velocity to indicate that the plane is going in a particular direction, in this case northwest. But when the builder of the airplane says that it is capable of going 630 miles per hour *in any direction*, the builder is talking of the plane's speed, not its velocity. Then the 630 miles per hour has no arrow. But in nontechnical English, speed and velocity are virtually synonyms. Physics has extended the meaning of each—and made each meaning different.

So although this chapter includes descriptions of particular techniques of producing copy—for example, how to write isotopes in the branch of physics known as nuclear physics (Section 4.9.4)—most of this chapter concerns definitions.

Since we cannot define every term you will meet in physics, we have selected only the most important and most frequently encountered. Most of the rest can be found in the "Dictionary of Technical Terms" in Part 2.

This does not mean that most of the text you will work with will be in the form of narrative. The opposite is true; most of it will be in the form of scientific notation. But we have already treated that notation in Chapter 2.

4.1.2 Classical Physics, Modern Physics

Like music, physics is divided into *classical* and *modern*. *Classical physics* includes essentially everything physicists knew before the year 1900, and *modern physics* is what physicists discovered after (approximately) that date.

There is nothing magic about the year 1900, it was just that around that time, scientists began to take a new view of matter and energy. They began to consider matter not merely as the "stuff" that surrounds us and affects our senses but from the point of view of the particles, the atoms, of which matter is made. At the same time, physicists realized that energy, such as light from the sun, moves not only as rays like beams from a flashlight, but as waves like surf breaking on a beach. These two new concepts of matter and energy changed everything in physics and made modern physics and all corresponding written material more complete and more accurate, which means it made it more complex, rigorous, and mathematical.

Classical physics includes

Mechanics
Acoustics
Electricity
Magnetism

Optics	Crystallography
Hydraulics	Heat and thermodynamics

Modern physics includes

Biophysics	
Geophysics	Particle physics
Nuclear physics	Plasma physics
Cryogenics	Astrophysics
Spectroscopy	Relativity
Solid-state physics	Quantum mechanics

These two lists tell us one more thing about the difference between classical and modern physics. A case can be made that everything in classical physics can be experienced by our senses. Classical physics deals, for the most part, with what we can see, hear, and touch. Modern physics deals with things that, for the most part, cannot be seen, heard, or felt. Most of them can only be experienced in our minds—but of course that mental experience can be transcribed to paper, which is where you come in.

If you work in a physics laboratory, the chances are that the emphasis will be distinctly modern. Be prepared to work with many complex mathematical equations. If you find yourself dealing with classical physics, you probably are not working with physicists at all, but with mechanical engineers, civil engineers, chemical engineers, metallurgists, electrical engineers, etc. Their texts tend to be more discursive and less filled with equations. On the other hand, engineers can get almost as mathematical as modern physicists, and in such cases the two kinds of physics tend to merge, and what you will be working with in either environment may be very similar indeed.

4.1.3 Theoretical Physics, Applied Physics

Physicists themselves can be divided into two groups depending on which of two approaches they use to perform their work—*theoretical* or *applied* (or experimental).

Theoretical physicists work primarily inside their minds, where they contrive to lay out the universe as they see it so that they may deal with it, which means that they speculate on questions of "What if?" The point is, they do not work in the laboratory; they work with paper and pencil (and with the input keyboards of computers).

If you work for a theoretical physicist, you will receive written copy that describes speculations on the laws of physics. These speculations are like journeys. The theoretician arrives at a new destination, never before reached. And the text describes, step by step, how he or she got there.

What is important for you is that each step of the "step by step" will be another mathematical equation, numbered in sequence. There will more than likely be more equations in the copy than written sentences (but, of course, we know from the mathematics chapter that an equation is nothing but a kind of sentence containing an equals sign that says "is" or some other sign that can be translated into words).

The applied physicist "applies" the theory worked out by the theoretical physicist by performing an experiment.

If the theoretical physicist's theory is shown to be right—because the results of the experiment conform with the theory's prediction—we say that the theory has been *confirmed.* If, instead, the experiment gives a different result from the one that the theoretician predicted, we know that someone has made a mistake. Either the theory was incorrectly conceived or the experiment was incorrectly done.

4.2 MECHANICS

Mechanics is the subdiscipline of classical physics which treats the actions of forces on objects.

Most of the time, when an object is subjected to a force, the object moves. If movement does not occur, the internal condition of the object becomes more *stressful.*

All this has been within our immediate experience from our earliest childhoods. From the beginning we pushed things around. So we have all been "practicing" physicists in the field of mechanics all our lives. What is different for someone who goes into that practice as a professional—a physicist or a mechanical engineer—is that all the words and definitions that are employed become more rigorous and exact, and all discussion becomes quantitative and mathematical.

4.2.1 Machines and Structures

The branch of mechanics that describes the movement of objects subjected to forces is known as *dynamics* or *kinematics.* In contrast, the part of mechanics that treats things that stand still is called *statics.* Machines are usually dynamic; structures are usually static.

1. Machines

A *machine* is a dynamic or kinematic device. It is a system of parts designed in such a way that when force is applied, the motion that results is useful.

A bicycle is a machine. A can opener is a machine. Skis are machines, as are surfboards and roller skates. An automobile and an industrial robot are complex machines. Any mechanism that extends our capability to perform a function involving motion that our bodies alone cannot do as well—or at all—is a machine.

2. Structures

A *structure* is an object that absorbs stress. When forces are applied to a structure, motion usually does not result. The usefulness of a structure lies in the fact that it is capable of "coping with" stress. A structure that tends to remain in a state of rest as it resists force is in a *static condition.*

A building is a structure. A bridge is a structure. A table and a chair are structures. Any device that extends our capability to hold weight or resist pressure is a useful structure.

NOTE: When the structure can't resist the force, it changes from a static to a dynamic condition—as when we move a chair.

4.2.2 Objects in a State of Motion—Dynamics

1. Motion Words

The manner in which the physicist generates new words and new meanings for ordinary words can be illustrated by an example in which forces applied to objects cause motion.

Consider a playground. To the engineer and the physicist

- The slide is an *inclined plane.*
- The swing is a *pendulum.*
- The seesaw is a *lever* balanced on a *fulcrum.*
- The merry-go-round is a *wheel* rotating on an *axle.*

All these objects are subjected to forces the result of which is movement. Physicists and engineers have constructed words to particularize each object (as in the preceding list), each force, and each kind of motion. This terminology adds to the ability of the English language to describe. These terms will be familiar to you, but even if the words are familiar, their usage may be unfamiliar.

- In the case of the slide, the motion down the slide involves *speed, velocity, acceleration,* and *deceleration.* That motion is affected by the *angle* of the inclined plane (its *inclination*), by its *length,* and by the *friction* between the slide and the bottom of the slider. That friction in turn may be described quantitatively by a term known as the *coefficient of friction* (the larger the coefficient, the slower the sliding).
- The movement of the swing is *periodic* (periodic motion is motion that repeats), and the time or distance of each back-and-forth cycle is a *period.* Its motion is also of the sort known as *harmonic,* and it turns out that mathematically it makes use of the trigonometric sine of the angle (Section 2.2.2) through which the swing *oscillates.* So the motion

of the swing is normally described as *sinusoidal.* Since the swing periodically, harmonically, and reiteratively makes an angle (the engineer likes to say *describes* an angle), this sort of motion is known as *angular motion.* And when the swing or pendulum is made to go at a given velocity (by reiteratively applied forces of a given intensity), that is known as its *angular velocity.* When the velocity increases or decreases, that is known as *angular acceleration* or *angular deceleration.* The distance traveled by the swing (as distinct from the angle described) is known as the *amplitude.* The number of oscillations (periods) per unit time is called the *frequency.*

- The seesaw also experiences angular motion, and all the same terms apply as for the swing. There is the additional factor of *balance.* Children of equal weights sitting motionless at equal distances from the fulcrum balance—which means that the opposing forces are equal and cancel each other out, so there is no *resultant force,* and thus no force to cause motion. In this case the situation is static. Such a state of balance is known as *equilibrium.* A static condition of *unbalance* can also occur with two motionless children of unequal weight, suspending one child in the air while the other remains on the ground.
- As for the merry-go-round, that also moves with angular motion, a result of which is the generation of *centrifugal force,* a force that directs itself outward from the center and acts to fling the riders outward from the center. There is an opposite force, *centripetal force,* which draws objects inward toward a body to which they are attracted by gravity. When centrifugal and centripetal forces balance, things move in fixed curved paths known as *orbits.* We encounter orbits in the study of *astronomy* and *celestial mechanics* and when we consider space vehicles, missiles, and satellites in the field of aerospace engineering.

The examples just dealt with illustrate how even such a simple set of mechanical systems as a playground can serve as the basis of a *highly enriched vocabulary,* as utilized by the scientist or engineer who specializes in the field of mechanics. Had some other example been chosen, such as a bicycle or a can opener, a similar set of facts would have been considered, and this would have resulted in other similar new definitions of words.

Some of the objects of mechanics are illustrated in Table 4-1.

2. Mathematical Treatment of Motion

All motion words can be represented by symbols—*algebraic symbols*—which can be plugged into mathematical equations and manipulated mathematically using the techniques of algebra, geometry, trigonometry, or calculus described in Chapter 2. From these equations, the engineer or scientist can construct graphs and sketches.

This is another way of saying that the techniques that must be employed when working with text concerning mechanics (once the special vocabulary is understood) are essentially the techniques of mathematics.

TABLE 4-1
SOME OF THE OBJECTS OF MECHANICS

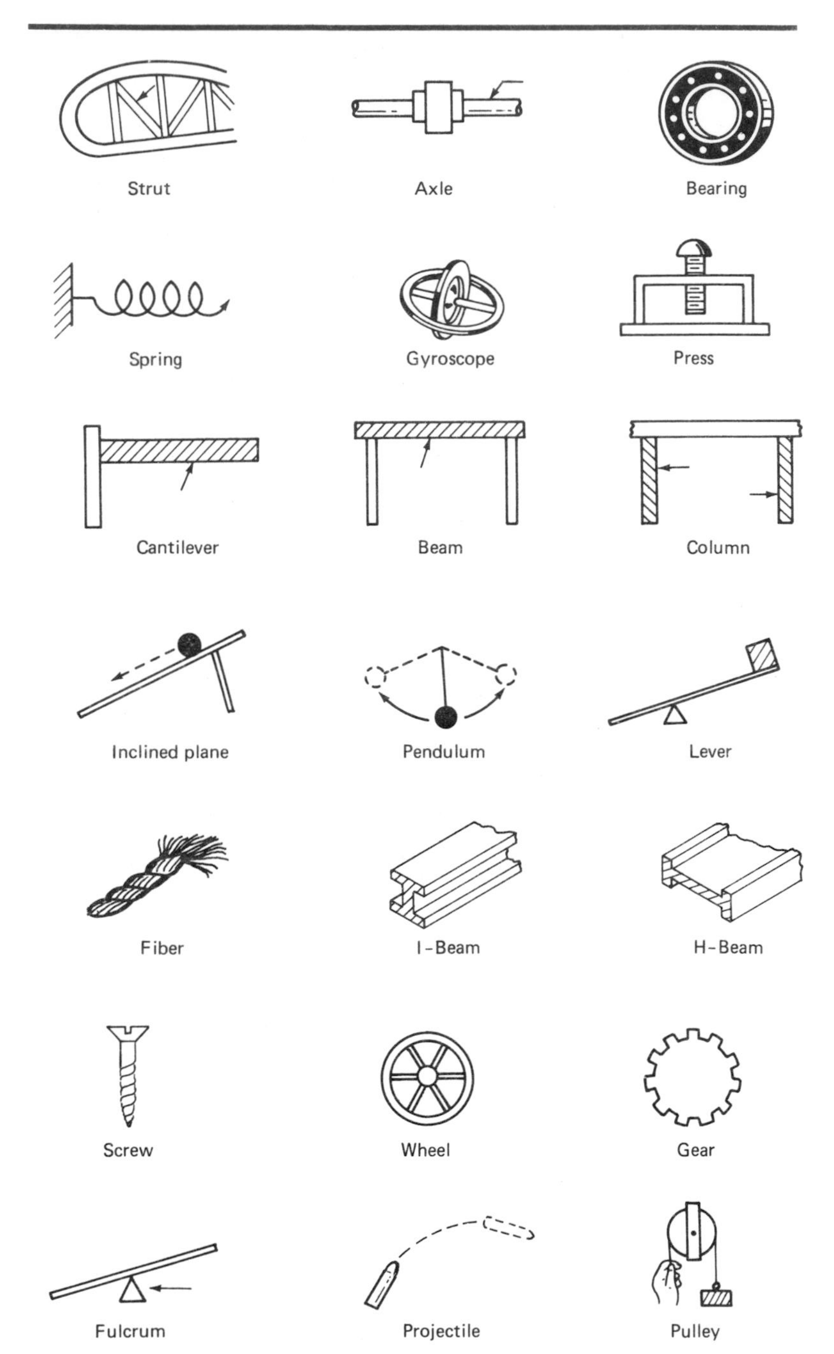

3. Force, Energy, Power

When forces in one direction exceed forces in the other direction, action results. (When they balance, action does not occur, but internal stress may result; see Section 4.2.3.)

The action may be in many forms: sliding, rolling, turning, flying, colliding, shattering, exploding, etc.

Do the same things happen to objects when *energy* or *power* is applied? The answer is yes; but force, energy, and power have different shades of meaning to the scientist and engineer, even though they may have about the same meaning in ordinary English. And the situation is further complicated because for the physicist another very ordinary English word, *work,* often has almost the same meaning as energy.

- *Force* describes the amount of push given to an object.
- *Work* describes the amount of push given to an object over a given distance.
- *Energy* describes the *capacity* to do that work. *Energy* is expended to do *work*.
- *Power* describes how much work is done during a given time interval.

Anyone who lifts a 100-pound weight must invest 100 pounds of *force*. If the weight is lifted from the floor to a table 4 feet high, then there has been an expenditure of 400 foot-pounds (ft·lb) of *energy* (4 feet times 100 pounds). If this act is completed in 1 second, then we say there has been an investment of 400 foot-pounds per second (ft·lb/sec) of *power*.

4. Units Applicable to the Mechanics of Motion

a. Two groups of concepts All that has been said thus far about mechanics confirms that it is a discipline that deals with objects and their reactions to force. The physicist describes these objects and their reactions by combinations of two sets of concepts:

- Force, energy, and power
- Mass, space, and time

These concepts can be quantified by defining units for each; then each particular factor can be given a numerical value.

b. Two systems of units There are two different systems of units, the English (pounds, feet; abbreviated lb, ft) and the metric (kilograms, meters; kg, m). The International System of Units (Section 2.1.6), known as the SI system, provides a rigorous standard for using metric units with internationally agreed upon names, values, prefixes, and abbreviations. That has not caused people to do away with the

English system, however, and both the English and the metric systems continue to be used. Eventually the international standard will prevail, and every year the SI system gains more ground.

Glossary M lists the most commonly encountered units—both SI and English—with their symbols.

4.2.3 Objects in a State of Stress—Statics

Here is a short "course" on what happens when we subject materials to forces in a static condition. The student of mechanics who later becomes a physicist, mechanical engineer, or metallurgist spends several years studying statics. But we will cover it, albeit superficially, in a few pages.

1. Applying Force to an Object

- Pulling something from two opposite directions at once is *tension.*
- Squeezing something together is *compression.*
- Twisting something is *torsion.*
- Sliding the top of something forward while the bottom is kept from moving is *shear.*
- Hitting something with a sudden blow is *impact.*

Table 4-2 illustrates these force terms.

2. Stress and Strain

In everyday language, *stress* and *strain* are often used to mean the same thing. Both refer to the psychological condition that often results when things are not going right. In the language of mechanics, the difference between the two terms is significant. Because of the application of force—as described in the previous section and in Table 4-2—the structure is said to be under a condition of *stress.* Only then does strain occur. *Strain* is the consequence of stress.

a. Stress *Stress* is force applied to an area.

Mathematically it can be expressed as F/A where F is the force and A is the area. The units of stress are in the English system *pounds per square inch* (psi or lb/in^2) or in the SI system *grams per square centimeter* (g/cm^2).

NOTE: Any combination of force words divided by area words can also designate stress, for example, *tons per acre* or *kilograms per square meter.*

b. Directional stress *Stress* is also force applied in any direction.

TABLE 4-2
SOME FORCE TERMS

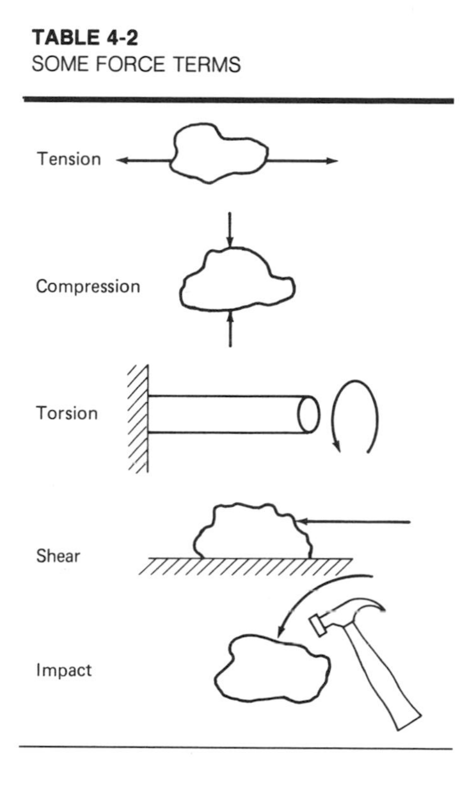

Thus it is possible to have *tension stress, compression stress, torsion stress, shear stress, instantaneous stress* (in any of the above forms) due to impact, or any combination of all these. Stress is a vector quantity (see Section 2.3.5).

c. Strain *Strain* is the *deflection* or *distortion* caused by stress. Any stress applied to a material causes it to change its dimensions (however slightly) or its shape (however slightly).

Mathematically, strain can be expressed as $\Delta L/L$ where ΔL is the change in length of a unit length L. The units of strain are *inches per inch* (in/in) or centimeters per centimeter (cm/cm) or any other ratio of length units.

It is interesting to note that some strain *always* results from stress. For example, if you touch any massive structure—like the wall of the Empire State Building in New York City—with the tip of your little finger, you will distort it, or cause it to deflect. However, the building will tilt by such a small amount that no one will notice it, although instruments exist that will show that deflection.

3. The Stress-Strain Diagram

The *stress-strain diagram* is a graph of stress vs. strain. It shows how much additional (incremental) strain results from each additional (incremental) bit of added stress.

This is the most important diagram for engineers who deal with the effects of forces on materials. If you work for such engineers, you will encounter different stress-strain diagrams over and over again. You may even be asked to draw them.

Figure 4-1 is a typical stress-strain curve for a metal. Note that the stress is always drawn on the vertical coordinate, the y axis, and the strain is always drawn on the horizontal coordinate, the x axis. Each axis is labeled with suitable units.

a. The proportionality of stress and strain For a large group of commonly encountered materials such as metals, glasses, and ceramics, stress and strain start off being linearly *proportional* to each other.

This means that if a doubling of stress causes a doubling of strain, a tripling of stress causes a tripling of strain, and so on. Thus, as shown in Figure 4-1, the first portion of the stress-strain graph is a straight line.

b. The nonproportionality of stress and strain It is possible for a stress-strain diagram not to be linearly proportional. That is, the early part of the stress-strain graph may be curved rather than straight.

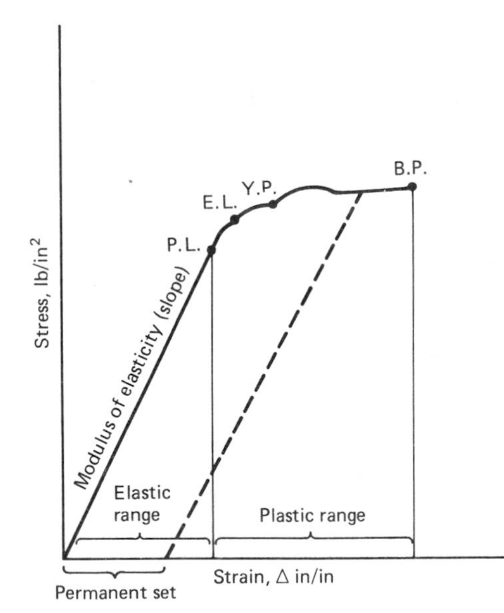

FIG. 4-1 A stress-strain curve of a metal.

Organic materials like rubber or plastic behave in this manner and have such curved stress-strain graphs, like that of Figure 4-2.

Figure 4-2 is a stress-strain curve for rubber. What it shows is that it takes very little force to stretch rubber when one first starts stretching it. Later, after the rubber has been stretched somewhat, it gets harder to stretch it further, and the stress-strain curve gets steeper.

c. The proportional limit As stress is increased on a material, a point is reached at which proportionality ceases. This is known as the *proportional limit (PL)* (Figure 4-1).

d. Elasticity Elasticity means reversibility. That means that when stress is relaxed, the material relaxes too. It does so by merely moving back down the same stress-strain curve until it reaches its initial state, at which with no stress there is no strain (the zero point on the graph).

A rubber band (correctly named an *elastic band*) is elastic. Pull it and it gets longer; relax the pull and it goes back to its original length. The direction of its change in length has reversed.

NOTE: It is possible for certain materials to be elastic (length changes are exactly reversible) without being linearly proportional. The rubber of Figure 4-2 is as elastic in its nonproportional way as the metal of Figure 4-1 in its proportional way.

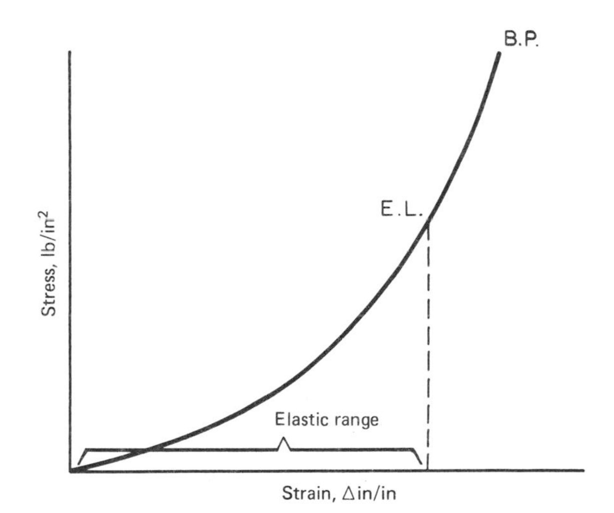

FIG. 4-2 A stress-strain curve of a rubber or plastic.

NOTE: Region to the elastic limit E.L. is elastic (reversible) but not linear.

e. The elastic limit and elastic range Although most metals behave elastically with small stresses (changes in length are exactly reversible), the same materials are not always elastic when stresses get very large. There is a limit to the amount of stress which causes a reversible change in dimension. The particular stress on a stress-strain curve at which elasticity ceases is known as the *elastic limit (E.L.)*, and the amount of dimensional change of any material at the elastic limit (and therefore the amount of dimensional damage that is reversible) is known as the *elastic range*. These may be seen in Figures 4-1 and 4-2.

f. The modulus of elasticity The slope (steepness of the line; see "Plotting Graphs" in Section 2.3.4) of a stress-strain curve in its proportional region, if it has one, is known as the *modulus* (plural, *moduli*) or the *modulus of elasticity* of the material being examined (Figure 4-1). The steeper the slope, the stiffer the material (Figure 4-3).

Stiffness is a measure of the resistance of a material to distortion under stress.

A high modulus (steep slope) means that a lot of stress results in only a small amount of strain—as is true of tungsten, a stiff material (Figure 4-3). A low modulus (shallow slope) means that a small amount of stress imparts a large amount of strain—as is true of the very unstiff metal lead (Figure 4-3). Lead is so soft (unstiff) that you can make a dent in it with your fingernail.

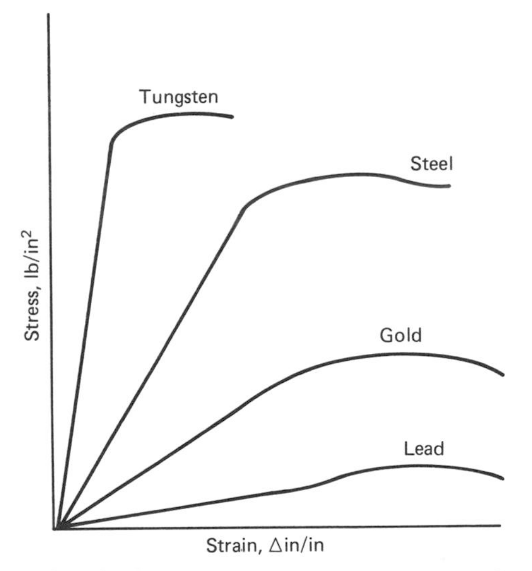

FIG. 4-3 Stress-strain curves for some metals of different stiffnesses (moduli of elasticity).

g. The permanent set As stress is further increased beyond the elastic limit, the material elongates further, but no longer elastically, that is, if stress relaxes and "lets go" entirely, the material does not come back to its previous shape or length.

Instead the curve follows a line (the dotted line in Figure 4-1) which has the same slope as the elastic line (the same modulus), but there is now an offset on the x axis.

The material is now longer than it was originally, and will remain so "permanently," even if all stress is removed. That offset is known as the *permanent set*.

h. Plasticity and the plastic range The portion of the stress-strain curve beyond the elastic limit within which permanent set can occur is known as the *plastic range* (Figure 4-1).

Plasticity is the opposite of elasticity. Plastic deformation is irreversible; elasticity is reversible.

NOTE: Although many classes of materials may show permanent deformation, the terms *plasticity* and *plastic range* are usually employed only when metals are being described.

i. The yield strength The materials engineer finds it useful to define a place on the stress-strain curve where permanent set can easily be observed to be beginning. That point is known as the *yield strength (YS),* and it is the stress that produces a defined amount (usually around 0.2 percent) of permanent set.

The yield strength is always slightly bigger than the elastic limit. Its value to the engineer is that it gives the information that from that point on, the material is surely in its plastic range and all further elongation will result in further permanent set.

NOTE: Although it is possible to define a yield strength for many classes of materials, it is usual to employ it for metals only.

j. The breaking point If stress is further increased beyond the yield strength, a level of stress is reached at which strain no longer increases and further elongation does not occur. At this point the material breaks. Logically enough, this is known as the *breaking point (B.P.)* (Figures 4-1 and 4-2).

k. Each material has its own stress-strain diagram The previous treatment, including the definitions of terms describing what happens to a material under increasing stress, showed ways of describing a material and how it differs from other materials. Every material has its own characteristic stress-strain curve, with a different modulus, proportional limit, elastic limit, yield strength, and breaking point.

We can think of some materials as *stiff* or *soft* or *strong* or *weak* or *resilient* or *ductile* or *springy* or any of dozens of strength adjectives. These characteristics can all be "picked off" the stress-strain diagram by the specialist in the field of *strength of materials,* which is a part of the study of mechanics.

4. Structural Analysis

Knowing how stresses may be applied to an object (see "Applying Force to an Object" and "The Stress-Strain Diagram" in Section 4.2.3) and how these stresses may cause strains in materials all the way to their breaking points (see "The Breaking Point" in Section 4.2.3), a specialist in the field of mechanics is in a position to design structures that will not fail. The crucial question always is: Will the bridge or building or structure fall down?

By analyzing the situation mathematically, the engineer can actually determine the amount of each stress at each point in any element of any structure. Then not only can he or she predict with confidence whether the structure will survive or fail, but the engineer can also say—in exact numbers—how far that point is from failing. From this it is possible to determine a factor of safety.

The *factor of safety* is a multiple of the breaking point. Thus if a girder will have to absorb a stress that will never exceed 30,000 pounds per square inch and it is made so large that it can absorb a stress of 300,000 pounds per square inch (the breaking point is 300,000 pounds per square inch), then the factor of safety is 10. (And factors of safety of 10 are common, so you can breathe easy.)

The techniques and notation of structural analysis are the same as those of algebra and geometry.

Particular algebraic symbols in common use in mechanics are listed in Glossary G.

a. Structural elements Most structures, like most living things (which are also structures), have skeletons. These skeletons resist applied forces and as a result have strength.

Vertebrates (human beings and all warm-blooded animals are vertebrates) are living things that have skeletons inside their bodies. The bone of the skeletons is stress-resistant; the flesh that covers the bones is far less stress-resistant. Many buildings are like vertebrates. They have a steel or a wooden-beam skeleton, which is strong, covered by insulation, sidings and shingles, which are weak. Invertebrates, such as lobsters and clams, have their skeletons outside their bodies. They are strong because their surfaces are strong. A man-made structure that has a similar skin

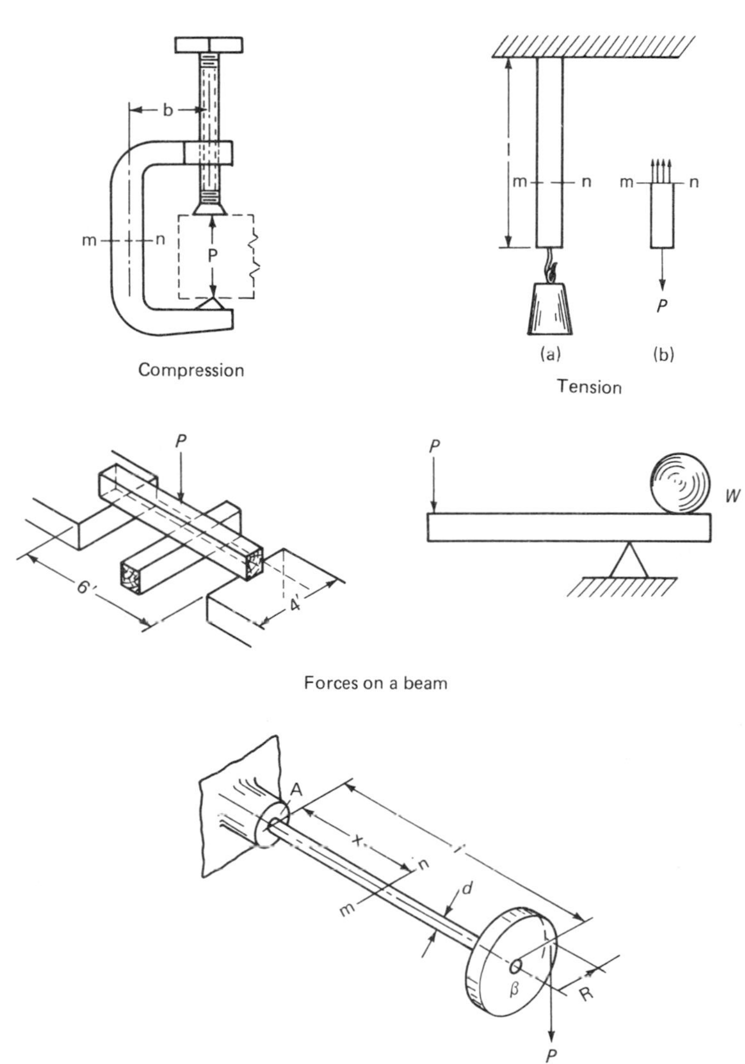

FIG. 4-4 Nature of stresses within structural elements as a result of force P.

strength is a canoe. Another is an igloo. Another is the Houston Astrodome.

The skeletal portions of structures, which bear all the weights and resist all the pressures, are known as the *structural elements.*

The beams and columns in Table 4-1 are structural elements. They behave like skeletons when they are inside the walls of buildings.

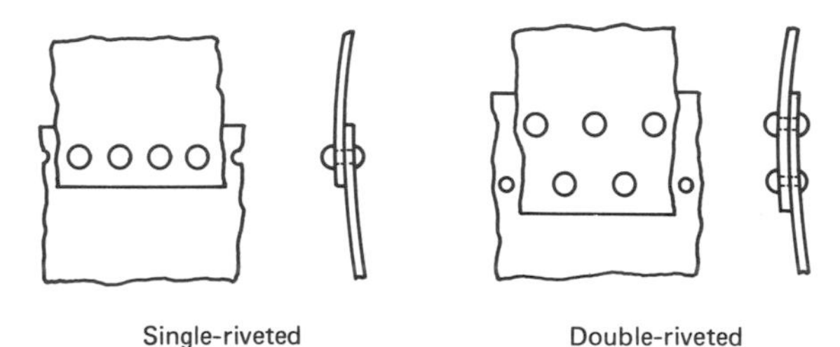

FIG. 4-5 Fasteners and fastening methods.

The structural elements are designed so that applied stresses may be absorbed within them without failure (Figure 4-4).

b. Fastening structural elements together Just as our bones are held together at joints, so must structural elements be held together. Joining can be accomplished by welding or gluing, but there is a variety of specially designed fasteners as well. As these fasteners hold the elements of a structure together, they too are subjected to stress, which they too must resist.

Figure 4-5 illustrates some of the more common fasteners and fastening methods.

4.3 HYDRAULICS

Hydraulics is a subdivision of mechanics. In fact it is the mechanics of fluids and is thus also known as *fluid mechanics*. (A *fluid* is anything that flows, and that includes water and other liquids, and air and other gases.) It has its own kind of statics, *hydrostatics*, and its own kind of dynamics, *hydrodynamics*. Hydraulics is the special province of civil and mechanical engineers, but it can appear in many other fields, even medicine.

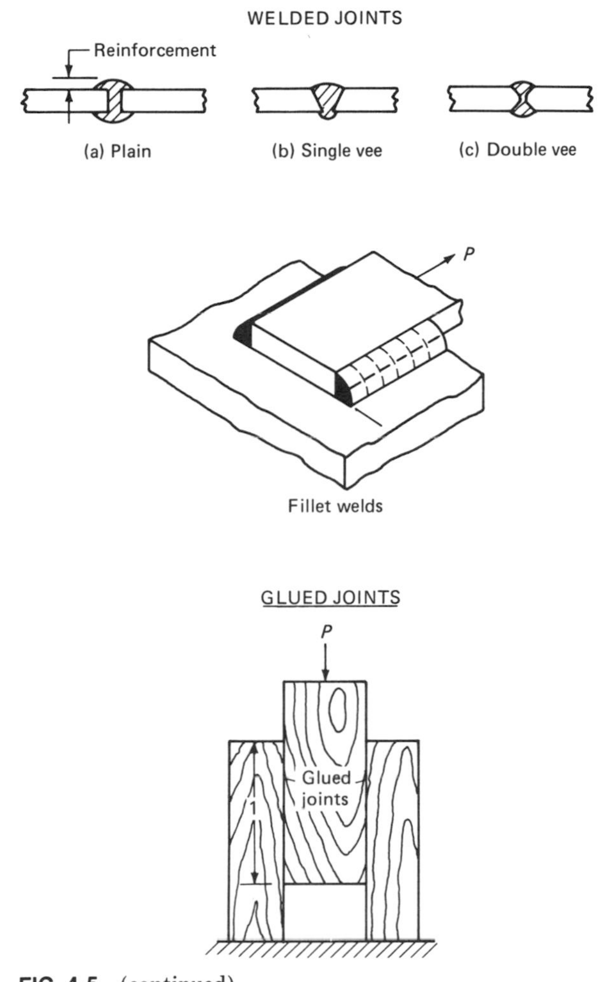

FIG. 4-5 (continued)

For example, the bloodstream in the human body is a hydraulic system. The biologist can describe the bloodstream in the same terms and units as the civil engineer uses to describe a system of rivers. And in both cases the fluid (blood, water) can be made to change direction and proceed with momentum in that new direction by a mechanical "pusher" (the heart, a pump).

The notation and the techniques of presentation on the printed page resemble those of mechanics whether the fluid is blood in the body or water in a river.

4.3.1 Hydrostatics

As in mechanical statics, in hydrostatics forces exerted upon objects are in equilibrium—or balance—and nothing moves.

A *dam* is a hydraulic device that is in a substantially static condition. The reservoir of water that it holds back presses on the dam with the force of the water's weight; the dam presses back with the same force. Nothing moves. The system is static.

The water faucet on your kitchen sink is a miniature dam. Until the faucet is turned, the water stands still in the pipes. But the water pushes on the

TABLE 4-3
SOME WORDS AND CONCEPTS OF HYDRAULICS

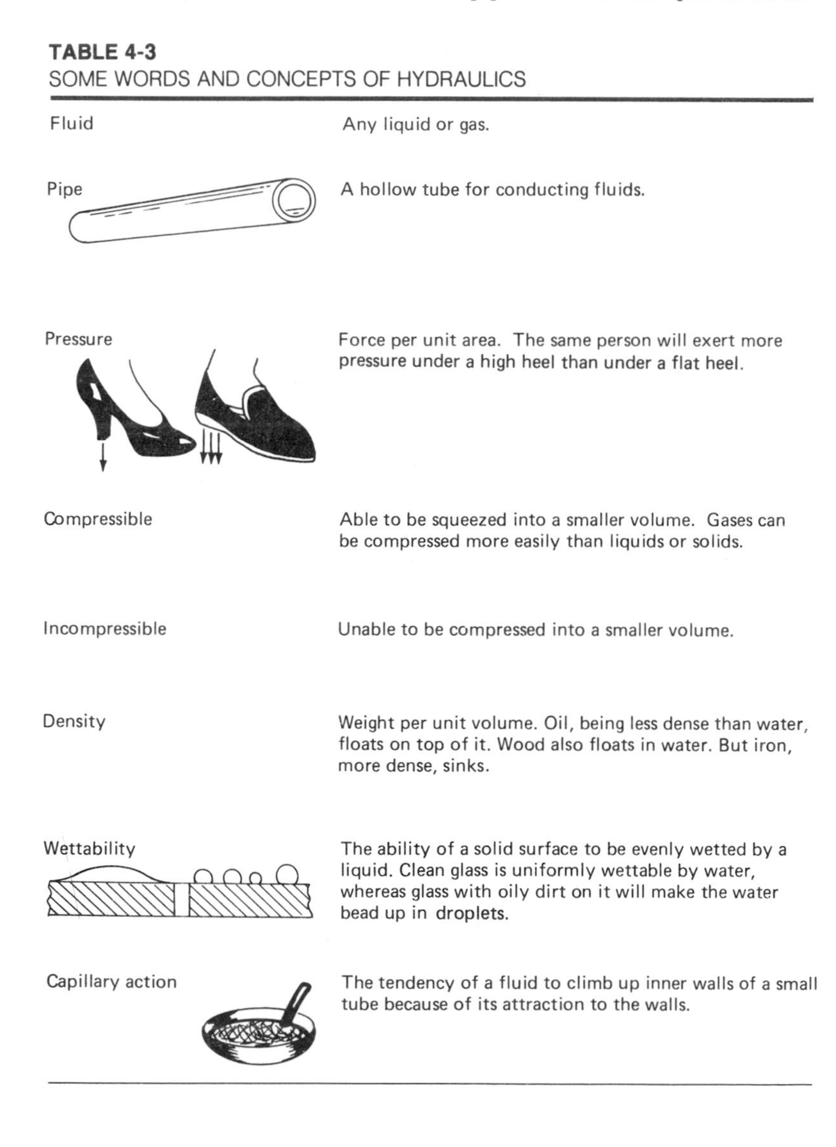

Fluid	Any liquid or gas.
Pipe	A hollow tube for conducting fluids.
Pressure	Force per unit area. The same person will exert more pressure under a high heel than under a flat heel.
Compressible	Able to be squeezed into a smaller volume. Gases can be compressed more easily than liquids or solids.
Incompressible	Unable to be compressed into a smaller volume.
Density	Weight per unit volume. Oil, being less dense than water, floats on top of it. Wood also floats in water. But iron, more dense, sinks.
Wettability	The ability of a solid surface to be evenly wetted by a liquid. Clean glass is uniformly wettable by water, whereas glass with oily dirt on it will make the water bead up in droplets.
Capillary action	The tendency of a fluid to climb up inner walls of a small tube because of its attraction to the walls.

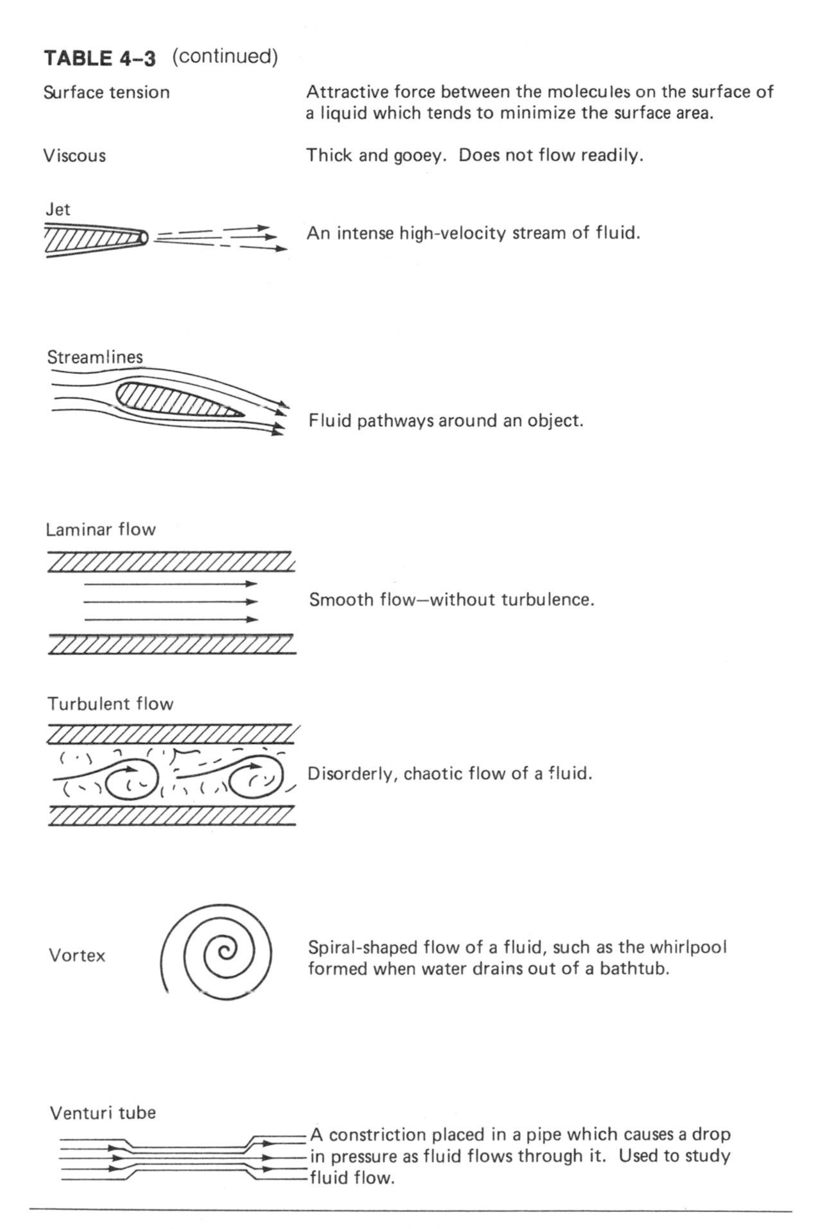

TABLE 4–3 (continued)

Term	Description
Surface tension	Attractive force between the molecules on the surface of a liquid which tends to minimize the surface area.
Viscous	Thick and gooey. Does not flow readily.
Jet	An intense high-velocity stream of fluid.
Streamlines	Fluid pathways around an object.
Laminar flow	Smooth flow—without turbulence.
Turbulent flow	Disorderly, chaotic flow of a fluid.
Vortex	Spiral-shaped flow of a fluid, such as the whirlpool formed when water drains out of a bathtub.
Venturi tube	A constriction placed in a pipe which causes a drop in pressure as fluid flows through it. Used to study fluid flow.

faucet with the *pressure* provided by the municipal water system. Faucets, and other such turn-on/turn-off devices, are known as *valves*.

4.3.2 Hydrodynamics

As in mechanical dynamics, in hydrodynamics forces exerted on objects cause them to move. There is imbalance.

When the sluice gates of a dam are lowered, the water—now unopposed by the equal and opposite force of that portion of the dam—rushes forth. It may turn a waterwheel, which in turn may cause an electric generator to rotate to provide electric power.

When the water faucet in the kitchen is opened, the same thing happens. The water rushes forth.

NOTE: Not all fluids have to be liquid. Gases are also fluids. They behave similarly to liquid fluids, and the laws governing their behavior are covered in separate disciplines of statics (similar pressures are exerted) and dynamics. The latter is *aerodynamics* and is the basis of aeronautical engineering. Jets and propellers push against the air, and airplanes derive *thrust* from this, and move.

NOTE: The movements (dynamics) of the air and the oceans are also the bases of *meteorology* and *oceanography,* as well as *naval architecture* and *marine engineering.*

Table 4-3 lists and illustrates some of the more common devices and actions of hydraulics. Further listings may be found in the "Dictionary of Technical Terms" in Part 2.

4.4 SOUND AND ACOUSTICS

In Section 4.1.2 it was pointed out that classical physics is that branch of the subject that deals with what can be experienced with our senses. *Acoustics* is certainly part of classical physics, and we experience it with our ears. It is the science of sound. It has many similarities with *optics,* the science of light, which we experience with our eyes.

4.4.1 "Aud-" and "Son-"

There are two Latin roots that form the basis of many acoustic words. They are "aud-" from the Latin *audire,* which means "to hear"; and "son-" from the Latin *sonus,* which means "sound."

Examples are *audio frequency, audiometer, sonar,* and *ultrasonic.*

4.4.2 The Movement of Sound

Sound travels. It starts at one location (where it is generated) and is gathered in (received) at another. It travels through materials (there can be no sound in the absence of materials, as in vacuum), and it goes through different materials at different speeds.

1. Generators and Receivers of Sound

Sound moves through a material from its source or *generator* to a *receiver.*

A bell struck by a hammer is a generator. So are clouds when there is a lightning flash accompanied by a thunderclap. Our ears are receivers.

Man has contrived machines for the generation of sound through material. The loudspeakers in your stereo set are such generators. Man has also contrived receivers of sound. A microphone is such a receiver. When machines generate unwanted sound, we call that sound *noise*.

NOTE: Noise is a collection of audible sound at random pitches and intensities and without rhythm. It is usually unwanted because it obscures meaningful sound, such as music or speech, which has particular organized nonrandom pitches, intensities, and rhythms.

2. Media

Sound travels through materials. A material that conducts sound is known as a *sound medium*.

Any material—solid, liquid, or gas—can be a sound medium, and a sound medium must be present in order for there to be sound. When no materials of any sort are present, there can be no sound. Since there is no air in outer space, it is utterly soundless.

We are most accustomed to sound's traveling through air, but it moves through water too (the movement of sound through water enables porpoises to communicate with each other). We make use of sound in water when determining water depth. In fact this method is known as *depth sounding*. We also use the movement of sound in water to detect submarines and ships at a distance. Such a system is called *sonar* (from *so*und *na*vigation *r*anging), and it is used by all the navies of the world.

Have you ever placed your ear against a tabletop and then drummed on the surface with your fingers? The loudness level is surprisingly high. Not only does sound travel through solids, but the more solid the medium, the better, and faster, sound travels.

NOTE: The same pertains to heat. Solids conduct heat better than liquids, and liquids conduct heat better than gases.

3. The Speed of Sound

It requires time for sound to travel through a medium. The more solid the medium, the faster sound goes through it.

- The speed of sound through air is 1086 feet per second.
- The speed of sound through seawater is 5022 feet per second.
- The speed of sound through wood (oak) is 14,600 feet per second.
- The speed of sound through steel is 18,000 feet per second.

In a thunderstorm, the flash of lightning and the clap of thunder actually occur at the same instant. However, since the light flash travels at a much faster rate than the sound of the thunder, we perceive them as happening at different times. The time interval is then a measure of the distance we are from that flash of lightning. If 3 seconds pass between the time we see the flash and the time we hear the thunder, we know that the flash took place 3258 feet away (since sound moves through air at 1086 feet per second).

4. Sonar

Measurement of the time of sound travel through a medium is a common way for engineers and scientists to measure distances. It is the basis of *sonar,* which by measuring time intervals necessary to detect echoes of sounds emitted—and then detected—by the sonar equipment can determine how far away the ocean bottom is and how far away an enemy submarine is.

NOTE: Since light travels at a speed of 186,000 miles per second, measurement of the time it takes for it to move from its source to its receiver may be used in the same way. The only difference is that the times are so short, compared with the times of sound travel, that measurement is harder and requires more complex electronics. This technique is known as *radar,* and is the analogue, using light, of sonar, using sound.

4.4.3 Vibrations

When a medium is made to *vibrate,* sound is generated.

The vibrating medium causes the air around it to vibrate too, in the same way (at the same number of vibrations per second and at the same intensity). Then that air becomes the medium and the sound travels through it. When sound reaches our eardrums, they also vibrate, and we hear the sound.

1. The Periodic and Harmonic Nature of Sound Vibrations

When something vibrates, it actually moves. That movement turns out to have much in common with the other motions we have encountered, such as those of the swing and the seesaw in Section 4.2.2. In "Motion Words" in that section, it was pointed out that repeated oscillations such as those that a swing or a seesaw make are called *periodic, harmonic oscillations.* A sequence of vibrations of a medium is similar to a sequence of swings of a swing and may be described in the same way.

Thus

- Vibrations are *harmonic* (can be described by trigonometric functions).
- Vibrations are *sinusoidal* (see "The Trigonometric Ratios" in Section 2.2.2).
- Vibrations are *periodic* (or reiterative—which means they repeat, and keep repeating in the same way as long as we hear the same sound).
- Vibrations have characteristic *frequencies* (which means so many periods, or cycles per second; see below).
- The time to complete one vibration is a *period.*

- One complete vibration is a *cycle*.
- The maximum distance traveled by the medium during a cycle is the *amplitude* (just as the maximum distance reached by the swing before it changes direction is the amplitude).

2. Waves

When vibrations occur in a fluid medium, it is easy to accept the fact that they are occurring in *waves*. If we could see them, they would look like waves at a beach—they are periodic (repeating), each wave occurs at a given frequency, occurs for a given period (duration), and has its own height (amplitude). The distance from the tip of one wave to the tip of the next wave is called the *wavelength*.

Exactly one wavelength occurs in each period. If the period is short, the frequency is high. For short periods (high frequencies), wavelengths are short; for long periods (low frequencies), wavelengths are long.

There is no limit to the size of wavelengths. They can be hundreds of feet long, or they can be millionths of a millimeter long.

4.4.4 Music

Music is a category of sound. It differs from noise in that it is organized sound. Noise is random sound (Section 4.4.2).

Speech is also organized sound. It differs from music in that it occurs at a narrower range of frequencies (pitches) and loudness (amplitudes).

1. Words and Concepts of Music

Both the musician and the physicist use the same concepts, although the words they use to designate these concepts differ.

- The physicist's *frequency* is the musician's *pitch*.
- The physicist's *amplitude* is the musician's *loudness* or *volume*.
- But the physicist's *noise* is very much the same as the musician's *noise*.

2. Phase

When two frequencies occur simultaneously, as when a musician plays two different notes at the same time, they may either *harmonize* with each other or *interfere* with each other. *Harmony* means that there is either a concurrence or a partial concurrence of the peaks of the sound waves, which means that the peaks of the sound waves at those frequencies occur at the same time.

Certainly if two violins play the exact same note at the exact same time, there is concurrence (harmony).

NOTE: Harmony (concurrence in time of the peaks of the sound waves) has nothing to do with the relative loudness of the two sounds. A sound peak of high amplitude (loudness) can be harmonious with one of low amplitude. They both produce the same frequency, and thus the same pitch. Then every peak of amplitude from violin 1 occurs at the same time as every peak of amplitude from violin 2. This is what happens in the violin section of a symphony orchestra. But what about a note played by the cello? It harmonizes too. Yet it is a lower note (lower frequency, lower pitch). The fact is that the peaks of the cello notes concur with the peaks of the violin notes every other time (one octave difference; see next section) or every third time (two octaves difference), and so on. Even if the notes are not exactly the same, they have a strong relation with each other—and that is harmonious.

When frequencies have some measure of concurrence with each other, we say they are *in phase*.

The acoustician's *out of phase* is the same as the musician's *discord*.

3. Intervals and Octaves

Musical sounds are in phase or harmonize if the ratios of their frequencies are small whole numbers. When the ratio is 1:2, 1:3, etc., the intervals are called *octaves*.

For example, a note and the same note one octave higher have frequencies in the ratio of 1:2.

- The "standard A" to which orchestras tune, the forty-ninth note on the piano, has a frequency of 440 cycles per second.
- The note A that is one octave higher in pitch has a frequency of 880 cycles per second.

NOTE: Nature makes harmonious sounds. Bird songs are usually harmonious, which means that songbirds tend to sing notes at harmonious intervals (the ratios of the frequencies are small whole numbers).

NOTE: Nature makes noise too. The ratios of the frequencies of the sound of a waterfall, or of the creaking of a tree in the wind, are random. Thus they are unorganized, and most people do not think of them as musical.

The harmonious sound made when notes at *in-phase* intervals sound simultaneously is called a *chord*.

4.4.5 Ultrasonic and Subsonic

Some sounds vibrate at frequencies that are too high or too low for our eardrums to detect.

NOTE: An eardrum is a thin membrane of skin that in fact resembles the skin of a drum. We experience hearing when that skin vibrates with the same frequency as the sound that hits it. But the eardrum has limited capability to vibrate. If the vibrations are too rapid, it isn't flexible enough to follow them; if they are too slow, it isn't sensitive enough to know they are there. In both cases, the eardrum merely "sits there," without vibrating at all, and even though the sound exists, we do not hear it.

1. Ultrasonic Sounds

Sounds that are too high for us to hear are described by the adjective *ultrasonic* and the noun *ultrasound.* We make use of ultrasound in sonar and in medical diagnostics, where it is now a tool of the radiologist, along with x-rays.

One reason to use such high frequencies of sound is to receive such an accurate echo of the object off which the sound is reflecting that we can construct a picture of it. The higher the frequency, the shorter the period, that is, the shorter the wavelength. The wavelengths of low-frequency sound are often as large as the object being viewed; in that case the sound image is just a vague blur. It is only when the wavelengths are short compared with the size of the object being viewed that we can observe details from each part of the object. Then the image of the object can be seen with *definition,* which means without fuzziness.

NOTE: The first time we get a "baby picture" is when we look at an ultrasound echo trace of a fetus in a womb. Such an ultrasonic picture is known as a *sonogram.* We could also see the fetus with x-rays, but x-rays can be harmful. Sound, however, does no damage that we know of.

Ultrasound is also the basis of a kind of cleaning machine used by the optometrist. And some dentists make use of an ultrasonic tool to aid in drilling and in blasting away hard deposits called tartar.

2. Subsonic Sounds

Sounds that are too low for us to hear are called *subsonic* or *infrasonic.* There are no known uses for subsonic sound, but such sounds do occur in nature.

NOTE: The eardrums of some birds and animals are far more sensitive than ours, and these creatures can therefore hear farther into the subsonic and ultrasonic frequencies than we can. Indeed, dog whistles make an ultrasonic sound to which dogs can be trained to respond, but which humans cannot hear.

4.4.6 Absorption and Reflection of Sound

In order to make any medium vibrate—and thus be a generator of sound (whether musical or noisy)—energy is required; force and power (see "Force Energy, Power" in Section 4.2.2) are also required.

You can sometimes see the vibration caused by the energy.

- A hammer hits a bell, and you hear the clanging of the bell. If you look at the rim of the bell, you will see a blur caused by the vibration.
- The fingernail of a guitarist plucks a string, and you hear the twanging of the string. If you look closely at the string, you will not be able to make out that it is a string because you will see only the blur caused by the vibration.

1. When Sound Energy Hits a Surface

When the energy of the sound "hits" something, one of two things happens to it. Either it rebounds—*reflects*—to continue sounding (in which case we experience what we call *echoes* and *reverberations*) or it is *absorbed* (in which case we experience immediate quietness the instant the initial sound stops).

Hard surfaces—like metal or a stone wall—reflect sound. Soft surfaces like people and their clothes absorb sound. That is why the same music sounds different in a concert hall that is empty and one that is filled with people. That is also why a room gets more cozy when you hang curtains. Not only does it look better (it has *visual harmony,* which is discussed in the section on color in Section 3.5.1), but your ears are not assailed by acoustical reverberations or echoes—because sound has been absorbed, not reflected.

2. Attenuation of Sound

Materials that absorb sound are referred to as providing acoustical *attenuation* (partial absorption) or acoustical *impedance.* Acousticians and architects who design concert halls are skilled in handling the attenuation of sound.

3. Anechoism

When the material that lines the walls of a chamber is extremely effective in absorbing sound, all the sound energy will be absorbed. Then there will be none left over to strike our ears, and no sound will be heard at all. Such a chamber is *anechoic* (no echo).

If you add curtains to a room, you make it more anechoic.

NOTE: The concept applies also to other waves that involve other kinds of vibration caused by other kinds of energy than mechanical, such as microwave electric radiation or light rays. They too can be reflected or absorbed (see "The Absorption and Reflection of Light" in Section 4.5.1 and "Blackness" in Section 4.6.3). Anechoic chambers and surfaces can be devised for them too.

CAUTION: There is a common spelling error associated with the concept of absorbing:

Absorption is CORRECT.

Absorbtion is INCORRECT.

4.4.7 Units of Sound

Sound intensity—loudness, volume—is a function of vibration amplitude. The greater the amplitude, the greater the loudness. The loudness can be measured quantitatively. It is then expressed in the sound unit known as the *decibel* (abbreviation dB).

The basic unit is the *bel* (after the inventor of the telephone, Alexander Graham Bell). But it is rarely used. Decibel, which is a tenth of a bel, is the commonly used unit.

NOTE: Zero decibels is defined as the threshold of hearing. A quiet office typically has a noise level of about 35 decibels; a noisy office has a noise level of about 60 decibels. A boiler factory—with hammers banging and machines clattering—will be at about 100 decibels. And a rock band in a small nightclub room, with all its amplifiers turned up to to maximum settings, will blast its clients with 135 decibels, which is at the threshold of tolerance of the human eardrum.

4.5 LIGHT AND OPTICS

Optics is the science of *light*. Like other areas of classical physics, it involves phenomena that can be experienced by one of our senses. Just as we are aware of sound because of our ears, we are aware of light because of our eyes. Light is also part of modern physics in that, like sound, it is a wave phenomenon. Thus like sound, light can be characterized by its frequency, amplitude, period, wavelength, and phase. It has, in fact, many similarities with sound, and reference should be made to the preceding sections on acoustics for a treatment of those features which are similar. They can be described as follows.

4.5.1 Comparison with Acoustics

1. "Opto-" and "Photo-"

Just as "aud-" and "son-" form the basis of many acoustical terms and words, so the prefixes "opto-" and "photo-" form the basis of many optical terms. Examples are *optometry, fiber optics, photosynthesis,* and *photomultiplier.*

2. The Movement of Light

Just as sound can be generated and received, so light can be generated (by a light bulb, for example) and received (by our eyes, or camera film, for example). And just as sound travels through space with the speed of sound, so light can travel through space with the speed of light (which is almost a million times faster than the speed of sound).

a. Media Light travels best in the *absence* of media. But it can travel in media too, for example, in glass or in air.

Thus in this aspect light differs from sound. Sound exists only when it causes materials to vibrate, and thus *requires* media in order to exist.

An example of the transmission of light in the absence of a medium is the transmission of light from the sun and the stars. Those rays of light reach us, through unimaginable distances of outer space, virtually unattenuated. The distances are so great that even at the enormous speed of light—186,000 miles per second—it takes years for starlight to reach us: *light years.*

b. Radar Radar is the analogue, in electromagnetic rays, of sonar (see "Sonar" in Section 4.4.2). By measuring the time it take for microwave beams (a part of the invisible spectrum of sunlight; see "The Wide Range of Possible Frequencies" in Section 5.1.5) to traverse a distance, we can calculate how great that distance is.

Radar equipment is essentially a system that measures those times, as sonar equipment is essentially a system that measures the times given sounds require to traverse given distances.

3. Frequency

The frequency of light waves in optics is analogous to the frequency of sound waves in acoustics.

a. Color In optics, frequency determines *color.* The higher the frequency, the more violet the color; the lower the frequency, the more red the color (see "Color" in Section 4.5.1).

In acoustics, the frequency determines *pitch.* The higher the frequency, the higher-pitched the sound; the lower the frequency, the more bass the sound.

b. The electromagnetic spectrum The range of light frequencies is part of the *electromagnetic spectrum.*

The spectrum includes not only visible light but invisible radiation—which is also a form of light—such as radio waves, microwaves, x-rays, cosmic rays, infrared radiation, and ultraviolet light.

NOTE: All these forms of rays are known as *electromagnetic radiation* because they all contain an electric and a magnetic component. They differ only in their frequency. This means that light, heat, electricity, etc., are all related.

In acoustics, similarly, there is a sound spectrum. It too is a range of frequencies and includes the audible region and the inaudible subsonic (infrasonic) and ultrasonic regions.

4. Art

Just as sound gives us music, light gives us visual art. The artist uses the whole visible color spectrum to create a painting on a canvas.

The "ingredients" of light which the artist manipulates are *hue* (color), *lightness* (or *brightness*) and *saturation* (the percentage, or vividness, of the hue).

a. Color When the artist chooses a particular color (the names of which—blue, green, red, etc.—are known as *hues*), he or she chooses a particular light frequency.

As in any wave situation, frequency determines the time for one oscillation or cycle or wave (the period), which in turn determines the length of that wave. (The slower the frequency, the longer the wavelength.)

- The wavelength of red light is 6600 angstroms.
- The wavelength of yellow light is 5800 angstroms.
- The wavelength of blue light is 4600 angstroms.

NOTE: An angstrom is a measure of length. It is equal to 10^{-10} meters. The abbreviation is Å; however, the SI system (see Section 2.1.6) does not use the angstrom, preferring the nanometer instead (10 angstroms = 1 nanometer). If your authors continue to use angstroms, you might ask if they would like to convert the measurements to nanometers (abbreviated nm).

b. Complementary colors A glass prism or a rainy sky can break white light into all the colors of the rainbow: red, orange, yellow, green, blue, and violet. Picture these colors laid out around the edge of a white circle so that the purple end of the rainbow and the red end are joined. Any two colors that lie directly opposite each other across the circle or any three colors that lie equidistant from each other along the edge of the circle are called *complementary colors.*

The word *complementary* means that when the colors are combined, they *complete* a whole. In the case of light, combining three complementary colors will produce white. In the case of paints, dyes, and inks, complementary colors produce a neutral gray or "shadow" color.

- In general, blues are complementary to yellows, reds to blue-greens, and greens to red-purples.
- Yellow, green-blue (cyan), and red-violet (magenta) are three complementary colors used in printing. All the other colors are made from various combinations of these three.
- To paint the shadow on some blue-green pines, an artist will mix with the color of the pines some of its complementary color, red.

c. Saturation versus lightness When white is mixed with colors of different hues, the lightness changes, but the hue does not. Thus brown becomes beige, blue becomes powder blue, and red becomes pink. But these are really the same as light brown, light blue, and light red, respectively, so the color remains what it was—brown, blue, and red. *Saturation* indicates the strength of a color, the amount of hue in that color. The saturation can be increased by adding more of the color (for example, more paint or dye or ink) so that the same color becomes more vivid and intense.

d. Brightness The height of the peak of a wave, its *amplitude,* is a measure of its strength, its intensity. This is so for light or sound or any wave. In the case of light, amplitude is a measure of *brightness.*

Thus sound waves with high amplitude are *loud* and those with low amplitude are *soft*. Light waves with high amplitude are *bright* and those with low amplitude are *dim*.

5. Visible and Invisible Light

Just as we hear sound in a limited range of frequencies (Section 4.4.5), there is a limited range of frequencies to which our eyes can react. There is much light, at higher and lower frequencies, which we do not see.

a. Below visible Engineers have built devices that can both generate and "see" light which our eyes cannot see. We can take pictures using light frequencies below those of red, which we call *infrared,* using infrared cameras and special infrared film.

Infrared light is generated by any hot surface. Light at this frequency is the same as radiated heat (see "Generators of Heat": Section 4.6.3). It is analogous to subsonic sound. If we go somewhat lower in frequency than infrared, we encounter *microwaves* and *radio waves,* which are "seen" by the antennas of our television sets and our FM and AM radios.

b. Above visible *Ultraviolet light* exists just at the top limit of our capability to see. It is analogous to ultrasonic sound.

Light that has even higher frequencies than ultraviolet light is *x-ray* radiation. The radiologist uses x-ray cameras and films as generators and receivers.

6. Absorption and Reflection of Light

Like sound, light can be reflected or absorbed, or partly absorbed *(attenuated).*

A *mirror* is an example of a light-reflecting surface.

Curtains that absorb (muffle) sound also absorb light. Different textiles have different reflectivities and absorptivities. Satin is somewhat reflective, though not so good as a mirror. Velvet is a highly absorbing fabric. Its "velvety" look comes from the fact that virtually no light at all reflects from it.

Frosted glass is attenuative. Some light goes through it, but some is absorbed.

4.5.2 Light as Radiation

1. The Nature of Light

So far we have dealt with the similarities between sound and light. This has been helpful because the same explanations apply to many of the phenomena. The remainder of the sections on optics describe matters that are unique to fields involving radiated energy in wave form, including optics, and have no relation to acoustics.

The major difference between sound waves and light waves is that sound is mechanical energy and light is electromagnetic energy.

You can tell that sound is mechanical energy (and thus really part of mechanics) because it causes media to move—to vibrate (Section 4.4.3). The movement is so real that you can feel it in your bones when low tones are sounded or see it in the blurred look of a plucked guitar string. But light is not mechanical and has no such mechanical effect on materials. When light passes through glass, it does not cause the glass to vibrate. The essential thing about light that distinguishes it from sound is that light is *radiation*. It *radiates* from a source (in *rays*).

2. The Generation of Light

a. Emission Both sound and light start at a source, known as a sound generator or a light generator. A bell is a sound generator when it is struck by a hammer. A light bulb is a light generator when electricity is run through it. The resulting radiation is known as emitted light. The object from which it is generated is called an *emitter*. A material which generates light easily is said to have good *emissivity*.

b. Different kinds of light emisison There are many ways to cause light to be emitted. All result from applying some sort of energy to the material of the light source.

Here material *is* required—all sources of radiation are made of something. (It is only for *light transmission* that a medium is not required.)

Sources of radiation include

- Incandescence
- Luminescence
- Fluorescence
- Plasma emission
- Laser emission

3. Incandescence

a. Dependence on temperature When materials are heated—and as a result they experience a rise in temperature—they become emitters of light. The process of producing that light by heat excitation is known as *incandescence*. The material that has been heated becomes *incandescent*.

This means that the material is emitting rays of electromagnetic energy in the frequency of visible light as a result of having attained a particular (high) temperature.

NOTE: Do not confuse *heat* with *temperature*. Heat is energy. A temperature rise is caused by the application of heat energy. First there must be heat; then there is an increase of temperature.

b. Each temperature has a color Incandescent materials emit light at different wavelengths (colors) depending on how hot they get.

If you were able to look into a self-cleaning oven while it was in its cleaning cycle, you would see that its walls were glowing (incandescing) a dull red shade. Self-cleaning of ovens takes place at about 950°F, and that temperature—of any material—always causes that exact color to be emitted.

NOTE: The relation between color and temperature is so exact that instruments exist for comparing colors of hot surfaces; the colors tell the temperatures of those surfaces. These instruments are known as *optical pyrometers*. Color is measured, but the dial on the face of the meter reads in degrees of temperature.

At an outdoor barbecue you might notice that the coals are a bright orange. This means that the coals are giving off (incandescent) light at the frequency and wavelength of orange. Anything that is hot enough to emit orange light is in the temperature range of 1200 to 1300°F, which means that the surface of the meat being barbecued should sear and blacken nicely.

If you ever visited a steel mill and looked into a furnace in which molten metal was seething and bubbling, you must remember that the emitted light from the molten metal had a color. It was yellow. Any material that approaches 3000°F emits yellow light.

We are all familiar with an example of a material that is hotter than yellow, and which we use every day. This is the tungsten filament in an electric light bulb (also known quite correctly as an *incandescent bulb*). The filament gets to about 6000°F, and at that temperature the light emitted is white.

The most important radiator we know also gives off white light—in this case called *sunlight*. In fact, it emits unimaginable amounts of every kind of radiation in addition to visible light. It is the *sun,* and it is an emitter at one million degrees Celsius.

4. Luminescence

There are many kinds of light that can be generated in the absence of incandescence. Such nonincandescent light is known as *luminescent* light, or *luminescence*. Most (but not all) of these types of light are emitted at cold temperatures.

a. Fluorescence *Fluorescent* light is light emitted when a substance is hit with some sort of external radiation.

The energy of the radiation is converted into light. The radiation then can be said to *stimulate* the light emission. When the radiation ceases to be applied, the light emission ceases.

b. Phosphorescence Phosphorescence is similar to fluorescence, but phosphorescent light continues to be emitted after the energy source is removed.

c. Other types of luminescence These include light emission stimulated by chemical and biological phenomena.

Several examples include the firefly, and the luminescent foam in the wake of a ship at sea, which is produced by glowing marine animals.

d. Terminology Words have been constructed to designate different forms of stimulated light. They consist of a prefix which is the name of the stimulator and the suffix, which is always "-luminescence."

Thus the kind of fluorescent light which results from stimulation by other light is *photoluminescence.* The prefix "photo-" means "light" (the stimulator). An example of photoluminescence is a road sign. At night your automobile headlights shine on the sign with white light, and the sign is stimulated to shine back to your eyes in green light with the message that this is, for example, Route 128.

Another example of photoluminescence is the light that results from shining ultraviolet light at a person's teeth. Living teeth respond with a rather ghastly yellow glow (photoluminescence), but if any false teeth are present, they won't glow at all.

Another example is the *cathodoluminescence* of the screen of your television set. Electrons from the cathode of the picture tube strike the coating of chemical particles that covers the inside surface of its glass, which you are viewing. These particles respond by fluorescing, giving off a light known as cathodoluminescence.

5. Plasma Emission of Light

Under certain conditions, certain gases glow when electric energy is applied

a. Definition What happens is that the molecules that make up the gas ionize, just as some molecules do in water solution (Section 6.3.2). Then the gas consists not just of molecules but also of positive ions and negative electrons. Such a gas is a *plasma* (see "Plasma Physics" in Section 4.8.3).

b. Light emission due to transition of electrons to different energy states Not all plasmas give off light just because the gas is ionized. It is necessary for some of the electrons remaining in the ions to make transitions into different energy states (see "Plasma Physics" in Section 4.8.3). But if the voltage conditions are right, that will happen, and different gases will emit radiation at different frequencies. Some of these frequencies will be in the visible range, and we will see the radiation as light. The process is *plasma light emission.*

An example of plasma emission we have all seen is the emission of light in a thunderstorm. Lightning is emitted from plasma.

So is light from the neon bulbs used in advertising signs.

So is the light of an arc in your automobile engine; the arc is made by the spark plugs. All sparks are plasmas.

So is the light of the arc of a welder's torch, another plasma, since all arcs are plasmas.

6. Laser Light

This is a kind of light never seen in the world until its conception in the 1950s and invention in the 1960s. It must have been a good idea since it won a Nobel Prize for its inventors.

a. Laser The word *laser* is an acronym which stands for "light amplification by stimulated emission of radiation."

Thus like fluorescence and luminescence, laser light is *stimulated,* which means one kind of energy is inputted and another kind is emitted.

NOTE: There is also the *maser* (microwave amplification by stimulated emission of radiation). This is a device which confines the radiation it emits to microwave frequencies, so a beam of microwave energy is emitted, rather than visible light.

The word *laser* has given birth to a large family of terms which incorporate it as a modifier to indicate a large variety of devices and functions. Such terms include *laser drill, laser gyro,* and *laser fusion.* And there is even a verb form—*to lase.*

b. Coherency Laser light is special, different from ordinary light. The difference is that it is *coherent.*

Coherent radiation is *monochromatic* (only one frequency) and has definite phase relations (the peaks of the emitted waves coincide).

For our purposes it is sufficient to know that coherent radiation allows the engineer and physicist to carry out certain operations and measurements not possible with ordinary light, which is *incoherent,* or contains many frequencies (colors) and has no definite phase relations.

c. Intensity Another quality of laser light is that it can achieve enormous intensities in very narrow beams.

The energy in such a beam—a beam only a few thousandths of an inch in diameter—is capable of melting any material in the world in a matter of fractions of seconds.

d. Laser devices Lasers have given birth to hundreds of optical devices which operate in unique ways because of the unique qualities of coherent light beams and because of the extreme intensities that these beams can achieve.

Two examples:

Holography uses coherent light to achieve what appear to be three-dimensional images from markings on flat sheets of film.

The *laser photocoagulator* directs bursts of laser light through the lens of a human eye to burn selected regions on a detached retina. These burns later heal into tiny scars that "weld" the retina back into position.

7. Units

- The unit of light energy is the *photon* (abbreviated *ph*).
- The unit of light intensity is the *lumen (lm)*.
- The unit of wavelength (wavelength is usually represented by the Greek letter lambda, λ) is the *nanometer (nm)*.

4.5.3 The Refraction and Reflection of Light

1. The "Bending" of Light

A distinctive property of light is that it is possible to cause it to change its direction. This occurs when *rays* or *beams* of light pass from one medium to another. They change direction, and that direction change is known as *refraction*.

NOTE: Anything that radiates gives off rays. And they are emitted *radially* (Figure 4-6), which means that the emitted energy, in this case light, goes off in all directions in straight lines, like the radii of a circle. The rays remain straight (unbent) until refracted.

a. Angle of refraction Refraction occurs at the *interface,* the boundary, between two media both conducting the same light. Refraction then causes the straight ray of light to change direction. Now the ray continues, but in a new direction.

This is shown in Figure 4-7. The light ray emitted from point K travels along the line KO through medium 1 and hits medium 2 at an angle α_1

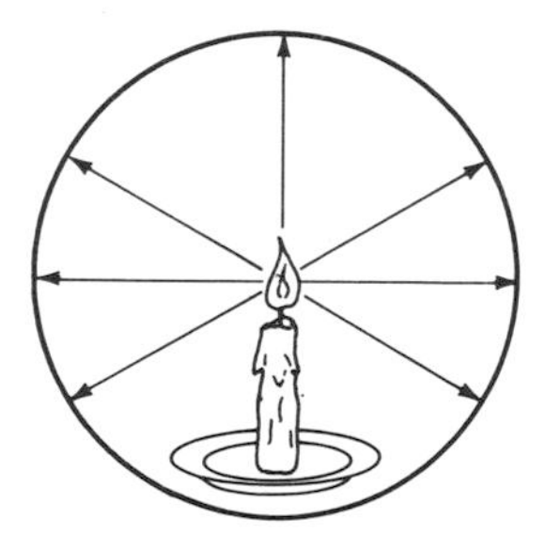

FIG. 4-6 Radial rays of light.

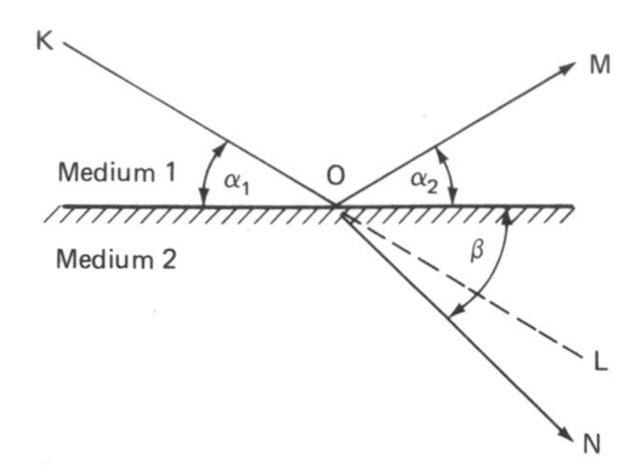

FIG. 4-7 The reflection and refraction of light.

with the interface. It enters medium 2 and *changes direction.* The ray now proceeds along line *ON,* which is at an angle with the interface. And angle β is larger than angle α_1. (If the ray had proceeded without refraction, it would have stayed on the same path in the same direction and would have headed for point *L*. But it didn't do that because it had been refracted.)

An example of the refraction of light can be observed when you go swimming. When you look at objects that are partially in and partially out of the water, you see that these objects seem to bend at the waterline. This can apply to your own body—your own arm or leg can seem bent. But these objects are not bending; the rays of light are bending.

b. Index of refraction The amount of bending depends on how different the two media are from each other. The extent of bending is described by a term known as the *index of refraction.* The larger the difference between the media (air and water are quite different), the larger the amount of refraction.

NOTE: The *index of refraction* is defined as the ratio of the speed of light in a vacuum to the speed of light in a particular substance. Thus strictly speaking, the important factor is the difference between the speed of light in one medium and the speed of light in the other.

In Figure 4-7 the angle β is large for large indices of refraction and small for small indices of refraction.

A *prism* is a transparent medium, like glass, shaped to refract a beam of light so widely as to separate the color elements which may be mixed together to form that beam.

This happens because each color refracts at a slightly different angle. What emerges from the other side of the prism is a set of rainbow beams of different colors (Figure 4-8).

2. Lenses

The most important use of refraction is in lenses such as are used in eyeglasses. A *lens* is a device made of a transparent medium—usually

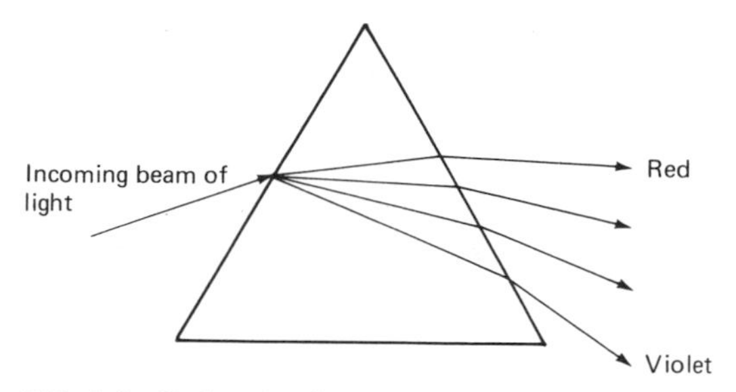

FIG. 4-8 Refraction by a prism.

glass or plastic—with a known index of refraction and with a carefully shaped pair of surfaces. The combination of the shape and the index of refraction causes a desired amount of bending of the light.

a. Convex and concave lenses A *convex* lens looks like a pair of parentheses facing each other—( ). A *concave* lens looks like the same parentheses, but back-to-back—) (.

Figure 4-9 shows such lenses and light paths through them. One can remember which is which by imagining being small enough to get into or onto a lens. The concave lens has a gentle depression, like a cave, to go into. The convex one has a round surface one could slide off of—which would be vexing.

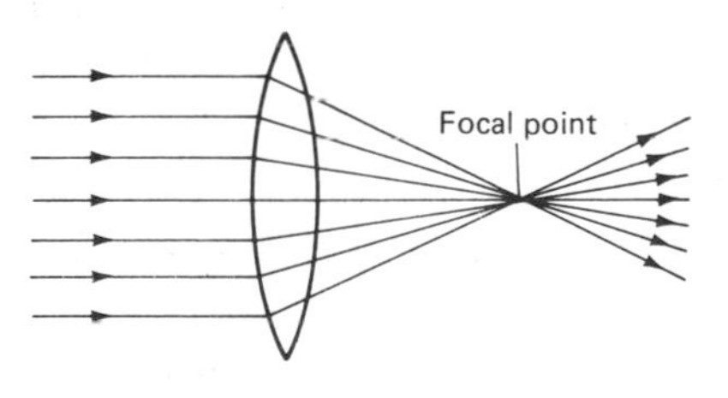

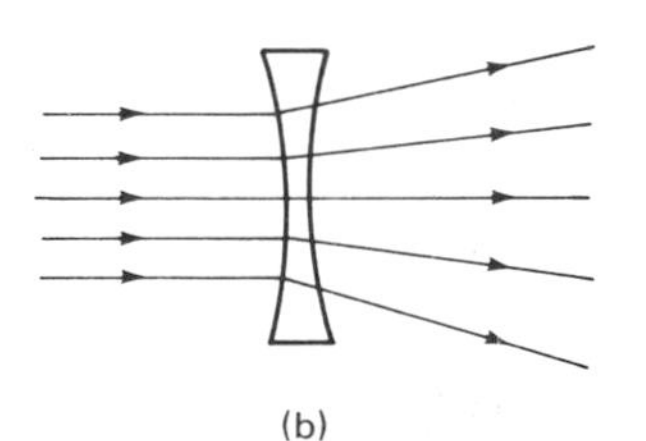

FIG. 4-9 Lenses: (*a*) Converging (convex), (*b*) diverging (concave). The arrows represent rays of light.

b. Converging and diverging lenses Lenses can cause rays to come together, *converge,* or spread apart, *diverge.*

Figure 4-9 shows the refractive actions of converging and diverging lenses.

c. Focus Convergence results in some point—the *focal point*—at which all the rays come together. The distance of the focal point from the lens is called the *focal length.* Beyond the focal point the rays spread apart again, but now the images are upside down.

Figure 4-9*a* illustrates the focus phenomenon.

d. Magnification, minification It should now be clear how lenses can be used (in eyeglasses, telescopes, and microscopes) to make images of objects appear larger or smaller by making the light rays spread out (diverge) or crowd together (converge). When arrays of lenses are arranged to make the image look larger, that is known as *magnification.* Lenses can also be arranged to refract light rays so as to make images appear smaller. That is *minification.*

If you are *nearsighted* or *farsighted,* that means your eyes are the wrong shape so that magnification and focus are wrong. Thus objects appear "out of focus" (blurred) and the "wrong" size. By adding lenses of the right shape and index of refraction (in your eyeglasses), the optometrist changes the overall refraction and thus "corrects" your vision.

e. Units of magnification The measure of the amount of magnification achieved by a lens, or the power of the lens, is the reciprocal of the focal length. The larger that number, the more powerful the lens. The unit of power is the inverse meter (m^{-1}), called a *diopter* (abbreviated D).

3. Angle of Incidence and Angle of Reflection

Not all the light from medium 1 in Figure 4-7 is refracted when it arrives at the interface. That which enters medium 2 is refracted. But some of the light is reflected, since the surface of almost any medium has some of the qualities of a mirror. The magnitude of the *angle of reflection* is unchanged from the angle of first contact (known as the *angle of incidence*). This is a well-known law in optics: The angle of reflection equals the angle of incidence.

This is illustrated in Figure 4-7; the angle of incidence α_1 is the same size as the angle of reflection α_2.

4.6 HEAT AND THERMODYNAMICS

Like sound and light, *heat* can be sensed by our bodies. Just as sound is heard by our ears and light is seen by our eyes, heat is felt by our skin.

There are other similarities. Heat is a form of energy; it moves in waves like light and sound; it can be absorbed or reflected, or both partially absorbed and partially reflected; it is radiated through space like light, or it can be conducted through a medium like sound. In fact, light and heat are virtually the same thing in that both are electromagnetic radiation; they differ only in that the frequencies and wavelengths differ. When radiated, heat is called *infrared radiation,* and it is similar to light in that case but is at a lower frequency and longer wavelength. The portion of physics that deals with heat is the science of *thermodynamics.*

4.6.1 "Thermo-"

The essential prefix that is used in most heat-related discussion is "thermo-," which means "heat."

Some examples are

- *Thermometer*—a measurer of the effect of heat
- *Thermal*—an adjective meaning "pertaining to heat"
- *Thermostat*—a control mechanism that regulates temperature
- *Thermos bottle*—a device that reduces the amount of heat that enters or departs from a particular space
- *Thermocouple*—a temperature-measuring device which generates voltage that varies with the temperature

4.6.2 Temperature

Temperature is not the same as heat. It is the consequence of heat. When heat is applied, the temperature rises; when heat is taken away, the temperature decreases.

The appliances in a kitchen illustrate this. If you want to heat a pan of water, you place it on a burner on your stove. If the burner is an electric coil, it *conducts* heat into the pan; if it is a gas burner, the hot flame *convects* heat into the pan. Or if you place the container of water into a micro wave oven, the oven *radiates* heat into the water. (See Section 4.6.3 for a fuller explanation of these terms.) *Then* after the passage of time in all three of these cases, it becomes evident that the water is experiencing an increase in temperature. The stoves and the microwave oven are *generators of heat.* The heat is absorbed by the water—and the consequence is the temperature increase of the water.

If the container of water is placed into a refrigerator, the movement of heat takes place in the opposite direction. Instead of heat energy going into the water, heat is removed from the water. The refrigerator is a device for the *absorption* of heat, which causes what you put into the refrigerator to become the *source of heat. Then* after the passage of a period of time, as a consequence of that absorption, we notice that the water has experienced a decrease in temperature. Again, this is a consequence of the movement of heat.

1. The Measurement of Temperature

When we measure temperature, we do not measure heat; rather, we measure the result of the history of heat changes in a system.

But knowing temperature changes enables the physicist to calculate how much heat was involved. Those calculations that you deal with will be in the form of rather conventional algebra and calculus notation such as was described in Chapter 2. Calculations involved with the mathematical treatment of heat and its effects are *thermodynamic* calculations.

2. Units of Temperature

Temperature is measured in *degrees*. But there are at least a five different commonly used *temperature scales,* all named after the scientists who first defined them.

a. Starting point A temperature scale is always defined as starting at a particular zero. Then the size of the increments of increase or decrease of temperature from that zero are defined. Those increments—or intervals—are called *degrees.*

No matter what the size of the interval, it is always called a *degree.*

The most logical temperature scale is the *centigrade scale,* also known as the *Celsius scale.* It sets zero at the freezing point of water, and 100 at the boiling point of water. Then it divides the temperature range between these points into 100 equal divisions (or degrees).

NOTE: The centigrade system is a metric system (in that it divides a specified range of physical phenomena into decimal units). It is part of the SI system (Section 2.1.6).

Some different temperature scales and the symbols with which they are designated in text or mathematical notation are shown in Table 4.4.

b. The symbol The symbol for *degrees* is usually a small circle written as a superscript (°) followed by an abbreviation for the kind of degrees involved.

Thus "one thousand and eighty-three degrees Fahrenheit" is written

1083°F or 1083 °F

"Seventy-four degrees Celsius" is written

74°C or 74 °C

But the kelvin temperature scale is an exception. The word and symbol for degree are both omitted. Thus the term *degree Kelvin* is not used and is replaced by the unit *kelvin* (abbreviated K).

Thus "Nineteen kelvin" is written

19 K

TABLE 4-4
THE DIFFERENT TEMPERATURE SCALES*

Unit	*Symbol*
degree centigrade (or Celsius)	°C
degree Fahrenheit	°F
kelvin† (or absolute)	K
degree Rankine	°R
degree Reaumur	°R

*There is a scale used to designate one of the properties of liquids (their specific gravity, which describes whether they are heavier or lighter than water and by how much). It happens that the scale resembles these temperature scales, but it is not a temperature scale. It is the Baumé scale, and degrees Baumé are written °Bé.

†The small circle symbol is not used for the Kelvin scale. The unit the *kelvin* replaces the older expression, the *Kelvin degree*.

4.6.3 The Movement of Heat

Like sound and light, heat moves. Also like sound and light, it has a starting point, a *generator*. And when heat leaves the generator, there are verbs that describe that leaving; these include *emission* and *radiation*. Then the heat travels, and the means by which it does so are described by the verbs *conduct, convect,* and *radiate*. Heat then arrives at its destination—a *receiver*—where it is *absorbed*.

1. Generators of Heat

a. Anything can generate or emit heat It is only necessary to inject energy into something and for that energy to have no way to "get out" in some way for the (confined) energy to be converted to heat energy (and subsequently to manifest itself as temperature rise).

If you push on the pedals of a bicycle with your feet, you cause movement. There is mechanical *energy transfer*. And there is no perceptible rise in the temperature of the gears. But if the bicycle chain is broken and you still keep rotating the gears of the bicycle around each other for a protracted period, they will increase in temperature (heat up) just because of the *friction* of rubbing together.

b. Many forms of heat The energy that may be transformed into heat may be in any form.

Some examples:

- ***Mechanical energy:*** In this case heat is generated because of *friction*. The Boy Scout trick of rubbing two sticks together builds up more heat the longer the rubbing goes on.
- ***Chemical energy:*** Heat is generated because of chemical reactions (Section 6.2.3). An example of such a reaction is burning. Just light a match. The flame is the visible result of the chemical reaction between the gases that are given off. The chemical reaction is a heat-generating reaction (known as *exothermic*).
- ***Atomic energy:*** The reactor in a nuclear plant creates heat because of the radioactivity of the fuel (Section 4.8.2). The temperature in the reactor rises and causes a similar increase in temperature in a network of surrounding pipes containing water. That water becomes steam and turns a steam turbine which generates electricity for our homes. Another, less pleasant device that generates heat from atomic reactions is the atom bomb.
- ***Electric energy:*** Such energy pushed through a small wire causes a kind of electrical friction. You can see it in the wires of your kitchen toaster or in the filament of an electric light. That friction is transformed into heat (with a consequent temperature rise).

2. Receivers of Heat

Anything can be a target for heat. Any material on which heat energy impacts is a *receiver* of heat.

It is only necessary for the heat to be directed in that direction. Once the heat hits the target, the target may be considered to be a receiver of the heat energy, and it will undergo changes because it has absorbed that energy.

3. Absorption and Reflection of Heat

Sometimes the body receiving the heat can reject it, can resist absorbing it.

This is the case for heat reflectors such as sheets of aluminum in the walls of your house. They help save fuel by reflecting heat back into your house instead of absorbing it.

a. Reflectors When heat bounces off a surface without being absorbed (and therefore without causing a temperature rise), that surface is known as a heat *reflector*.

Heat reflectors are analogous to light reflectors in their actions. Light reflectors, of course, are mirrors.

b. Absorbers That heat which is not reflected is absorbed. The receiver of heat energy in that case is called an *absorber*.

Black surfaces are the best absorbers. A black cloth placed on snow on a sunny day will absorb enough heat from the sun to cause the snow under it to melt, even if the temperature of the air is below freezing. But the snow beyond the black cloth will remain unmelted.

4. The Three Methods of Transmission of Heat

The previous sections have described how heat is generated at sources and absorbed or rejected (reflected) by receivers or targets. But what happens in between? What causes the heat to travel from the generator to the receiver? There are three methods—or mechanisms, or phenomena—which must function, either singly or in combination, to cause heat to be transmitted. These are *conduction, convection,* and *radiation.*

All three have already been mentioned in the example in Section 4.6.2 which illustrates temperature increase and decrease in a pan of water.

5. Conduction

Like sound and light, heat can travel through a medium. Any material—solid, liquid, or gas—can act as a path or channel for the movement of heat. When it does, that material is known (logically enough) as a *conductor* of heat, and the process of heat transmission through it is known as *conduction.*

What happens is that the heat energy causes the atoms of which the material is made to vibrate. The more heat, the more vibration. And each atom which is vibrating impacts against the atoms that lie adjacent to it, causing *them* to vibrate in the same way. And they do the same thing to the next layer of atoms, and so on, until the heat that comes in at one side of an object emerges at the other side. In Figure 4-10, heat coming in at side *A* emerges at side *B*. If you touch a point on *B*, it feels hot, even if the heat is generated far away, at *A*. Conduction of the heat has taken place.

a. Temperature gradient What is it that tells the heat to go from point *A* to point *B*? Why does it not rather go (get conducted) from point *B* to point *A*? The reason is that the temperature at point *A* is higher than the temperature at point *B*. Such a temperature difference

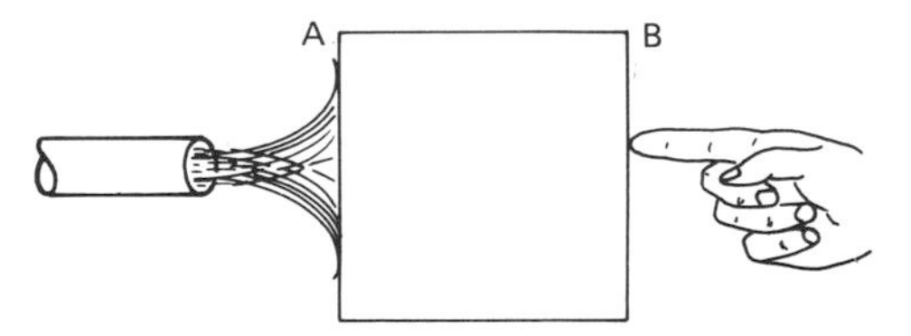

FIG. 4-10 The conduction of heat.

is known as a temperature *gradient*. Gradients push heat flow in a particular direction, from high temperature to low temperature.

Since the temperature at A is higher, this means that there is more heat energy residing at A than at B. It is like a round stone sitting at the top of a hill. Give it a push and it rolls down. The hill represents a gradient of height, which is analogous to a gradient of temperature. If the stone were at the bottom of the hill, it would not roll up.

b. Thermal conductivity We have already shown (see "The Speed of Sound" in Section 4.4.2) that some materials are better sound conductors than others. (Sound moves faster and louder through steel than through wood, for example.) The same pertains to heat. Some materials conduct heat more effectively than others.

Physicists can measure the effectiveness of heat conduction through various materials and can then label the materials with regard to whether heat is transmitted easily or with difficulty through them. This is the quality of *thermal conductivity.*

Coming back to the kitchen, consider that a kettle of water has been placed on the burner to heat up. The heat enters the bottom of the kettle and is conducted through the bottom to the water inside. At the same time the heat is conducted up the walls of the kettle, raising their temperature. Then the heat is conducted into the handle of the kettle. That handle can be quite far from the burner yet become warm to the touch, because of conduction.

Materials with high thermal conductivity are known as *conductors.* Materials with low thermal conductivity are known as *insulators.* Materials in between run the spectrum from *moderate conductors* to *moderate insulators.*

Silver is the best conductor. It is about 5 times as conductive as iron. It is about 400 times better as a conductor than glass, and glass is considered an insulator. But there is a worse thermal conductor than glass: wool felt is about 60 times worse (or to put it another way, wool felt is 60 times better as an insulator than glass).

6. Convection

Convection is the primary means of transmission of heat through fluids. Both gases and liquids are fluids.

a. Resemblance to heat conduction Convection has some resemblance to the conduction of heat through solids (see "Conduction" in Section 4.6.3).

For example:

- Just like solids, gases and liquids are media and thus can conduct heat. (In the same way, they are media for the conduction of sound and light; see "Media" in Sections 4.4.2 and 4.5.1.)

- As conductors of heat, gases and liquids are on the poor side. Thus they are insulators.
- The activating force and the determination of the direction of heat flow in a gas or liquid are set by temperature gradients for convection as for conduction.

b. Thermal expansion The difference in heat transmission in a gas or liquid from that in a solid is that (convected) *heat rises*. This is caused by *thermal expansion*.

Once some heat is conducted or radiated into a *fluid* (gas and liquid are fluids), the atoms of which the fluid is constituted increase their rate of vibration. This causes them to hit each other and bounce off each other, and the result is that the atoms tend to separate from each other. Separation means that they are occupying a greater volume of space, that the fluid has expanded.

NOTE: Solids expand because of temperature increases just as fluids do. But the expansion of solids for every degree of temperature increase is less dynamic than that of fluids, because the atoms and molecules of solids are held together more rigidly by intermolecular bonds (Section 4.7.1) than those of fluids.

- The expansion rate for solids is given by the *coefficient of linear thermal expansion,* expressed as an increase in length per unit length of the material for every degree of rise in temperature. This coefficient is commonly designated by the Greek letter α and may be written in English units in any of the four following ways:

$$\frac{\text{in/in}}{°\text{F}} \quad \text{or} \quad (\text{in/in})/°\text{F} \quad \text{or} \quad \text{in}/(\text{in}\cdot°\text{F}) \quad \text{or} \quad \text{in}\cdot\text{in}^{-1}\cdot°\text{F}^{-1}$$

In SI units, the coefficient is expressed in centimeters per centimeter per kelvin, using any of the same styles as for USCS units, such as (cm/cm)/K.

NOTE: A multiplier of 10^{-6} (one millionth) is usually present, which indicates that the amount of thermal expansion for solids is very small—not detectable by the human eye. For example, the coefficient for aluminum is 13.3, which means that an inch of aluminum will increase in length by 13.3 thousandths (0.0133) of an inch for a temperature increase of one degree Fahrenheit. And aluminum expands at a much faster rate than most other solids. The coefficient for Pyrex glass is 1.8, which means that with the same one degree temperature increase, one inch of it will expand by only 1.8 thousandths (0.0018) of an inch.

c. Density Once a material has expanded it has less *density*. Density is a measure of the amount of space (volume) occupied by a standard weight or mass of any material.

- In English units, density is expressed in pounds per cubic inch:

$$\frac{\text{lb}}{\text{in}^3} \quad \text{or} \quad \text{lb}\cdot\text{in}^{-3}$$

TABLE 4-5
DENSITIES OF SOME SELECTED MATERIALS

Material	*British Units, lb/in³*	*SI Units, g/cm³*
Gold	0.7	194
Lead	0.41	113
Copper	0.32	89
Steel	0.28	78
Ceramics	0.14	39
Aluminum	0.10	28
Glass	0.09	25
Plastics	0.05	14
Silk	0.045	12.5
Plastic foams	0.002	0.55

- **In SI units, density is expressed in grams per cubic centimeter:**

$$\frac{\text{g}}{\text{cm}^3} \quad \text{or} \quad \text{g}\cdot\text{cm}^{-3}$$

Table 4-5 lists the density of some selected materials.

Because of the force of gravity, materials stack in the order of their density if the materials are easily movable, as fluids are.

So the heavier—denser—fluids move closer to the earth (where the gravity resides), and the lighter—less dense—fluids move to positions stacked on top of the heavier ones. And that is the (mechanical) driving force—derived from heat gradients—of convection.

d. Currents This process of stacking, in which the less dense fluids move up and the more dense move down, involves *movement.* The movement occurs in *currents (convection currents).*

Ocean currents are caused by these phenomena—thermal expansion and the consequent attempts by the fluid (water) to stack, with the denser water sinking and the less dense rising. If you work with text written by *oceanographers,* you will encounter much discussion of water currents.

Where does the energy of this movement come from? From the sun, which radiates heat to the oceans.

Air currents are caused by these same phenomena—thermal expansion and the consequent attempts by the fluid (air) to stack, with the denser air sinking and the less dense air rising. If you work with *meteorologists,* you will encounter much discussion of air currents.

Where does the energy of this movement come from? As with ocean currents, from the heat of the sun.

e. Convection depends on movement of fluids Movement of fluids because of heat means that the heat moves too.

The heat is in the atoms and molecules of the fluid—it got there in the first place by conduction or radiation—but now that it is there, it moves in currents, carried by the fluid currents. Heat transfer by means of these heat currents is known as *convection.* It is heat transfer by virtue of the movement of fluids.

In Figure 4-11 the heat that was on a surface *A* has been conducted into the gas above it and has been carried upward with the upward gas currents. The heat now resides, in part at least, in the top layer of the gas—region *B*. It has been transferred there by convection.

Convection is the primary way in which we heat our homes. Either the convection begins with the baseboard heating unit of Figure 4-11, or a fan pushes warm air into the room from a duct opening. Once the warm air is in the room it *convects* in the same way, as though it came from a baseboard heating unit.

One more example of convection heat transfer is in cooking. When you steam something, you place it on a kind of platform in a closed cooking vessel. Water is at the bottom of the vessel; when you heat the water, some of it changes into a gas—known as steam. That gas, being hot, is less dense than the air in the vessel, and the steam rises until it meets the food you have placed on the tray inside. It then transfers its heat to that food (by conduction) and causes it to cook. We call the process *steaming.*

7. Radiation

We have already dealt with the radiation of light (Section 4.5.2). Heat is a form of electromagnetic energy and is like light, but its waves oscillate at a lower frequency, the *infrared* frequency.

So infrared light is heat.

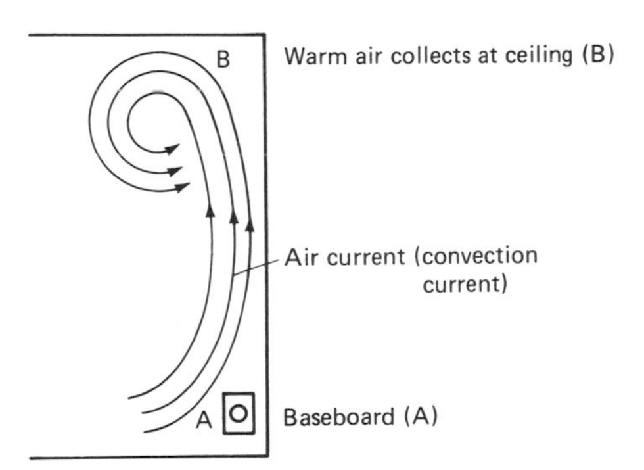

FIG. 4-11 Convection in a room heated by a baseboard heating unit.

a. The third method of heat transfer Although energy can be conducted in media by means of the heat-transfer mechanisms of conduction and convection (see "Conduction" and "Convection" in Section 4.6.3), media are not required for the transfer of heat. Energy can be *radiated*—transferred through space without the participation of media. It is the third method of transmission of heat.

The heat we get from the sun is an example. It reaches us through millions of miles of space. And space is not a medium, since it contains no materials.

NOTE: Most heat is not transmitted by only one of the three mechanisms. Rather, in the real world, heat is usually transmitted by a combination of all three—all occurring at the same time. But in each situation, one mechanism usually predominates.

b. Generator and receiver All that is required for heat transmission by radiation is a *generator* of infrared radiation and a *receiver* so placed that it detects the radiation from the generator.

Once the radiated heat is received, it can be reflected or absorbed or any combination of these two. When the receiver acts as an *absorber,* the result will be what always happens when something accepts heat: its temperature will go up, it will expand, etc.

c. Blackness As was pointed out in "Absorbers" in Section 4.6.3, the darker the surface of the absorber, the more readily it absorbs heat. In this regard the effect of the nature of the surface of an object is quite independent of the nature of the material under that surface.

- A surface that absorbs all the radiant heat energy that falls on it (is *incident* on it) is known as a *blackbody.*
- A surface that absorbs none of the radiant heat energy that falls on it is a *reflector.* A *mirror* is a reflector—of both light and heat.

Reflectors and blackbodies are the extremes. Perfect reflectors and perfect blackbodies do not exist. All surfaces in the real world are someplace in between.

- The amount of absorption of a surface is designated by its *emissivity.*

Emissivity is a decimal fraction that ranges between zero and 1. Perfect reflectors have an emissivity of zero. Perfect absorbers have an emissivity of 1. Real materials that we often encounter have the following approximate emissivities:

- Newly fallen snow has an emissivity of 0.15.
- Coal dust has an emissivity of 0.9.

- A white shirt or blouse has an emissivity of 0.2.
- A wool sweater in some dark color, like purple, has an emissivity of 0.85.

d. Emissivities work both ways A surface that absorbs heat readily emits it just as readily. A surface that is a good reflector is a poor heat emitter. Thus the term *emissivity* applies equally to generators and absorbers of heat.

Engineers design the surfaces of all the products you buy and use to have emissivities (degrees of blackness, degrees of shininess) that work best for a particular product. They do this by specifying

- Surface roughness (shininess, dullness)
- Color

Then they achieve these specifications by designing techniques of producing metals and plastics or by devising different paints, enamels, porcelains, dyes, etc. You will deal with these matters if you work for a paint chemist or with engineers in any of the fields of material finishing.

4.6.4 Quantities of Heat

The physicist and engineer must describe thermal phenomena quantitatively. The units of temperature (the manifestation of the effect of heat) were treated in Section 4.6.2. The units of heat are described in the following sections.

1. Heat Units

The basic SI unit (see Section 2.1.6) of heat is the *joule* (named after a British scientist who lived in the nineteenth century, James Joule). A unit in the English engineering system—the *British thermal unit*—is also often used. The third heat unit in common use is the *calorie.*

a. The British thermal unit One British thermal unit (abbreviated Btu) of heat will raise the temperature of 1 pound of water 1°F.

If we heat a pan containing 2 pounds of water and the temperature of the water rises from 70 to 100°F, that means that 60 British thermal units of heat were pumped into the water (2 pounds times an increase in temperature of 30 degrees equals 60 British thermal units).

NOTE: People frequently write the abbreviation as BTU or B.T.U. These are WRONG. Btu is CORRECT.

b. The joule A *joule* (abbreviated J) is a unit of mechanical energy while a Btu is a unit of thermal energy. But one kind of energy is convertible into another (for example, heat runs engines). So 1055 joules is equivalent to 1 Btu.

c. The calorie The *calorie* (abbreviated cal) is the SI equivalent of the Btu. It is the amount of heat required to raise 1 gram of water 1°C. It takes 4.184 joules to make a calorie, and it takes 252 calories to make a Btu.

NOTE: There are actually more exact definitions of the British thermal unit and the calorie. A British thermal unit is actually 1/180 of the heat required to raise the temperature of 1 pound of water from 32 to 212°F; and a calorie is actually the quantity of heat required to raise the temperature of 1 gram of water from 14.5 to 15.5°C. But for most purposes, the "cruder" definitions are quite adequate.

2. Conversion Factors

Certain numbers (4.184 joules in a calorie, 252 calories in a Btu, etc.) allow us to convert from one unit to another. They are called *conversion factors.*

NOTE: There are other units of energy which are not thermal or thermochemical as British thermal units and calories are. But like joules, they are convertible into heat units. These units include the units of mechanical energy and electric energy—*ergs, foot-pounds, watt-seconds, kilowatt-hours,* etc. (see Glossary M).

4.7 CRYSTAL PHYSICS

The materials of the universe, whether living or dead, organic or inorganic, solid, liquid, or gas, are made up of atoms, of which there are only somewhat more than 100 kinds (the chemical elements; see Section 6.1.1 and Glossary H). Most atoms don't exist free and uncombined but rather as a variety of kinds of combinations with other atoms. Chapter 6 describes how atoms attach themselves to each other with chemical bonds to form molecules and chemical compounds. Section 6.4 describes a particular kind of compound, the organic compound, which can become very complex. In crystals, the arrangements of atoms are also very complex and extensive. As a result of these complex atomic arrangements, crystals can be beautiful. Diamonds, rubies, and emeralds are all inorganic crystals.

When atoms or molecules are arranged in a regular three-dimensional pattern that repeats and persists throughout a solid structure, that structure is known as a *crystal.* A solid consisting either of a single crystal or of many microcrystals fused together is called a *crystalline* solid. Such solids are studied by *crystallographers, crystal physicists, mineralogists, mining engineers, metallurgists, ceramists,* and others. The study of crystals is called *crystallography.*

4.7.1 Crystallography

1. Crystal Morphology

Although atoms are extremely small, the patterns of crystals are large enough to be visible. It is possible to differentiate crystals from each other by recognizing their characteristic shapes. The classification of crystals by their shape is known as *morphology,* which comes from the Greek root "morph-," which means "shape."

NOTE: All other materials—those that are not crystalline and therefore comprise molecules that are not arranged in regular patterns—have no regular characteristic shape. They are *amorphous,* which means "having no shape" (see "Amorphous Materials" in Section 4.7.1).

a. Facets The regular arrangements of molecules are always in straight rows and columns, which then stretch out into layers, or planes. So all crystals have surfaces which are in regular layers, or planes—or faces. We call these flat faces *facets* (Figure 4-12). It is the shape and arrangements of these facets that allow us to classify crystals morphologically.

From ancient times, people recognized the beauty of crystalline minerals, particularly the ones we know as precious and semiprecious stones. A major part of their beauty results from the facets. These are at angles to each other and reflect and refract light (Section 4.5.3) in complicated ways (which is what we refer to when we say that certain stones used in jewelry *sparkle*).

Crystals occur in nature, and we can recognize them by the fact that they have facets which may cause them to sparkle. Many naturally occurring crystals look crude and lumpish, with no visible facets. But the facets are there; their corners have been worn off by rolling around in the world for a few million years.

Examples of crystals in nature are particles of rock salt or quartz (which you might see on a beach) or any mineral outcropping in the landscape.

b. Isotropy, anisotropy Some crystals are *symmetrical.* They present the same facets no matter how we turn them in our hands. These are called *isotropic.* Nonsymmetrical crystals are called *anisotropic.*

A crystal which is in the shape of a cube is isotropic. As shown in Figure 4-13*a*, it is evident that all facets of a cube are equal squares.

A crystal with an elongated rectangular form as shown in Figure 4-13*b* is not symmetrical, since some faces, those along the length, are much longer than others, the faces at the ends.

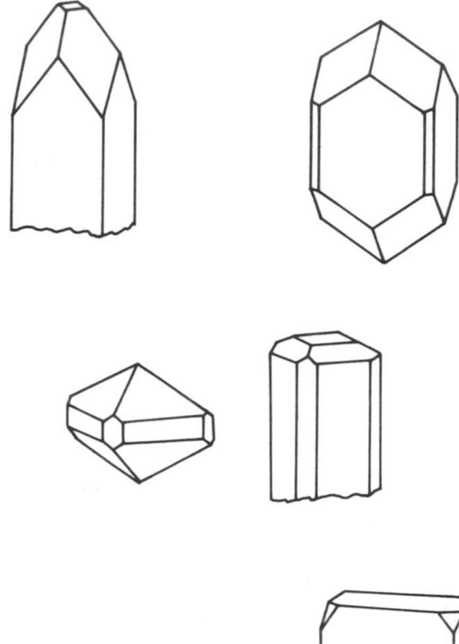

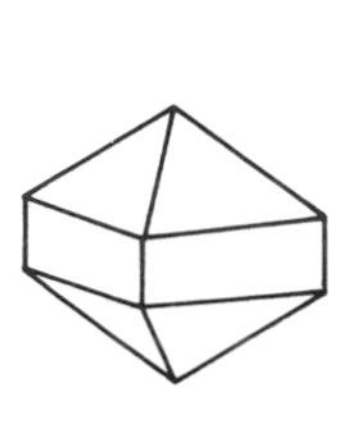

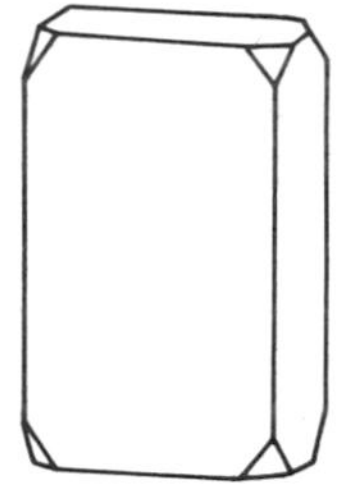

(a)

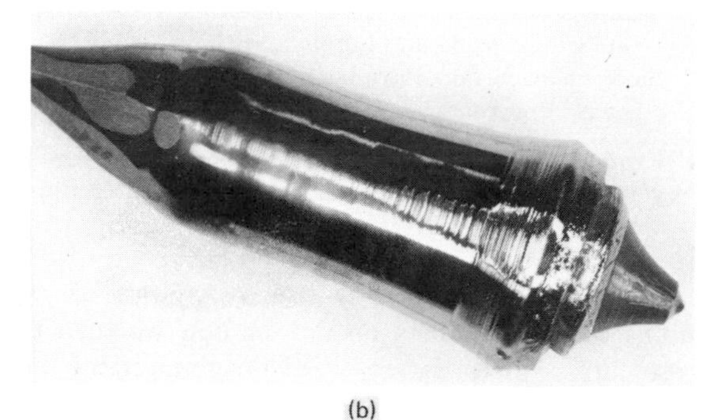

(b)

FIG. 4-12 Crystals: (*a*) A selection of crystal forms (habits) showing various shapes of facets, (*b*) a single crystal of silicon as grown from the melt. *(Photograph courtesy of Manlabs, Cambridge, Mass.)*

NOTE: Facets are visible at left in final portion grown.

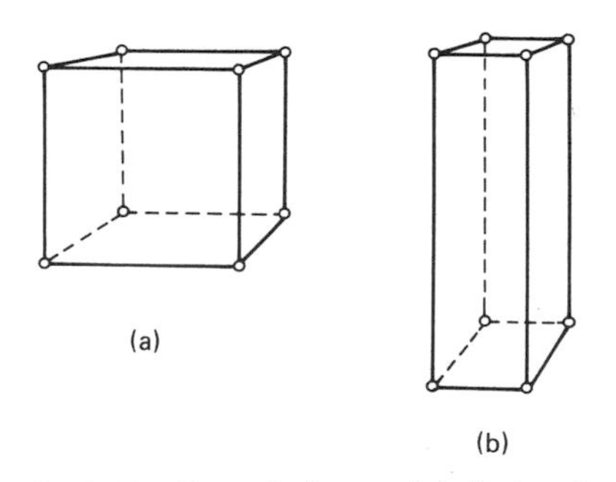

FIG. 4-13 Crystal forms: (*a*) Isotropic, (*b*) anisotropic.

Some examples:

Every particle of rock salt is a cubic isotropic crystal.

The mineral *galena,* which is the principal ore of lead (chemically it is lead sulfide, PbS), when taken from the ground (mined) is in the form of beautiful, shiny, metallic-looking isotropic cubic crystals.

The mineral *asbestos* as it occurs in nature is very anistropic. The crystals are so elongated, so much longer than they are wide, that it is in the form of fibers, or threads.

The mineral *mica* is anisotropic in the opposite way. Its crystals are so much wider than they are long that it is in the form of sheets that can be lifted apart from each other like layers of thin paper.

NOTE: The easy separation of layers of crystals from each other, as in the case of mica, is known as *cleavage.* The separation occurs along the *plane of cleavage.* The crystals most easily cleaved are the most anisotropic.

c. Miller index A scientist named William Hallowes Miller in the nineteenth century devised a logical means for describing crystal morphology with sets of three-digit numbers. His system is the *Miller index* (plural, *indices*).

Figure 4-14 illustrates a crystal with its facets labeled with sets of Miller indices.

These are most commonly grouped in parentheses, but brackets or braces are sometimes used:

- (001) Most common
- [001]
- {001}

Occasionally, the crystallographer has need to call attention to one of the digits. This is done with an overline:

$(1\bar{1}0)$

$(\bar{2}01)$

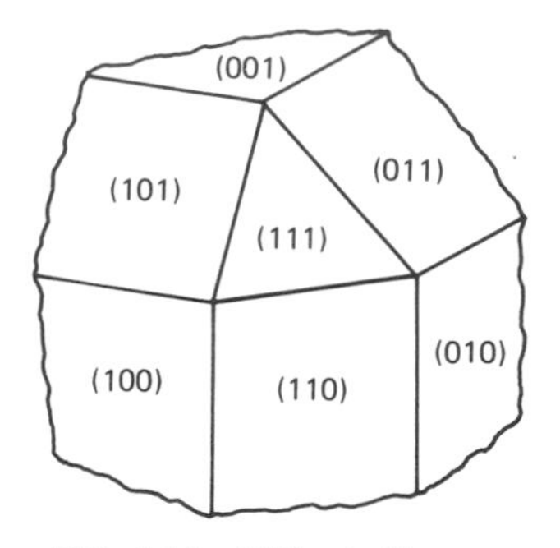

FIG. 4-14 Miller indices.

d. The crystal classes Any crystal that occurs in nature or that can be synthesized in the laboratory—or that can be imagined—can be classified; and for the crystallographer, just knowing the name of a crystal within the classification system means that the shape of the crystal is automatically known too. These crystal classes are essentially nothing but designations of geometrical shapes similar to those used in solid geometry (Section 2.2.1); for example, cubes, parallelepipeds, pyramids, and so on. But the crystal shapes are sometimes more complicated than these simple forms, so a more complicated system of designating crystal shapes is needed. The system is based on the fact that three-dimensional objects made up of flat facets that intersect in straight-line edges can exist in no more than 32 possible combinations. These are the 32 *crystal classes.*

NOTE: Using a different set of ground rules, some crystallographers say that the number is 48. The actual number is not important. What is important is that only a limited number of crystal configurations are possible.

The names of the 32 crystal classes are given in Table 4-6.

NOTE: The different crystal classes are also known as crystal *habits.*

2. The Growth of Crystals

We get crystals by growing them. A number of factors are common to all crystal growth:

- Crystals grow from *seeds.*
- Crystals take on a form predetermined by these seeds.
- The longer we let crystals grow, the larger they get.
- Crystals such as minerals grow in nature, but, like many things that occur naturally, most of the same crystals can be grown in the laboratory as well.
- During growth, crystals can divert into *twins,* which is the name given to crystals that result from a change in direction of the crystal. The

TABLE 4-6
THE 32 CRYSTAL CLASSES

System	*No.*	*Class*
Triclinic	1	Asymmetrical
	2	Pinakoidal
Monoclinic	3	Sphenoidal
	4	Domatic
	5	Prismatic
Orthorhombic	6	Rhombic disphenoidal
	7	Rhombic pyramidal
	8	Rhombic dipyramidal
Tetragonal	9	Tetragonal dispheroidal
	10	Tetragonal pyramidal
	11	Tetragonal scalenohedral
	12	Tetragonal trapezohedral
	13	Tetragonal dipyramidal
	14	Ditetragonal pyramidal
	15	Ditetragonal dipyramidal
Hexagonal with rhombohedral subsystem	16	Trigonal pyramidal
	17	Rhombohedral
	18	Trigonal trapezohedral
	19	Ditrigonal pyramidal
	20	Hexagonal scalenohedral
Hexagonal with hexagonal subsystem	21	Trigonal dipyramidal
	22	Ditrigonal dipyramidal
	23	Hexagonal pyramidal
	24	Hexagonal trapezohedral
	25	Hexagonal dipyramidal
	26	Dihexagonal pyramidal
	27	Dihexagonal dipyramidal
Isometric	28	Tetrahedral
	29	Gyroidal
	30	Diploidal
	31	Hextetrahedral
	32	Hexoctahedral

twinned region is known as a *twin band* (Figure 4-15). When crystals are polished and examined with a microscope, twin bands may also be apparent (Figure 4-21*a*).

a. Nucleation The initiation of crystal growth from seeds is known as *nucleation*. Each seed is a *nucleus* (plural, *nuclei*).

NOTE: This is not the same as the nucleus of an atom, although both in atomic physics and in crystal physics the word means "a center" of something.

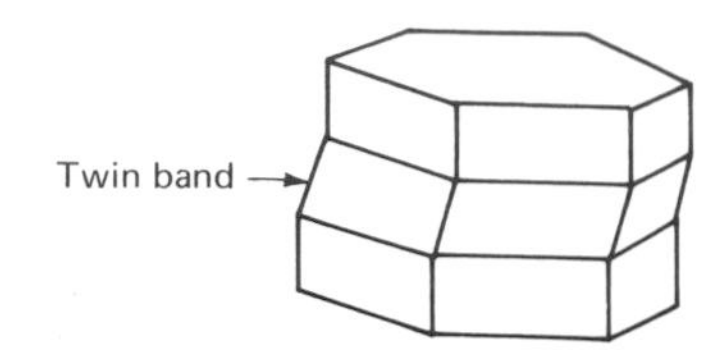

FIG. 4-15 Crystal twin (with twin band).

Snowflakes are crystals of solid water (ice). Each flake starts (nucleates) on a tiny mote of dust or on a microscopic speck of ice—the seed—and grows so large that it can no longer be supported in the air (can no longer float), and as a result it falls.

b. Change in physical state All crystal growth involves a change in physical state (see Section 6.2.1).

All crystals are solid, but they can be formed from either liquids or gases that change into the solid state. They may also be formed from other crystals, solid to solid.

- Crystals made by inserting a seed in the form of a thin rod into a crucible containing a liquid and *pulling* are said to be grown by the *Czochralski technique*. As the rod is pulled out of the liquid, some of the liquid—eventually all—solidifies (crystallizes) on it (Figure 4-16*a*).

  The silicon crystal of Figure 4-12*b* was grown by the Czochralski technique. Note the seed, the small projection at the lower right.

- Crystals made by inserting a seed into a liquid pool and *drawing them out* at the bottom are grown by the *Bridgman technique* (Figure 4-16*b*).
- Crystals made by deposition onto a seed from a gas (from a *vapor phase*) are grown by *chemical vapor deposition* (abbreviation CVD). This is also known as the *epitaxial* method of crystal growth (Figure 4-16*c*).

NOTE: Engineers and scientists in this field often refer to this process as the *epi* process and the layers of crystalline material deposited in this manner as *epi layers*.

Snowflakes are examples of crystals that grow by CVD.

NOTE: All semiconductors used by the electronics industries (see "Semiconductor Diodes and Transistors" in Section 5.3.3) are made of crystals which are grown by either the Czochralski or the epitaxial technique.

3. Unit Cells

It has already been pointed out that we can recognize and classify crystals by studying their surfaces—their facets (see "Facets" in Section

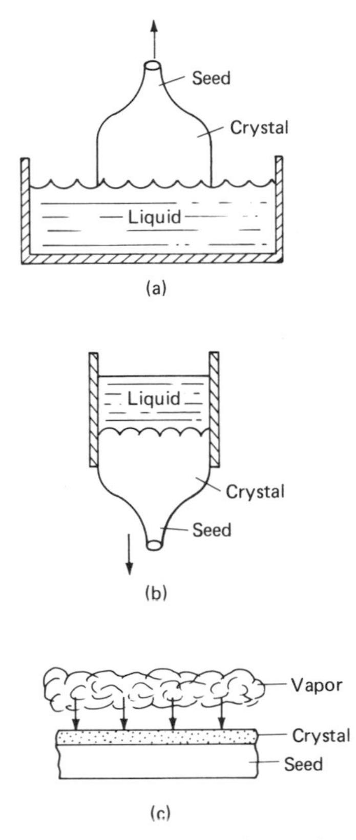

FIG. 4-16 Crystal growth: (*a*) Czochralski method, (*b*) Bridgman method, (*c*) epitaxial method.

4.7.1). But these facets are only the result of a basic arrangement of atoms which is repeated over and over again trillions of times to form the solid hunk of crystal which has grown large enough for us to hold in our hands. That basic arrangement (different for each class of crystal) is called a *cell,* and in more common usage, a *unit cell.*

a. The unit cell A unit cell is a solid geometrical figure with atoms at each corner and often with additional atoms either in the faces of the figure or inside the figure.

Figure 4-17 illustrates the unit cells of two different forms of iron (different crystals of iron; described in "Phases" in Section 4.7.2), and a unit cell of diamond (consisting of carbon atoms).

b. The x-ray analysis of crystals It is possible to "look inside" a crystal and identify the structure of its unit cell. This is done with x-rays, with which a distinctive picture may be made.

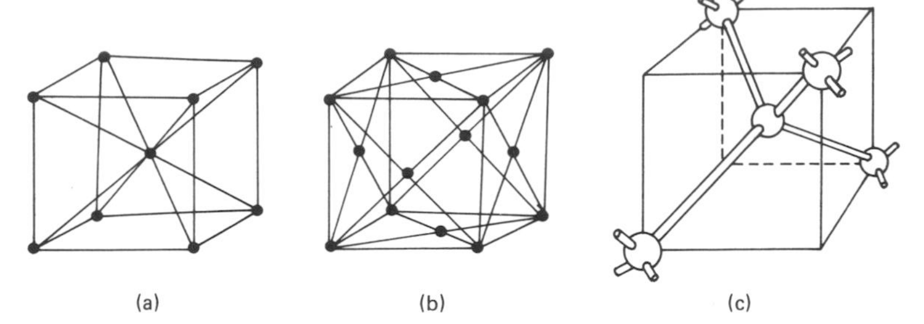

FIG. 4-17 Unit cells: (*a*) Alpha iron (known as *ferrite*), a body-centered cubic unit cell, (*b*) gamma iron (known as *austenite*), a face-centered cubic unit cell, (*c*) carbon, a diamond unit cell.

A fine beam of x-rays is directed into the crystal by the crystallographer. As the radiation encounters the trillions of unit cells, it finds "easy paths" through the crystal while other paths are blocked.

Imagine walking past a large cornfield. You can see a long distance down certain rows at certain angles, but at other angles your view is blocked (Figure 4-18*a*). The x-ray beam has a similar experience with the rows of atoms which make up the unit cells of the crystal.

The x-ray beam passes through the spaces between the rows of atoms and emerges from the crystal in a set of new beams determined by the shape of the unit cells. These beams are recorded on x-ray film (such as the radiologist in a hospital uses), and by measuring and analyzing the patterns on the film, the crystallographer can determine the exact structure of the unit cell.

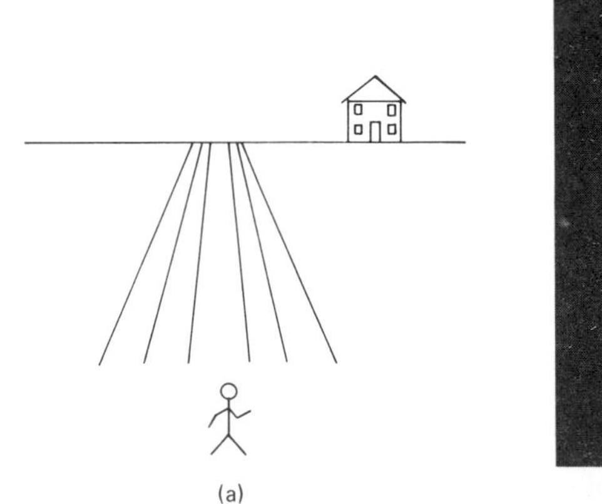

FIG. 4-18 Laue pattern: (*a*) Straight lines in a cornfield, (*b*) Laue picture of a cubic crystal in (100) direction. *(Photograph courtesy of Manlabs, Cambridge, Mass.)*

Such pictures are known as *Laue* (rhymes with *Howie*) pictures. Figure 4-18*b* is a typical Laue picture.

A value of x-ray analysis is that a crystal cannot hide its secrets from the physicist just because its corners have been knocked off and its facets destroyed. The physicist—with x-ray analysis—can look below the surface.

4. Amorphous materials

Not every material in the world is crystalline. The noncrystalline materials of the world are called *amorphous*.

Since *amorphous* means "without shape" (and in this case what is being referred to is molecular arrangements), such materials have no unit cells, and their molecules are in no set pattern. They are in random arrangements. And their surfaces have no facets.

a. Fluids All fluids—liquids, gases, gelatins, greases, etc.—are amorphous.

b. Solids Some solids are amorphous. Rubber is an example of an amorphous solid.

c. Living materials The flesh of our bodies is amorphous (but the calcium and sodium compounds of our bones and teeth are crystalline).

4.7.2 Physical Metallurgy

Physical metallurgy is the engineering discipline that deals with such physical properties of metals as strength, ductility, and resistance to wear, and how to control these properties.

The properties of metals may be changed and controlled by creating different metal mixtures (alloys) and also by using different heat treatments or mechanical treatments.

Metals are crystalline. Heat treatments and mechanical treatments cause the crystals of any metal to change shape or basic nature (from one crystal form to another). It is these crystal changes that result in certain property changes.

Because of the dependence of the performance of metals on the crystals of which they consist, physical metallurgy is actually a branch of crystal physics, which is discussed in the previous section.

1. Alloys

The verb *to alloy* is the metallurgist's way of saying *to mix*. And the noun *alloy* is the metallurgist's way of saying *mixture*. The only thing that is special about the word *alloy* is that it refers specifically to a mixture of two or more different metals.

a. Liquid alloys Because most metals melt at high temperatures, the mixing must generally be done at high temperatures, often above red heat. At the moment of mixing, when all the metallic ingredients have melted, and in fact intermelted, the alloy may be said to have formed. Since it is then in the liquid state, it is a *liquid alloy.*

b. Solid alloys When the liquid alloy is allowed to cool, it freezes. It becomes a solid. But it is still an alloy, a *solid alloy.* The important thing to remember about solid alloys is that they consist of metallic crystals, millions all frozen together.

2. Phases

In any metal, pure or alloy, there is one or more variety of crystal.

- Each crystal variety in a pure metal or alloy is known as a *phase.*
- Each crystal variety in a pure metal or alloy has its own characteristic *unit cell,* which is the basic form in which the metal atoms are arranged in the metal crystal.

Section 4.7.1 describes how the atoms of crystals are fixed in orderly geometrical patterns. The smallest collection of atoms that still maintains such a pattern is called the *unit cell* (see Figure 4-17).

Most pure (unalloyed) metals have only a single solid phase. That means they are in only one single crystal form at all temperatures from the coldest that can be imagined all the way to their melting point.

When crystals melt, they cease to be crystals. Unit cells cease to exist, and the resulting arrangement of the metal atoms is the loose and random one associated with the liquid state. Then the material has gone from an ordered condition to an amorphous condition (see "Amorphous Materials" in Section 4.7.1). The term *phase* is used with liquids as well as with crystalline solids. Thus it is correct to refer the *liquid phase* of a metal.

Certain pure metals have more than a single crystal phase. These phases exist at different temperatures.

Thus iron, unalloyed, is in one crystal form (known as *alpha iron*) up to 910°C; then between 910 and 1390°C it changes to another crystal form (known as *gamma iron*), and from 1390°C to its melting point at 1534°C, it reverts back to alpha iron. So it has two crystal phases, which exist in three temperature ranges, and one liquid phase.

When two or more metals are mixed to form an alloy, there are two forms in which the atoms may be arranged in the unit cells of the metal crystals which make up the alloy:

- Solid solution
- Intermetallic compound

a. Solid solutions When two or more metals of an alloy share a unit cell over a range of different proportions, *dissolving* has taken place, and the resulting crystals are called a *solid solution*. There are two ways in which the atoms of alloys dissolve into each other:

- Substitution
- Interstitial alloying

When metal B replaces one or more atoms from the unit cell of a crystal phase of metal A, it is clear that A and B are sharing that unit cell. In such cases the alloy process is known as *substitution*.

Figure 4-19*a* shows atoms of metal B (black) replacing atoms of metal A (white). Thus B and A have formed a *substitution solid solution*.

When metal B enters the unit cell of metal A without substituting itself for any of the A atoms, but rather by "tucking itself" into available spaces in the unit cell, the alloy process is known as *interstitial alloying*. Such alloying usually takes place when metal B is much smaller than metal A.

Figure 4-19*b* shows atoms of metal B (black) occupying spaces that surround atoms of metal A (white). Thus B and A have formed an *interstitial solid solution*.

NOTE: The spaces in a crystal lattice into which interstitial atoms may fit are known as *interstices*.

Solid solutions, whether subsitutional or interstitial, do not differ from liquid solutions in many basic aspects, although there is no question that the solid mixture is stiffer and more rigid (more solid) than the liquid one. Thus as with all solutions, one is often interested in knowing the solubility of one substance in another, and this involves defining which material is the solvent and which the solute.

For example, Metal B has dissolved in metal A. Metal A (since it is present in a larger amount) is the *solvent* and metal B (which is present in a small proportion compared with metal A) is the *solute*.

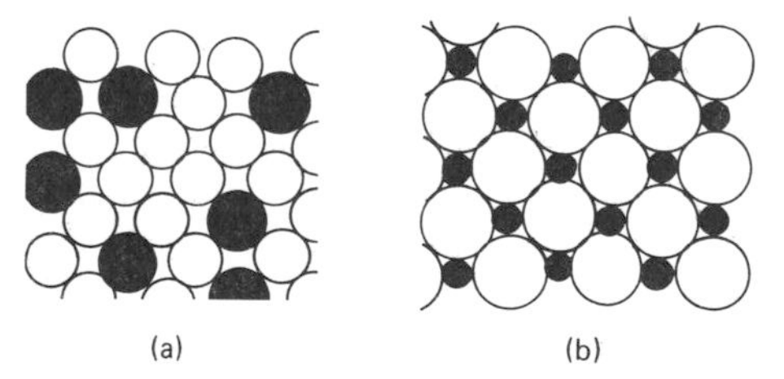

FIG. 4-19 Solid solutions: (*a*) Substitutional crystal lattice, (*b*) interstitial crystal lattice.

Solutions are described in Section 6.2.2. The same rules apply to all solutions, whether solid or liquid.

NOTE: We are used to dissolving things in liquid solvents—as when we dissolve sugar in tea. But there is no reason that the same actions cannot take place in solid solvents with solid solutes. A solution consisting of one solid dissolved in another solid is known as a *solid solution.* The degree to which one metal (B) dissolves in another (A) is known as the *solid solubility* (of B in A).

Solid and liquid solutions are also similar to each other in how they are affected by the temperature. In general, as the temperature increases, the solubility increases. But not always. Also, there may be combinations of materials and temperature at which no solution takes place and the materials merely exist side-by-side as mechanical mixtures. These relations of solubilities and temperature are conveniently shown in *phase diagrams* (see the description of phase diagrams in this section).

b. Intermetallic compounds When two or more metals of an alloy share a unit cell of a metal crystal at only a single proportion of these metal ingredients—and no other—the resulting crystal phase is known as an *intermetallic compound.*

For example, in an alloy of metal A and metal B, if the unit cell of the crystal phase is a cube and has atoms at each of its eight corners of which three are A and five are B, the resulting intermetallic compound is A_3B_5.

NOTE: Intermetallic compounds are written with subscript numerical coefficients as though they were chemical compounds (see Section 6.1.4). But they are not chemical compounds; they are crystal compounds. The difference is based on the fact that the metal atoms are bonded to each other differently and are arranged in a unit cell.

The relations of solubilities and temperature of intermetallic compounds, like those of solid solutions, may be represented in phase diagrams (see the description of phase diagrams following).

3. Phase Diagrams

The metallurgist has a graphical way to show for any alloy system all the crystal phases such as solid solutions, mechanical mixtures in which no dissolving has taken place, and intermetallic compounds. This is the *phase diagram,* also known as the *equilibrium diagram.*

Phase diagrams are the lifeblood of the metallurgists' work. Such diagrams are the primary way in which they communicate with other metallurgists. If you work with metallurgists, you may be expected to draw simple ones on occasion. The rules for drawing graphs pertain (Section 2.3.4).

NOTE: These remarks also pertain to *ceramists,* who use the same techniques as metallurgists.

A *phase diagram* is a graph of temperature vs. proportion of one metal in an alloy relative to the amount of the other metal in the alloy. It shows all the phases and their normal (equilibrium) amounts at each temperature for each proportion of the mix. Since it tells the temperature at which a solid alloy becomes a liquid alloy for different proportions of the metals, it is also a graph of melting points.

a. Plotting a phase diagram In a phase diagram, temperature is always the y axis; alloy proportion is always the x axis. Phases show as areas (that is a way to recognize them). They are always labeled with the appropriate Greek letters.

NOTE: Figure 4-20*a* (next page) shows a phase diagram for a typical alloy system—in this case all the possible alloys of copper and aluminim. It is common to show the proportion of the two alloy ingredients both in terms of atomic % (number of atoms of each) and in terms of weight %, with the atomic % numbers along the bottom x axis and the weight % numbers along the top. Where numbers are marked within the diagram to indicate values at specific points, those based on weight are written in parentheses.

b. Binary and ternary systems Alloy systems of two metals are known as *binary* systems. But there are many alloys that contain three or more metals. When the alloys are constituted of three metals, the systems are known as *ternary* systems. (Stainless steel is a ternary system, consisting of iron, nickel, and chromium.) Phase diagrams for ternary systems always have a triangular form. Figure 4-20*b* shows the ternary equilibrium (or phase) diagram for this system.

c. Showing melting points on a phase diagram Above the melting point the metal is all liquid. This is a phase too, the liquid phase. Liquid phases are not labeled with Greek letters. The liquid phase is either left unlabeled, as in Figure 4-20*a*, or given the label *L*.

In Figure 4-20*a* the liquid phase is the entire region above the top curve, from 1083°C on the left side to 660°C on the right side.

d. The eutectic—a special melting point In any alloy system there is always one composition at which the melting point is the lowest.

Usually, when metal A is added to metal B, the melting point of the resulting alloy is less than that of pure B. And as more A is added, the melting point continues to go down. In the same way, as metal B is added to metal A, the melting point of the resulting alloy is less than that of pure A, and as more B is added, the melting point continues to go down. There is one

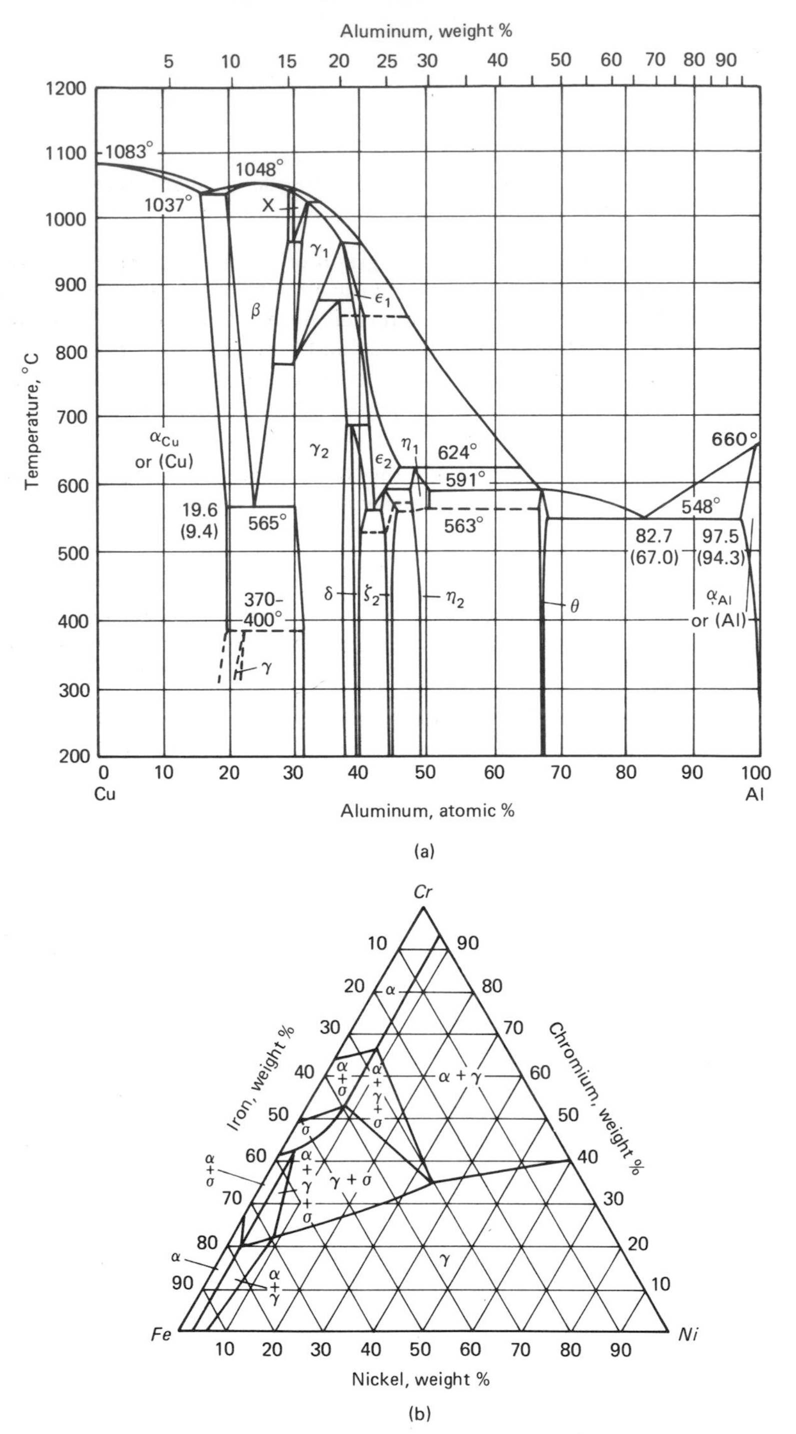

FIG. 4-20 Equilibrium diagrams. (*a*) Copper and aluminum binary diagram. (*From M. Hansen,* Constitution of Binary Diagrams. *2d ed., New York: McGraw-Hill, 1958).* (*b*) Iron, nickel, and chromium (stainless steel) ternary diagram, phases at 650°C. (*From* Metals Handbook, *1948 Edition, Cleveland: The American Society for Metals, 1948.)*

particular alloy composition at which the melting point line from metal A intersects the melting point line from metal B. That is where the melting point is the lowest.

That is known as the *eutectic composition* or *eutectic alloy,* and it occurs at the temperature known as the *eutectic point.*

In Figure 4-20*a*, the eutectic alloy is shown to contain 67 percent aluminum and 33 percent copper, and the eutectic point (which is the melting point at that composition) is seen to be 548°C.

4. Controlling the Properties of Metals

The ways in which a metal behaves in use—for example, whether it is hard or soft, strong or weak, capable of keeping a cutting edge or easily dulled, magnetic or nonmagnetic, yellow (like brass or gold) or white (like aluminum or silver), easily corroded (like iron in acid) or corrosion-resistant (like stainless steel in acid)—are called its *properties.*

We have already discussed mechanical properties (the strength of materials) in the section on stress and strain and have shown how these properties may be shown graphically in stress-strain diagrams (Section 4.2.3).

The metallurgist can design for and predict the properties of any metal because they depend on two major factors which the metallurgist can control:

- The choice of the metal alloy
- The choice of the crystal form of the metal alloy

a. Changing the properties of metals by the choice of alloy Knowing any alloy system (by examining its phase diagram), the metallurgist can control the conditions to create particular mixtures that are the best fitted for particular uses. In doing this, the metallurgist utilizes the knowledge that certain alloy compositions and certain kinds of crystals behave best (have properties closest to those required) under certain conditions to which they will be exposed in use.

For example, an almost pure iron with very little alloy addition works best for the skeleton structure of a skyscraper. Its crystals are tough, which means they will bend a little under pressure, but will not easily break. But such a metal is very soft and will not cut other metals. To make a saw which requires sawteeth that stay sharp even after much use, it is common to make an alloy with a few percent tungsten and carbon in it. That makes all the difference, because crystals of tungsten carbide will form, and they are very hard.

b. Changing the properties of metals by heat treatment Knowing any alloy system (by examining its phase diagram), the metallurgist can purposefully manipulate an alloy of some selected composition from one crystal phase to another. This is done by controlled procedures of heating and cooling. Those procedures are known as *heat treatment.*

NOTE: *Pressure treatment* can do the same thing, or some combination of heat and pressure treatment. But it is most common (and easier) to do the required job simply by using a furnace or a hot torch, that is, by heat treatment alone.

Surely you have seen blacksmiths or metalworkers plunging hot metal into a tub of water, the process being accompanied by a sizzling sound and much steam.

They do this because at high temperature steel forms a crystal that is harder and stronger than that formed at room temperature. When the metal is heated to red heat, the stronger crystal is formed, replacing the softer one that is "normal" at room temperature. And the harder crystal is then *retained*—brought down to room temperature, where it does not normally exist—by cooling it so fast that the atoms of that harder crystal do not have time to rearrange themselves back to the room-temperature crystalline form. This process of rapid cooling is called *quenching*.

NOTE: The literature of metallurgy abounds with legends and myths—all fascinating, although probably not true—of how ancient blacksmiths hardened sword blades by quenching them—red hot—into the bellies of slaves. (Water at body temperature would have done as well.)

We prefer the legend that the best swords were made by quenching them—red hot—into the urine of a red-headed boy.

In the above examples, the metallurgist would describe what had happened more scientifically by reference to the binary phase diagram for iron and carbon, since steel is an alloy of iron and carbon.

The diagram shows that the iron-carbon alloy system has different crystal phases, each with a different name and with different known properties. And in a text you might process if you work for a metallurgist, there might be discussion of a number of terms that designate these crystal phases. Here are some typical ones—familiar to the metallurgist but not to the average blacksmith (although the blacksmith works with the same phases and achieves the same results as the metallurgist).

- *Alpha iron,* which is written with the notation $Fe_{(\alpha)}$, or in text may appear also as α *iron,* is a crystal in which the iron atoms form the eight corners of a cube, but there is a ninth atom in the center, the body, of the cube (Figure 4-17*a*). So the alpha phase is a *body-centered cubic* crystal (in notation: bcc). When the alpha-iron phase contains carbon, and is thus a form of steel, it has a particular name—*ferrite*.
- *Gamma iron,* which is written with the notation $Fe_{(\gamma)}$, or in text may appear also as γ *iron,* is a crystal in which the iron atoms form the eight corners of a cube, but in each of the six faces of the cube there is an additional atom of iron (Figure 4-17*b*). So the gamma phase of iron is a *face-centered cubic* crystal (in notation: fcc). When the gamma-iron phase contains carbon, and is thus a form of steel, it has a particular name—*austenite*.

Coming back to the crafting of the sword blade, what the sword maker did was to heat the steel to a temperature at which austenite was the equilibrium crystal phase. The sword maker then quenched it to retain the austenite at room temperature. By virtue of this heat treatment, a metal phase was created in the sword blade that had the properties required for use as a sword.

NOTE: Normally austenite exists in steel only at red heat. But quenching red-hot steel in water *retains* the austenite at room temperature, where normally only ferrite would exist.

Under most conditions, austenite is far stronger, more wear-resistant, more difficult to indent or scratch, of higher tensile strength than ferrite, so it makes a better sword blade.

5. Metallography

In addition to phase diagrams, stress-strain diagrams, heat-treating furnaces, and quenching baths, the metallurgist has one more important tool: the microscope. When microscopes are used to examine the crystal structure of specially prepared metal specimens, that procedure is known as *metallography*.

NOTE: Metallography is as important to the metallurgist as microscopic procedures are to the biologist and the bacteriologist. The difference is that the metallurgist looks searchingly at tiny metal crystals, while biologists and bacteriologists look just as hard at living cells and microbes. And the more they look, the more they learn and understand.

Metal samples for metallography are prepared by cutting off a small sample of the metal to be examined, mounting it in a plastic mold or holder, polishing the metal until it is very smooth, and then etching it with a chemical *etchant* to bring out different reflections from each phase so the phases look different from each other, and then looking and taking pictures through the microscope. The entire procedure is known as *metallography*. The pictures—which you may have occasion to mount in reports as illustrations—are called *photomicrographs*. (They are also sometimes called "microphotographs," but that is not the preferred term).

Figure 4-21 shows two typical photomicrographs of metal alloys at magnifications of 200 times (×200). Figure 4-21*a* shows two different crystal phases occurring side-by-side. (We know they are different because they look different.) Since there are two simultaneous phases present, we call this a *two-phase* alloy system. Figure 4-21*b*, on the other hand, is obviously a single-phase alloy system.

NOTE: There are twinned crystals (Section 4.7.1) in both these photomicrographs. Twins are recognizable by their twin bands, sets of parallel lines within particular crystals.

4.8 MODERN PHYSICS

Modern physics is concerned primarily with the structure of matter and the interrelations between matter and energy. It includes—but is not

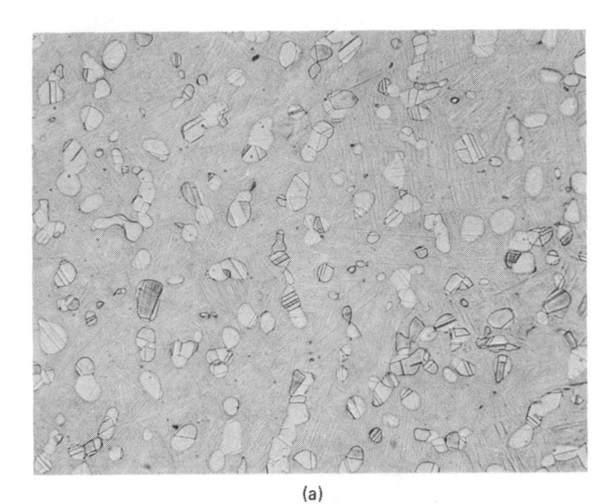

(a)

(b)

FIG. 4-21 Alloys. (*a*) Two-phase alloy, photomicrograph at 200 magnifications (200X). Note presence of twin bands in one of the phases. (*b*) Single-phase alloy (200X). *(Photographs courtesy of Manlabs, Cambridge, Mass.)*

limited to—solid-state physics, plasma physics, spectroscopy, nuclear physics, relativity, and quantum mechanics. It also embraces anything "theoretical." It delves one step deeper into all aspects of physics (and chemistry) than classical physics does.

Although classical studies attribute the behavior of materials to the fact that materials are made up of atoms, modern physics takes off from there and tries to understand what makes up atoms, and from this, how matter and energy overlap and become different aspects of the same thing.

4.8.1 The Structure of the Atom

In previous sections of this book, when an atom was represented in an illustration (Figures 4-17 and 4-19), it was shown as a circle, by which was meant some sort of sphere. But atoms are more complicated than that. In the first place, the atoms of each chemical element are different from the atoms of every other element. In fact, it is the atomic difference that causes the chemical difference. Secondly, atoms are in turn made up of several basic "ingredients," and these behave in very complex ways.

1. The Three Fundamental Particles

Atoms are arrangements—in different proportions and in different locations and orientations—of three other, smaller parts. These may be considered to be *building blocks* of atoms. They are called *subatomic particles* and are

- Electrons
- Protons
- Neutrons

NOTE: These in turn are made up of other, smaller subatomic particles. There is a large variety of these, and new ones are still being discovered or hypothesized.

All three may be described in terms of their electrical behavior.

a. Electrons Each *electron* has one negative charge.

Section 6.3.1 discusses how electrons contribute to the electrical nature of the chemical bond. In Section 6.3.2 it is shown how more or fewer electrons attached to an atom or group of atoms results in negative or positive ions. When electrons move in any direction, they cause an electric current—in fact they *become* electricity (Section 5.1.4).

b. Protons Each *proton* has one positive charge.

All atoms are balanced in terms of electric charge, which means that unless charges are added or taken away—as occurs when ions are produced—the total atom has no charge. This is because inside it there are always just as many protons as there are electrons.

c. Neutrons The electrical characteristic of the *neutron* is that it has no electric charge.

It is neutral electrically (thus the name—neutron).

But, in spite of making no electrical contribution to the atom, the neutron does participate in determining some of the qualities of each atom. Each atom has different numbers of neutrons.

2. The Nucleus

Neutrons and protons always associate themselves with each other in a kind of clump which always resides at the center of the atom. This clump is called the *nucleus.*

This word gives us the name of this field of physics—*nuclear physics.* It is the study of the qualities, behavior, and internal structures of the atomic nucleus.

NOTE: The constituents of the nucleus, the protons and the neutrons, are collectively known as *nucleons.*

3. Atomic Structure

For any atom, the three particles always arrange themselves in an array that may be considered spherical.

The nucleus is always in the center, and the electrons always surround the nucleus.

a. The fixed nucleus The nucleus may be considered to be *fixed in space*—although it does vibrate somewhat. But the electrons are much freer to move. Thus they move around the nucleus in specific paths.

The arrangement may be compared to our sun and the planets that surround it, with a nucleus functioning like the sun and each electron behaving like a planet—and with each electron (planet) remaining always in its designated orbital (orbit). Figure 4-22, which represents the structure of the sodium atom, also looks somewhat like a drawing of our sun surrounded by Mercury, Venus, Earth, and Mars.

b. The orbital—energy levels The path around the nucleus in which an electron may move is called the *orbital* of the electron. Each orbital represents a different *energy level.* Electrons at lower energy levels (closer to the nucleus) can leap to a higher energy level if they receive additional energy.

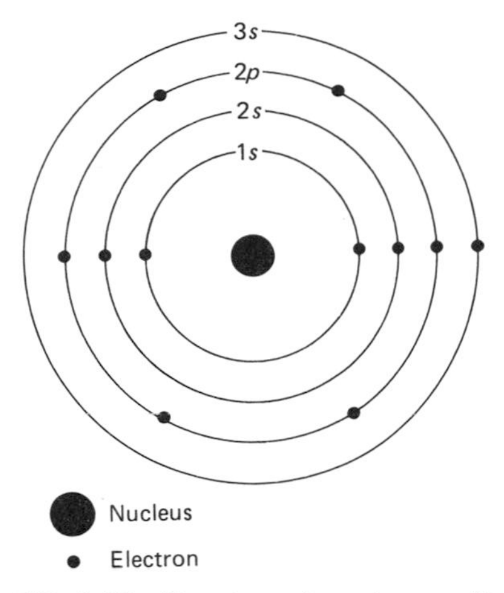

FIG. 4-22 Structure of an atom: sodium.

Electrons closest to the nucleus are tightly bound and cannot easily come free, but electrons already in higher energy states (levels) are easily dislodged and may leap to even higher energy states or from atom to atom.

It is the outer electrons that—by their ability to depart from the parent nucleus, even momentarily—change atoms into ions (Section 6.3.2) and at the same time provide the basis for electric current.

c. Constructing atoms Each atom species—which is the basis of each chemical element (see Section 6.1.1)—results from a unique combination of the three basic particles.

Three examples are hydrogen, copper, and gold.

- The hydrogen atom has 1 proton, no neutrons, and 1 electron.
- The copper atom has 29 protons, 34 neutrons, and 29 electrons.

NOTE: The 29 electrons balance the 29 protons electrically. The 29 electrons move in separate orbits (some of which are considered inner and some outer).

- The gold atom has 79 protons, 118 neutrons, and 79 electrons.

d. Atomic number, mass number Each atom is described by a numbering system as follows:

- The *atomic number* indicates the number of protons (and thus the number of pairs of electrons and protons). It is designated Z.
- The *mass number* is a measure of the size of the nucleus. It is the sum of the number of protons and the number of neutrons in any atom. It is designated A.

TABLE 4-7
ATOM CONSTRUCTION—ATOMIC NUMBERS AND MASS NUMBERS

Element	Atomic Number, Z	Mass Number, A	Structure
hydrogen	1	1	1 ● ; ⊖ 1
carbon	6	12	6 ●, 6 ○ ; ⊖ 6
aluminum	13	27	13 ●, 14 ○ ; ⊖ 13
copper	29	63	29 ●, 34 ○ ; ⊖ 29
gold	79	197	79 ●, 118 ○ ; ⊖
lead	82	207	82 ●, 125 ○ ; ⊖ 82

*Protons are ● ; neutrons are ○ ; electrons are ⊖

Table 4-7 shows how some atoms are constructed. It also gives the pertinent atomic numbers and mass numbers. Such a listing indicates the *atomic structure* of each atom, showing how each is different from all others. Of course, this table could be extended to include every one of the 107 atoms that are known to exist.

NOTE: You could also construct the table to include elements that aren't now known to exist. In fact, physicists did predict a number of atoms by extending this table beyond what was known at that time. Later it was shown that the predicted atoms do actually exist. Even now we can theorize what atom structure the atom with a mass number of 108 would have—and someday that atom might be discovered or synthesized.

The number of protons is indicated by a subscript numeral and the mass number is indicated by a superscript numeral, both preceding the symbol for the element.

For example:

$${}^{6}_{3}\mathrm{Li}$$

is lithium (three protons) with a mass number of 6.

$${}^{206}_{82}\mathrm{Pb}$$

is lead (82 protons) with a mass number of 206.

NOTE: When atoms are characterized in terms of both their number of protons and their mass number, they are known as *nuclides*. Isotope designations differ from nuclide designations in that only mass numbers are shown. Thus in the preceding example, the isotope lead-206 is designated ${}^{206}\mathrm{Pb}$.

4. Isotopes

A certain proportion of the atoms of some elements have a different number of neutrons in the nucleus while the number of protons remains unchanged.

a. Same atomic number, different mass number This means that the atoms of a given element with a given atomic number Z may have two or more mass numbers A.

b. Definition These different forms of atoms—different only because there is a different number of neutrons in the same element—are called *isotopes*.

For all elements there is a predominant isotope. The other isotope or isotopes, if they exist at all (some elements in fact have only one natural isotope), are normally present in very small proportion.

Hydrogen generally has one electron and one proton (giving it an atomic number of 1) and no neutrons (giving it a mass number of 1). But in nature, for approximately 1 hydrogen atom out of 5000 there is a neutron. The resulting atom is an isotope of hydrogen.

c. Isotope names Names have been given to a few isotopes. (Those that have no names have to make do with just the Z and A designations.)

The "minority" isotope of hydrogen is *deuterium.*

The isotopes of hydrogen thus include the predominant isotope,

$$Z = 1 \quad \text{and} \quad A = 1$$

which is symbolized

$${}^{1}_{1}\text{H}$$

and deuterium,

$$Z = 1 \quad \text{and} \quad A = 2$$

which is symbolized

$${}^{2}_{1}\text{H}$$

d. Chemical differences between isotopes For most isotopes of the same element there is very little different in chemical properties or behavior. But there are some exceptions, which become important in certain chemical compounds.

Water is such an exception. Here the effect of using the deuterium isotope of hydrogen to form the compound H_2O (which then can be written ^2H_2O or sometimes D_2O) is that the nucleus contains a neutron where usually there is none. This form of water is known as *heavy water* and is important in the making of atomic weapons.

But compounds of lead, which has four isotopes with mass numbers of 204, 206, 207, and 208, are very similar chemically, regardless of which isotopes are present in whatever proportion.

e. Radioactive differences between isotopes Isotopes may differ from each other very greatly in regard to their radioactivity (see Section 4.8.2 for a description of this subject).

Table 4-8 lists the natural isotopes of a number of elements. It is an extension of Table 4-7, which lists only the predominant isotope.

NOTE: New isotopes are always being discovered. And isotopes that do not exist naturally can be produced in the laboratory.

4.8.2 Radioactivity

Some types of atoms are *unstable*. Under normal conditions, they *disintegrate* (or *decay*), throwing out *(emanating) particles* and *rays*. This property of disintegrating is known as *radioactivity*, and elements (atoms) that exhibit radioactivity are *radioactive*.

TABLE 4-8
THE ISOTOPES OF SOME COMMON ELEMENTS

Element	*Atomic Number Z*	*Mass Number A*
Hydrogen	1	1
		2
		3*
Carbon	6	12
		13
		14*
Aluminum	13	27
Copper	29	63
		65
Gold	79	197
Lead	82	204
		206
		207
		208

*Occurs naturally in trace amounts.

Radium is such an element. (Its name comes from the Latin word for ray.) Some elements which are not normally radioactive have forms, *isotopes* (see "Isotopes" in Section 4.8.1), which are (for example, *tritium* is a radioactive isotope of hydrogen).

1. Radiochemistry

The changes that occur during radioactive disintegration of isotopes (while in the realm of physics) have chemical outcomes. These changes are studied and discussed as part of the field of *radiochemistry*.

NOTE: Radiochemistry has nothing to do with the radio which provides you with news and music—other than that in both cases energy is involved, which radiates.

2. Forming New Elements

In ordinary chemistry (see Chapter 6), no matter how many chemical reactions occur and how many chemical compounds are formed and unformed, the atoms never change: an atom of chromium remains an atom of chromium. In radiochemistry, on the other hand, one of the isotopes of an atom can break down (disintegrate) into some combination of one or more of the following:

- Other isotopes of the same element
- Isotopes of some other element
- Subatomic particles
- Electromagnetic radiation (Section 4.5.2)

3. The Role of Subatomic Particles

Subatomic particles are involved in two very important processes. These are *emanation* and *bombardment*.

a. Emanation When radioactive materials disintegrate, subatomic particles and electromagnetic energy are given off as *rays* or radiation. In radiochemistry anything given off in this manner is called an *emanation*.

b. Bombardment Such emanation may trigger or aid radioactive disintegration. This process is called *bombardment*.

Subatomic particles are directed to materials in beams or rays. When such beams impact materials, we say the materials are being *bombarded*.

4. Radioactive Emanations

The most common radioactive emanations are alpha particles (also called alpha rays), beta particles (also called beta rays), neutrons, positrons, and gamma rays.

a. Alpha particles These are made up of a clump of two protons and two neutrons. Thus they have a double positive charge. When emanated, they are known as *alpha rays* (also written *α rays*). In notation they are represented by α.

Alpha particles are equivalent to the nuclei of helium atoms. The helium atom consists of two protons, two neutrons, and two electrons, and can be represented by ${}^{4}_{2}He$ (see "Isotopes" in Section 4.8.1). When the two electrons are removed, only the nucleus, with its double positive charge, is left. And you can see that it is the same thing as the alpha rays. Thus in notation the alpha rays are sometimes represented by ${}^{4}_{2}He$.

b. Beta particles These are high-velocity electrons.

Like electrons, they have a single negative charge. When emanated, they are known as *beta rays (β rays)*. In notation they can be represented by β, e, e, e^-, or e^-, depending on the preference of the author or publication.

c. Positrons Positive electrons exist. They can be emanated or emitted, and this is known as positron emanation. Such electrons are called *positrons* and are designated by the symbol β^+, e^+, or e^+.

d. Neutrons Neutrons (see "Neutrons" in Section 4.8.1) are elementary particles having the same mass as a proton and bearing no electric charge.

When emanated, neutrons are known as *neutron rays*. In notation they are n or n.

e. Gamma rays Gamma rays (or γ rays) are electromagnetic radiation like light or x-rays.

They oscillate at a frequency higher than that of visible light or x-rays. In notation they are represented by γ.

NOTE: Any of the above may either bombard materials or be emanated by materials.

5. Radiochemical Reactions—Notation

The notation of these reactions resembles that of chemical reactions, but it differs in that special notation is added to indicate subatomic particles (emanating or bombarding) and isotopes. This notation is described in Section 4.9.6.

6. Fission

The process of disintegration of an isotope is known as *fission* or *nuclear fission.*

Materials that undergo fission are *fissionable* or *fissile.*

Why is fission so important? Because enormous amounts of energy are locked up in small amounts of material. And fission is the means by which this stored energy is released.

Only a few pounds of radioactive material are needed to make devastating atom bombs.

Only a few pounds of radioactive material are needed to make nuclear power stations capable of supplying all the electricity needed by a large city for several years.

Only a few milligrams of radioactive material injected into the human body can help us diagnose certain diseases using the techniques of *nuclear medicine.*

7. Half-Life

The fission process does not go on indefinitely, nor does it proceed at the same rate through time (it slows down). The fission time for different radiochemicals is measured by their *half-life.*

Half-life is the time required for one-half of the atoms in a radioactive sample to "decay."

8. Units

The basic unit of rate of emanation from radioactive materials is the *curie* (abbreviated *Ci*).

A thousandth of a curie is a *millicurie* (*mCi*).

NOTE: This unit is named after Madame Marie Curie, who pioneered the science of radiochemistry.

4.8.3 Quantum Mechanics

This is a branch of physics that treats the theory of matter from the point of view of the interactions of matter and radiation. It differs from the many branches of classical physics in that everything is considered to be based on atomic and subatomic phenomena rather than on the relatively large machines and structures of our everyday experience.

If you work with text that deals with quantum mechanics and related subjects, you will find that it is highly mathematical. There will be endless equations. The techniques of working with such material on the printed page have already been discussed in Chapter 2. Here we will define a few key concepts of quantum mechanics so that you will have an idea of the meaning of the text.

1. The Quantum Concept

The basic contribution of modern physicists to modern physics is the realization that—on an atomic or subatomic scale at least—things do not happen continuously; they happen in steps, in discrete intervals and quantities. Things do not flow smoothly; they jump. Also, no jump is divisible; there is no such thing as a "half jump." Every such indivisible increment is called a *quantum* (plural, *quanta*).

> It is like walking from the first floor of a building to the second. We go by steps. We have to go by steps because there are no half steps—thus every step is a full step: no more, no less. So a staircase is a sequence of *quantized* intervals.

2. Particles and Waves

In quantum mechanics, wave phenomena are reinterpreted as particle phenomena. This is consistent with the definition of the quantum concept, since the movement of a particle can be described using a *step* concept, as opposed to a wave, which requires a nonstep or *continuous* concept to explain its movement.

> So the quantum physicist deals with particles rather than waves. For example:
>
> - The indivisible particle of sound is the *phonon,* in quantum mechanics considered the unit of *vibration* of atoms in a crystal.
> - The indivisible particle of light is the *photon,* in quantum mechanics considered to represent all electromagnetic radiation, not just light.

3. Energy Levels

What is considered to move in quantum mechanics is particles. They do not so much move in space as into different levels or states of energy. What causes the movement, or results from it, is a change in energy.

a. Allowed levels For every atom, there are *allowed* energy conditions and *unallowed* energy conditions for particles. These are the *steps* of quantum mechanics previously mentioned. They are also called allowed and unallowed *energy levels.*

b. Upper levels, lower levels A particle may jump from a lower level to an upper level or fall from an upper level to a lower level.

c. Energy gaps The regions between allowed levels in an atom are known as *energy gaps.*

The wider the separation between allowed levels in an atom, the more energy is required for particles to leap the gap. Materials with wide gaps are hard to change. They are *stable.* Materials with narrow gaps are easy to change—with the investment of little energy—and are relatively *unstable.*

Photographic film (see "Photochemistry" in Section 4.8.3) utilizes a relatively unstable material, silver iodide. It takes relatively little light energy to activate the silver so it can be developed to give us a photograph.

d. Energy bands Sometimes the allowed energy levels clump; they are close to each other. Then it is hard to tell one allowed energy level from another within that clump. The clump is then called an *energy band.*

e. Energy—invested or contributed For a particle to jump up, energy must be invested or contributed from the outside. When a particle falls down, energy is given off. In both cases the energy is in the form of electromagnetic radiation.

4. Photochemistry

Chemical changes can be produced by various sources of energy. One source is light. Photography is an example of this. We make use of quantum jumps of particles to result in the chemical changes that give us images on film.

> Light hits the film. Quantum changes take place in the silver iodide crystals that are on the film. As a result, the silver iodide is in a higher energy state (less stable) and can easily be reduced by the developer chemicals to form silver and iodide ion. The resulting metallic silver is black, and that provides us with the negative, which we perceive as a (negative) picture.

5. Fluorescence

Fluorescent light (see "Luminescence" in Section 4.5.2) is an example of the investment of energy by means of particle impact on a material with the result that radiation emanates from the material.

> The screen of the picture tube in your television set is hit by electrons. This causes a leaping of energy gaps by particles within the material of the screen. When they fall down again, the result is that light waves (or

light particles—photons) are created for us to see, and we perceive the picture on the screen as a result.

6. Spectroscopy

Spectroscopy is a technique for identifying materials, which is to say for identifying which atoms the materials are made of.

This is accomplished by investing heat energy so that the material rises to such a high temperature that it vaporizes. Then it gives off light. The nature of that light is different for each atom, because each atom has different orbitals or allowed energy levels for its electron particles—and each gives off light at different frequencies (colors). By identifying the colors, the *spectroscopist* can identify the atoms (the elements) present. When the intensity of the color is measured, a quantitative measure of how much material is present can also be obtained. This is known as *spectrographic analysis.*

You can do some spectroscopy yourself. Just light a match. The hot vapor gives off a characteristic orange glow which we recognize as *flame*. It is due to the energy changes of electrons in the vaporized atoms that made up the wood or paper of the match. When they jump from one allowed energy level to another, they give off some orange radiation. Or light a fire in your fireplace. You get the same orange flame. But now throw in some salt. You will now see some yellow flame jets. That is because when hot the sodium of the salt *emits* radiation at the wavelength of yellow. That radiation is released from the sodium atom because of quantum changes within it.

NOTE: We are used to referring to the emission of visible light from hot bodies as *incandescence* (see "Incandescence" in Section 4.5.2). An example is the light from an electric light bulb, which we call an *incandescent bulb.* How does incandescent light differ from the light utilized for analysis purposes by the spectroscopist, which also is emitted from hot bodies? It doesn't really. It happens that when a body gets to higher and higher temperatures, many more orbital jumps occur than at lower temperatures—each giving off its own light at its own frequency (color)—and all the light emanations combine. When many colors combine, the resulting color begins to look white.

7. Plasma Physics

That branch of quantum mechanics which deals with gases is known as *plasma physics.*

Any glowing gas, like a flame or an arc, is a *plasma* (see "Plasma Emission of Light" in Section 4.5.2). Spectroscopy (previous section) is a branch of plasma physics.

8. Solid-State Physics

That branch of quantum mechanics that deals with phenomena within *closely packed* matter (liquids and solids as distinguished from gases) is known as solid-state *physics.*

The quantum jumping of electrons from one energy band to another in a controlled fashion within a particular kind of material (a *semiconductor*) is the basis of the electronic components known as *transistors, integrated circuits,* and *computer chips,* which in turn give us all of modern electronics, including computers. All these topics are discussed in the following chapter.

9. Relativity

No discussion of modern physics is complete without mention of Albert Einstein's *theory of relativity.* This theory deals with the motion of objects through space and the notion that defining the position of any object at any moment in time is dependent on the position of the observer. This theory has more practical significance than is apparent to the layperson, but it is beyond the scope of this book to discuss this very complex topic. If you are interested. ask your author or company librarian to recommend a book that attempts to explain relativity to the nonscientist. If you are employed by a physicist who works in this area, you will probably find that in form the texts that you work with will look familiar. A typical text will consist of mathematical equations and in that regard will not differ from other modern physics text described in this chapter.

10. Statistical Mechanics

When dealing with atoms and subatomic particles, the physicist never really knows what happens. What he or she knows is what *probably* happens. Thus quantum phenomena must be treated by the mathematical techniques of *probability* and *statistics.* (see "Probabilities" in Section 2.5.2). These techniques applied to quantum mechanics are known as *statistical mechanics.*

Physicists not only recognize what they can know and can never know, but they are confident enough of where they stand to define their uncertainty—with certainty. This definition is known as the *Heisenberg uncertainty principle* after its conceiver, the German physicist Werner Heisenberg. (It states that the greater the certainty of a particle's position, the greater the uncertainty in the particle's velocity. This is because with such tiny particles as atoms and electrons, any method used to measure velocity changes the position of the particle, and vice versa.) It is also known as the *indeterminacy principle.* Since it also indicates how unending the physicist's search must be, it is (probably) a good place to end this section on modern physics.

4.9 Physics Notation

Most of the text to be reproduced in physics will be in the form of scientific notation, and most of that notation has been covered in the chapter on mathematics. Following are a few techniques and notations more specific to physics.

4.9.1 Units

Both the English system and the International System of Units (the SI system) continue to be used in physics, although every year the SI system gains more ground, and whenever there is a question of which to use, the SI system should be considered to be the preferred one. Glossary M lists the most commonly encountered units—both English and SI—with their symbols.

4.9.2 Degree Sign

The degree sign is smaller than the lowercase letter o, but it is not usually found on typewriters or word processors, and you will thus have no choice but to use the lowercase o. Write it one-half index up and follow it with the capital letter that designates which temperature scale is being used.

Some authors and some publications use a space between the number and the degree sign; others don't. If your author uses both, ask which form you should use, and then use it consistently.

NOTE: The Kelvin scale does not use a degree sign.

Here are some examples of practice to follow when writing expressions containing degrees:

"One thousand and eighty-three degrees Fahrenheit" is written

1083°F or 1083 °F

"Seventy-four degrees Celsius" is written

74°C or 74 °C

"Nineteen kelvin" is written

19 K

Refer to Table 4.4, which illustrates usage of the degree sign for different temperature scales.

4.9.3 The Use of Miller Indices in Crystallography

The facets of crystals are shown using the system of Miller indices. These are sets of three-digit numbers. They are most commonly grouped in parentheses, but brackets or braces are sometimes used:

- (001) Most common
- [001]
- {001}

Occasionally, the crystallographer has need to call attention to one of the digits. This is done with an overline.

$(1\bar{1}0)$

$(\bar{2}01)$

4.9.4 Phases of a Material

Phases are designated by Greek letters, either as subscripts to symbols or as prefixes in text:

$Fe_{(\alpha)}$ or α iron or alpha iron

4.9.5 Isotope and Nuclide Notation

Isotopes are shown in equations and text by using the chemical symbol for the atom (see "Symbols for Chemical Elements" in Section 6.1.1) preceded by the mass number that is pertinent. The mass number is placed one-half turn up. In text, the name is also sometimes written out.

For the element uranium, for example, the isotope with a mass number of 235 is written

^{235}U or uranium 235

Isotopes may also be designated in terms of their atomic numbers as well as their mass numbers. In that case atomic numbers are shown as subscripts preceding the chemical symbol. When atoms are so characterized—when both mass number and atomic number are indicated—they are known as *nuclides*.

For one of the nuclides of tellurium, for example, the nuclide symbol is

$^{125}_{52}Te$

4.9.6 Subatomic Particles

Alpha particles when emanated are known as alpha rays or α rays. In notation they are designated α. They can also be designated using nuclide notation (see Section 4.9.4) as $^4_2$He.

Emanated beta particles are known as beta rays or β rays. In notation they can be represented by β, e, e, e^-, or e^-, depending on preference of authors or publications.

Positrons can be designated by the symbols β^+, e^+, or e^+, again depending on individual preferences. Remember, however, whichever notation an author chooses should be used consistently within one paper or book.

Neutrons, when emanated, are known as neutron rays. In notation they are n or n.

Gamma rays (or γ rays) in notation are represented by γ.

4.9.7 Radiochemical Reactions

The notation of these reactions resembles that of chemical reactions, but it differs in that special notation is added to indicate subatomic particles (emanating or bombarding) and isotopes.

1. When Bombarding

When subatomic particles are used to bombard other atoms, they may be shown as though they were atoms in their own right; they are simply "connected" to the other chemicals by plus signs.

This is the case for the n and the γ in the following example:

$$^{238}\mathrm{U} + \mathrm{n} + \gamma \rightarrow {}^{239}\mathrm{U}$$

NOTE: Bombardment is always understood to take place on the left side of the equation.

2. When Emanating

Subatomic particles may be shown in two ways when they are emanated:

- As though they were atoms in their own right; they are simply "connected" to the other chemicals by plus signs.

  For example:

$$^{239}\mathrm{Pu} \rightarrow {}^{235}\mathrm{U} + \alpha$$

  NOTE: Emanation is always understood to be taking place on the right side of the equation (as in the preceding example).

- By placement over the equals sign or arrow of the equation.

For example, the reaction used as an example in the preceding section actually continues as follows:

$$^{238}\mathrm{U} + \mathrm{n} + \gamma \rightarrow {}^{239}\mathrm{U} \xrightarrow{\beta} {}^{239}\mathrm{Np} \xrightarrow{\beta} {}^{239}\mathrm{Pu} \xrightarrow{\alpha} {}^{235}\mathrm{U}$$

Nuclide notation may be required and may be used in the same manner, as in

$${}^{15}_{8}\mathrm{O} \rightarrow {}^{15}_{7}\mathrm{N} + {}^{0}_{1}\mathrm{e}$$

4.9.8 Half-Life

Half-life is shown, in time units such as seconds, minutes, days, or years, in a reaction either over the arrow sign or below it. The latter practice is used when the position over the arrow is already occupied by an emanation symbol.

For example:

$${}^{239}\mathrm{U} \xrightarrow{\text{23 min}} {}^{239}Np + \beta$$

or

$${}^{239}\mathrm{U} \underset{\text{23 min}}{\overset{\beta}{\rightarrow}} {}^{239}Np$$

Electricity and Electronics

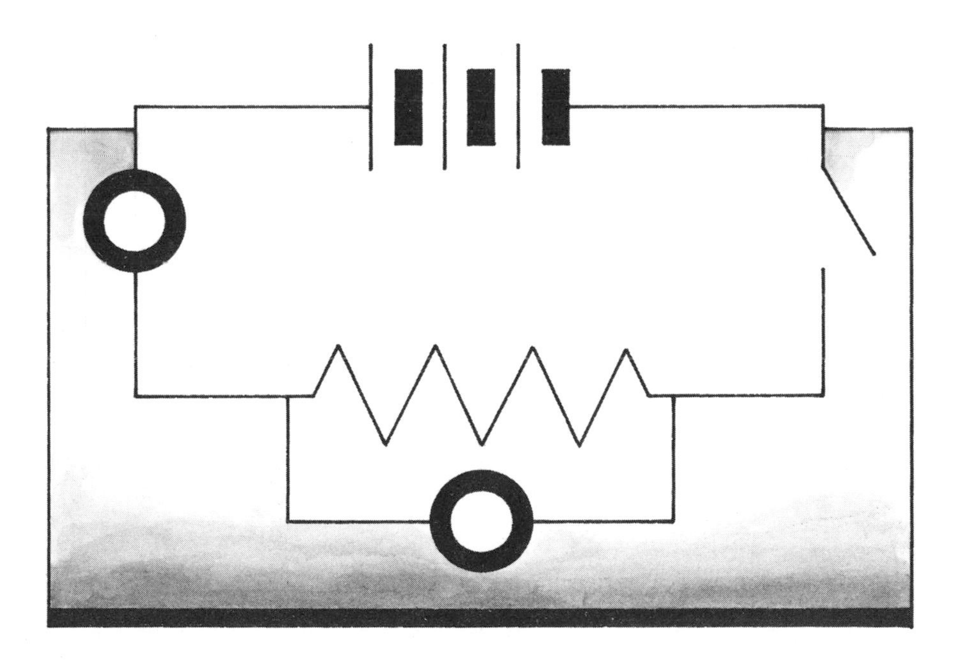

Basic Principles

When electrons, one of several kinds of subatomic particles, which are the fundamental building blocks of matter (Section 4.8.1), are subjected to external forces that free them from their atomic orbitals, motion results. That movement of electrons through a material is called *electric current* or *electricity.*

Electricity can be used to provide power, as in turning a motor; or to provide and process signals, as in modern communication methods such as television or radar; or to process data, as in computers.

Until the time of Benjamin Franklin, which is to say through most of the world's history, people did not know how to produce and harness electricity. But when he led a current of electrons (lightning) down a kite wire to earth—and nearly killed himself in the process—he changed all that. It then eventually became possible, after the inventors got busy, to replace horses with horseless vehicles (although we still rate motors by their *horsepower*) and the pony express by the telegraph, the telephone, and electronic mail.

We deal with electrons and electricity in other chapters. Electrons and attractions between positive and negative electric charges are the basis of chemical bonding (Section 6.4.1) and are responsible for ions and dissolving (Sections 6.3.1 and 6.3.2). They also are the basis of crystal bonding (Section 4.7) and indeed of atom structure (Section 4.8.1). Finally, the equivalency of particles and waves explains the various forms of electromagnetic radiation such as light and heat (Sections 4.5 and 4.6).

In this chapter the uses of electricity as studied and exploited by the *electrical engineer* and *electronics engineer* will be described.

As for the difference between these two, the electrical engineer deals mostly with the *power aspects* of electricity, using such components as motors, transformers, and circuit breakers; and the electronics engineer deals mostly with the *signal aspects* of electricity, using such *electronic devices* as electron tubes, transistors, and semiconductor chips to control these signals. But these two technologies have much in common since both depend on the manipulation of electrons in *electric circuits.*

Whether the engineer or author is dealing with the generation and use of power or the handling of signals, a central concept is that electrons are sent down complex paths, mostly made of metal wires, called *circuits.* The technical secretary and the technical writer processing text in this subject should understand (be literate in) a number of basic

5.2

concepts relating to how electrons move in circuits, what components and structures they encounter along the way, what effects they cause, and how to represent them in text and notation.

5.1 THE MOVEMENT OF ELECTRONS IN MATERIALS

How does it happen that electrons move through materials? Are not the materials which conduct electrons, such as copper wires, hard and impenetrable? The answer is that when things as small as electrons are considered, "solid" materials are very penetrable indeed.

Not only are there spaces between individual atoms (in crystals the atoms occupy only the corners or centers of geometrical figures—the regions between are empty; Section 4.7.1), but there are also large regions of emptiness—large compared with an electron's size at least—within any atom. These empty regions exist between the nucleus of the atom and the orbits (called orbitals) of the the electrons. They also exist at "forbidden" areas between orbitals, and even within orbitals wherever the electrons don't happen to be at any particular moment—just as the planets at any given moment occupy only a tiny part of their enormous orbits (see "Atomic Structure" in Section 4.8.1).

The fact is that most of what an electron confronts when it encounters a solid material is emptiness—vacuum. And electrons find it very easy to move (become the basis of electric currents) in vacuum. They need only a push, which *voltage forces* impressed on the solid material provide (Section 5.1.3).

5.1.1 Conductivity

Electrons move through some materials more easily than through other materials. The ease of movement may be described quantitatively by a measure called the *conductivity,* which has a characteristic value for each material. The symbol for conductivity is σ, the lowercase Greek sigma.

Substances are classified according to their conductivities, those with higher conductivity values being the more *conductive.*

The major classes of materials, grouped in terms of their relative conductivities, are conductors (through which electrons move most easily), resistors, semiconductors, and insulators (which are such poor conductors that they hardly conduct at all).

1. Conductors

The materials which conduct electrons best are called, simply, *conductors.*

Most conductors are metals.

2. Insulators

The materials which conduct electrons least, in fact essentially not at all, are called *insulators.*

Most insulators are ceramics or glass. Wood and air are also classed as insulators.

NOTE: Heat (thermal) insulators (Section 4.6.3) are usually also electrical insulators.

3. Semiconductors

The materials which conduct electrons with intermediate ease (between that of conductors and that of insulators) are called *semiconductors.*

Semiconductor materials are in the class known as *metalloids,* which means that they are somewhat like metals, but not entirely. *Silicon* is the most common semiconductor material; *gallium arsenide* and *germanium* are the next most common.

4. Resistors

Resistors are electronic devices (see "Resistors" in Section 5.3.2) which are designed to have an exactly prescribed level of *conductance* (ability to conduct electrons). Thus resistors are provide *just the right amount* of conductance for a particular use—no more and no less.

The name *resistor* comes from the fact that the conductance is "held down" to the value required by selecting its material, size, and shape. Thus it can be said that the resistor is a device that resists the flow of electrons. But it is essential to know that the electrons still flow (are still conducted) in a resistor. The circuit may be thought of as a broad river that suddenly becomes narrow. The narrow place resists (restricts) the flow of the water but still lets it flow.

Figure 5-1 shows some resistors. Note that they are of many different sizes (for different values of conductance described in the following section), but almost all are in the shape of small cylinders with one wire (one *lead*) at each end. (Leads are also called *terminals, connectors, contacts,* or *electrodes* depending on the context.)

NOTE: It is logical that resistors should have two leads. One is to lead the electrons in, and the other is to lead the (resisted) electrons out.

5. Conductance and Conductivity

The ease of movement of electrons is further affected by the size and shape of the object in which the electrons move.

This accounts in part for the different sizes of the resistors in Figure 5-1.

The longer the object, the more trouble electrons have moving through it from one end to the other; the larger the cross-sectional area of the object, the more easily electrons move through it from one end to the other.

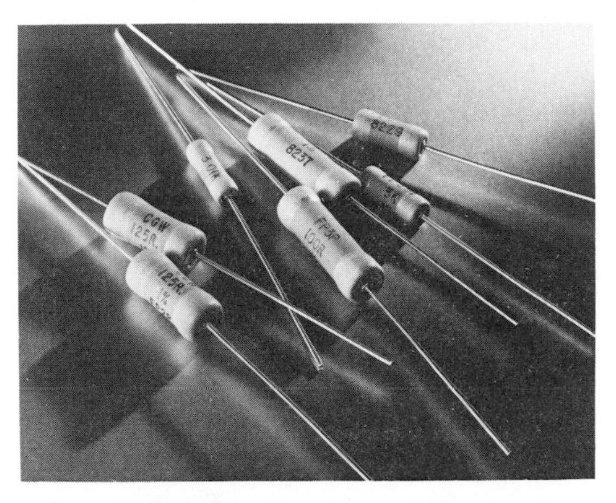

FIG. 5-1 Resistors. *Courtesy of Corning Glass Works, Corning, N.Y.*

This ease of electron flow that is affected by size and shape is called the *conductance* C and is related to the conductivity σ by an expression that takes the size and shape of the object into account.

This relation is expressed in equation form as

$$C = \sigma \frac{A}{L}$$

where A is the cross-sectional area of the object in units such as square inches (in^2) or square centimeters (cm^2), and L is the length in units such as inches (in) or centimeters (cm).

The unit of conductance in SI units is the *siemens*, S.

NOTE: The siemens is also known as the *mho*, for which the symbol is the uppercase Greek letter omega drawn upside down, ℧. Siemens is the preferred usage.

6. Resistance and Resistivity

The interference offered to the movement of electrons by a material may be described quantitatively as the *resistivity*, which has a characteristic value for each material. The symbol for resistivity is ρ, the lowercase Greek rho.

The interference with the movement of electrons is further affected by the size and shape of the object in which they move.

The longer the object, the more the interference; the larger the cross-sectional area of object, the less the interference.

This interference with electron flow that is affected by size and shape is called the *resistance* R and is related to resistivity by the expression

$$R = \rho \frac{L}{A}$$

where L is the length of the object and A is the cross-sectional area of the object.

The unit of resistance is the *ohm.* The symbol for ohm is the uppercase Greek omega, Ω.

NOTE: In the preceding example the symbol R designates a resistance. The magnitude of that resistance is measured in the number of resistance units (in ohms). Thus, for example:

$$R = 100\ \Omega$$

Or, expressed in words, R is a resistance of 100 ohms.

The ohm is as basic a unit in electricity and electronics as the pound or kilogram is in the measurement of weight or the mile or kilometer is in the measurement of distance.

NOTE: The ohm is named after a nineteenth-century German physicist, Georg Simon Ohm, who was one of the first people to understand the concept of resistivity.

NOTE: Since conductance is the opposite of resistance, the unit of conductance is the *mho,* which is *ohm* written backward. Mathematically, it is the *reciprocal*—which means "one over," in this case 1/ohm or ohm^{-1} (see "Fractional Exponents" in Section 2.1.4). Thus a resistor which has a resistance of 5 ohms (5 Ω) has a conductance of ⅕ mhos, or 0.2 mhos. In recent years the SI unit the siemens, S, has become preferred to the mho. If you see *mho* in a manuscript, ask if the author would like to update it to *siemens.*

a. Ohm's law The ohm is defined as the amount of resistance that allows a flow of electrons of 1 ampere (Section 5.1.4) when these electrons are subjected to a force of 1 volt (Section 5.1.3). This definition has its basis in *Ohm's law,* a very simple but very important equation which electrical and electronic engineers use very frequently:

$$E = IR$$

where E = voltage or electric potential in volts
I = current in amperes
R = resistance in ohms

Ohm's law states that for a given force pushing electrons from one end of an object to the other (for a given voltage), more resistance means fewer electrons moving (less current), and less resistance means more electrons moving (more current).

NOTE: When E = 1 volt and I = 1 ampere, R must equal 1 ohm, according to Ohm's law:

$$1 = 1 \times 1$$

which brings us back to the definition of the ohm given here.

b. Unlimited resistance Resistance of resistors is not limited. They may range from infinitely large to infinitely small. For example, the following units may be used to express resistance:

- A *megohm* (MΩ) is a million ohms.
- A *millohm* (mΩ) is a thousandth of an ohm.

NOTE: In the SI system, the prefixes for million and thousandth are "mega-" and "milli-" respectively. "Megaohm" and "milliohm" are awkward to say, however, so the spellings given here are used instead.

5.1.2 The Atomic Basis of Conductivity and Resistivity

The primary factor determining a material's conductivity is how tightly the nucleii of its atoms hold their electrons.

These are the electrons that move in orbits (called *orbitals*) around the nucleus of every atom (see "Atomic Structure" in Section 4.8.1).

In any atom the electrons farthest from the nucleus are less tightly held than the ones closer to the nucleus. Thus the outer electrons are the freest to move and to make their contribution to conductivity. Since conductivity depends on electron movement, good conductors have more "loose" outer-orbital (higher energy level) electrons than poor ones.

NOTE: Even though electrons move, there is always exact electrical balance. As electrons move away, they are replaced by other electrons. Thus the total negative charge, in any location, is always the same as the positive charge of the nuclei in that location (which are fixed and do not move).

1. The Electron Cloud—The Loosest Electrons

Sometimes the outer electrons are held so loosely that their orbitals are difficult to define. Then they are said to hover at the outer boundary of the atom in a kind of cloud—an *electron cloud*. Materials with electron clouds make the best conductors.

All metals have electron clouds, so metals are the best conductors (Section 5.1.1).

But

- The outer electrons of the atoms of insulators are tightly held; thus they have no electron clouds (no especially loose electrons).
- The outer electrons of the atoms of semiconductors are intermediate; a small proportion of them are loosely held, but most of them are tightly held.

Figure 5.2 illustrates the atom atructure of a material with an electron cloud.

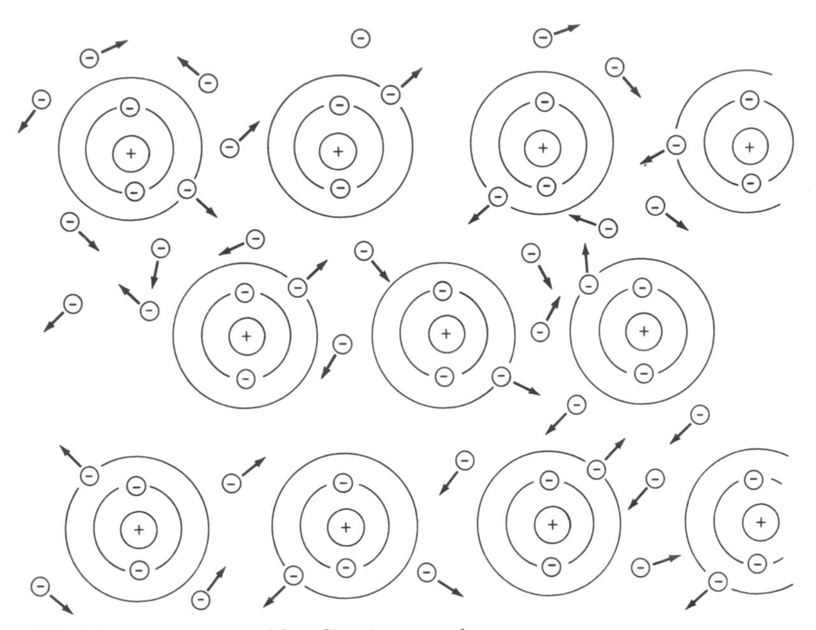

FIG. 5-2 Electron cloud bonding in a metal.

The electrons in inner orbit remain tightly bonded to their respective nuclei. The outer ones form a freely moving "cloud" of shared electrons.

2. The Metallic Bond—The Sharing of Electrons

The atoms in a crystal or in a chemical compound are held together by electric bonding forces (see Sections 6.3.1 and 6.4.1).

The metallic bond is a special kind of bond. It differs from other bonds in that some of the electrons involved in the bonding are in the electron cloud, and it is in the nature of the electron cloud that its electrons are shared by all the atomic nucleii.

It is not possible to identify which electrons of the electron cloud belong to any one atom in a unit cell of a metallic crystal. The electrons of the electron cloud are so loose that they can hop from atom to atom. Thus all the electrons of the electron cloud of a metal are shared by all the atoms.

In the hypothetical example of Figure 5-3, each nucleus contains three protons (each with one positive charge; see Section 4.8.1) and thus requires three electrons circling around it in order to have electrical balance. The two electrons in the two inner orbitals have stayed put, but the third electron in each case has been borrowed and "traded off" with electrons that once were associated with adjacent nuclei. What happened is *electron sharing*—but done in such a way that the electric charges are still balanced.

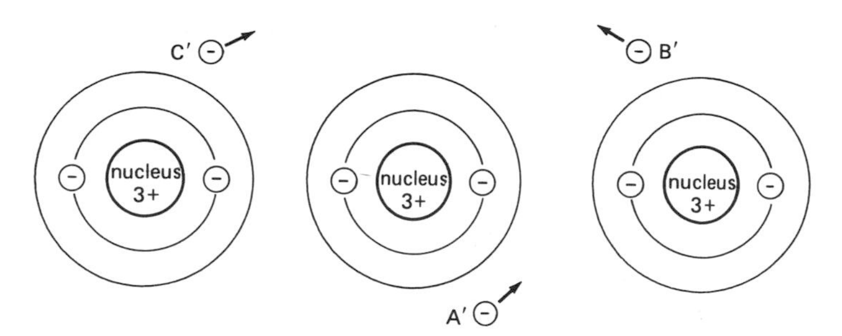

FIG. 5-3 Electron sharing in the electron cloud of a lithium metal crystal.

Electrons with prime superscripts have jumped from adjacent atom orbitals.

5.1.3 Pushing (and Pulling) with Voltage

The force which causes loose—and therefore available to move—electrons to hop from atom to atom, and thus to *flow,* is electric potential. It is measured in volts, the unit of electric force.

1. Volts and Voltage

a. The symbol for the volt is the uppercase V.

NOTE: The volt is named after the nineteenth-century Italian scientist Count Alessandro Volta.

b. Definition The volt may be defined as the amount of force necessary to cause 1 ampere's worth (Section 5.1.4) of electrons to flow through a 1-ohm resistor.

c. Infinite range of volts The number of volts may be any size, from infinitely large to infinitely small.

- A *megavolt* (MV) is a million volts.
- A *kilovolt* (kV) is a thousand volts.
- A *millivolt* (mV) is a thousandth of a volt.
- A *microvolt* (μV) is a millionth of a volt.

NOTE: It is common usage to refer to the number of volts as the *voltage.* Other terms used to express the same thing are *electric potential, potential difference,* and *electromotive force,* emf.

2. Positive and Negative Voltage

Voltage may be *positive* (indicated by a plus sign) or *negative* (indicated by a minus sign).

a. Positive voltage If the voltage is positive, electrons, each of which has a negative charge, are attracted or "pulled" toward that positive charge.

In electricity, opposite charges attract (see the discussion of chemical bonding, Section 6.3.1).

b. Negative voltage If the voltage is negative, electrons, each of which has a negative charge, are "repulsed" or "pushed away" from that negative charge.

In electricity, like charges repel each other.

Figure 5-4 illustrates the pulling and pushing that positive and negative voltages exert on electrons. In Figure 5-4*a* the directions of movement of the electrons in the cloud are random. In Figure 5-4*b*, where a voltage is imposed, most of the electrons tend to move in the direction of the positive charge.

3. Charge

The term *charge* refers to the number of volts, the *voltage,* which may be located at some particular location.

a. Terminals and electrodes The locations where charges are generally measured or indicated on circuit diagrams are at connection points between different major elements of a circuit (see the discussion

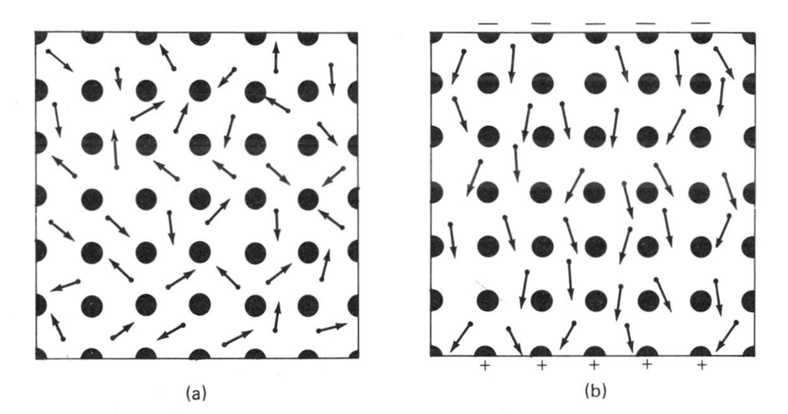

FIG. 5-4 The effect of pulling and pushing forces exerted by an imposed voltage on the electrons of an electron cloud. (*a*) With no imposed voltage, the electrons of the electron cloud move in random directions. (*b*) With an imposed voltage, the electrons of the electron cloud tend to move in the direction of the imposed positive voltage. The direction of the motion is no longer random, and most of the electrons are now headed toward the positive charges.

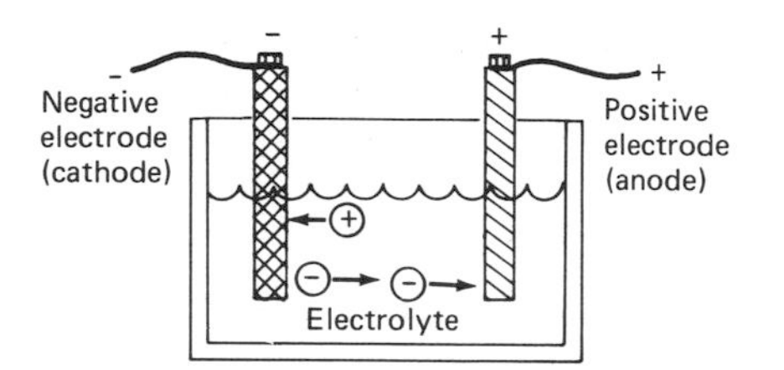

FIG. 5-5 Voltage charges at the terminals of a battery.

Electric current enters at the cathode and moves through the solution to the anode. Negative ions in the solution are attracted to the positive terminal (anode), positive ions to the cathode.

of circuit components such as resistors, capacitors, transistors, and batteries in Section 5.3.1). These connection points are known as *terminals* (or *electrodes* depending on the context).

> The terminals of a *battery* (see Section 6.3.4 and "Batteries" in Section 5.1.5) are commonly called electrodes (Figure 5-5). The charges that reside at these electrodes provide a pair of puller and pusher forces to move electrons around, as in Figure 5-5.
>
> NOTE: When a battery ceases to be capable of moving electrons around, it ceases to be useful. We then say it has "lost its charge." It has *discharged.* We can make it useful again by giving a charge back to its electrodes. This is known as *recharging* the battery, or "giving it a charge."

b. Wires are extensions Any time there is a connection wire to an electrode (Figure 5-5), that wire has the same charge along its entire length as the electrode has.

> Wire connectors are simply *extenders.* They are a means to bring a charge—unchanged—from one location to another.
>
> NOTE: Any wire with a charge on it is called *live.* If you touch it, you will regret it—you get a *shock,* which means that a flow of electrons will pass through your body.

4. Ground

Voltages exist only as they relate to other voltages. In fact the term *voltage* actually designates a difference between the amount of electric force at one location and the amount at another location. Unless otherwise stated, that other location is always *ground,* and is understood to be the spot where the voltage is—by definition—zero.

a. Earth (ground) as universal zero-voltage location The most convenient location on earth is the earth itself. The voltage of the earth is the same anywhere in the world. So driving a metal pole into the ground and making that a voltage reference understood to be zero allows all voltages in the world to be compared with each other.

That is the origin of the concept of *ground* and *grounding* of voltages. It means that at the point (terminal) shown as "ground," the voltage is considered to be at zero charge.

NOTE: Practice in the United Kingdom is to use *earth* and *earthing* instead of ground and grounding.

b. Voltage forces and gravity forces behave similarly There is another way to use the ground as a zero reference. Consider a falling object—or one about to fall—because of the force of gravity. The same sort of thinking pertains whether the forces are mechanical or electric.

In Figure 5-6*a*, a sack of stones weighing 100 pounds is positioned 5 feet off the ground. Should we push the sack off the platform, it will impact the ground with a force that combines the 100 pounds and the 5 feet.

When the same sack is placed at an elevation of 10 feet, as in Figure 5-6*b*, the force of impact will be much greater. It will be greater still when the height is 100 feet (Figure 5-6*c*).

When the sack is placed on the ground over a hole 5 feet deep (Figure 5-6*d*), the impact of the falling load at the bottom of the hole will be the same as that in Figure 5-6*a*. But now the force will be exerted *below ground*.

NOTE: The force never makes itself felt until the moment that the sack is pushed into space. Yet it is always present there at the edge of the platform. It is a *potential* force (see Section 4.2.2). It is present, but it causes no movement. It merely has a *potential* for causing movement.

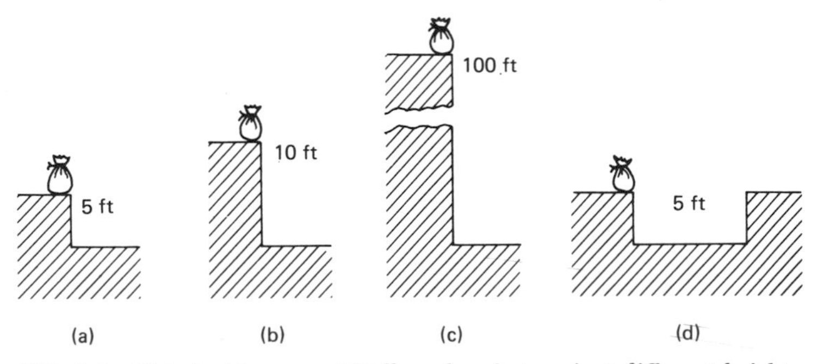

FIG. 5-6 Objects (shown as 100-lb sacks of stones) at different heights with respect to ground.

In each of the cases of Figure 5-6, the electrical analogue of the sack of stones is known as *voltage,* above or below zero, or above or below ground—and the term *ground* is used in both cases, whether the force is mechanical like the sack of stones or electric, and whether the ground is really the actual earth or the imaginary earth in an electric circuit.

c. Symbol The symbol for electrical ground in text, notation, or illustration is the

drawn freehand or with the aid of a template. Other (less used) forms are

and

Figure 5-7 tells the same story as Figure 5-6, but tells it about an electrical rather than a mechanical situation. Instead of sacks of stones on platforms, consider batteries that maintain different voltages across their terminals, with one terminal (electrode) of each battery connected to ground. This means that the other battery electrode is automatically at a voltage charge either above or below ground. Thus in Figures 5-7*a*, *b*, and *c*, it is at 5, 10, and 100 volts above ground; and in Figure 5-7*d* it is 5 volts below ground.

NOTE: The symbol for a battery (as shown in Figure 5-7) is

5. Voltage Potential

When an electrode (or any location in an electric circuit) carries a charge which merely "sits there," like the sack of stones (Figure 5-6),

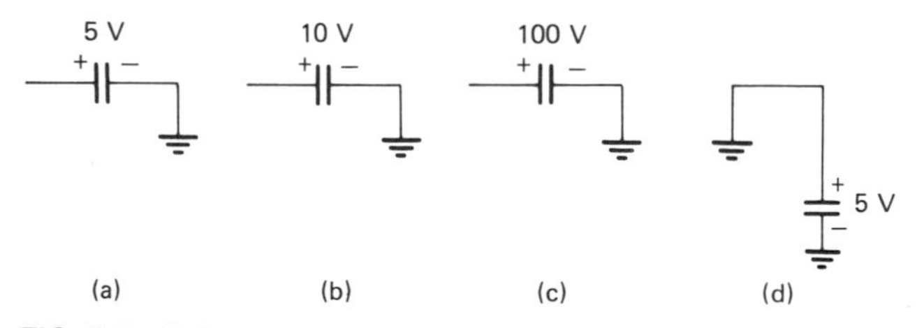

FIG. 5-7 Voltages with respect to ground. (Shown as impressed on the plates of a capacitor.)

the voltage is said to be a potential *force,* just as the sack of stones is up to the instant it is pushed off the edge of the platform. It is then referred to as *voltage potential* or *potential voltage,* or merely as *voltage* (with potential understood) or *potential* with voltage understood.

Example:

> The potential maintained at the high-voltage plate of the capacitor is 1500 volts. . . .

a. Electrostatics The field of *electrostatics* deals with potential voltages, voltages that potentially may cause electron motion—but for the moment do not.

Electrostatics is the electrical analogue of mechanical statics (Section 4.2.3) or hydrostatics (Section 4.3.1).

b. Electrodynamics The field of *electrodynamics* deals with active voltages, those which cause electron movement.

5.1.4 The Flow of Electrons—Amperes

The flow of electrons is called a *current*. The rate of flow of that current of electrons is measured in *amperes*.

When electrons move down the length of a conductor, or through a resistor, or through any circuit component, they are considered to flow, as the water in a river flows. The flow of water in a river is also called a current.

1. Definition of Ampere

An ampere of electron flow is defined as a coulomb's worth of electrons per second.

The rate of flow of water (its current) in a river may be measured in *gallons per minute*. The current in a wire may be measured in *electrons per second*. A coulomb is 6.25×10^{18} (more than 6 quintillion) electrons; that many electrons flowing in a second makes 1 ampere.

NOTE: The ampere is named after the nineteenth-century French scientist André Marie Ampère, and the coulomb is named after the eighteenth-century French scientist Charles Augustin de Coulomb.

2. Symbol

The symbol for ampere is the uppercase A. The abbreviation for ampere is *amp*. These are often used interchangeably, but A is preferred.

Here is a case in which the abbreviation is used in conversation and in text as a word in its own right (to mean the same as *ampere*). Thus it is common to talk of or write about a current of some number of *amps*.

Example:

> The current through the resistor network of the circuit reached an instantaneous peak of 1400 amps. . . .

3. Another Definition of Ampere

The ampere may be defined as the amount of electron flow (in coulombs per second) that results from applying 1 volt to a resistor of 1 ohm.

4. Infinite Range of Amperes

The number of amperes may be any size, from infinitely large to infinitely small.

- A *milliampere* (mA) is a thousandth of an ampere.
- A *microampere* (μA) is a millionth of an ampere.

NOTE: It is common usage to refer to the number of amperes as the *amperage.*

5.1.5 Two Kinds of Current—Direct and Alternating

When electrons flow consistantly in one direction—as a river does—that flow, that current, is called *direct current* (abbreviated dc).

When electrons rapidly and regularly and repeatedly (periodically) flow first in one direction and then in the reverse direction, that flow, that current, is called *alternating current* (abbreviated ac).

It is hard to imagine the water of a river doing this because water has bulk and weight, but it is not hard to imagine electrons reversing their direction easily, because electrons have no bulk and almost no weight. Although electrons are particles, they are so unsubstantial that they may also be considered as electric waves (see "Particles and Waves" in Section 4.8.3). It is easier to think of such waves as being easily diverted in direction than it is to think of a massive river flow being easily diverted.

1. Alternating Frequency

a. Units of alternating frequency The number of alernations per second, the *frequency,* is measured in *cycles per second.* One cycle per second (cps) is the *hertz* (abbreviated Hz).

Metric or SI prefixes (Section 2.1.6) may be used with hertz, for example, *megahertz,* a million hertz.

NOTE: The hertz is named for Heinrich Hertz, a nineteenth-century German physicist.

b. The wide range of possible frequencies There is no limit to the number of alternations (the frequency) an electric current may make in a given time. There may be only a few per second or billions of billions per second.

The voltages which cause these varying currents oscillate at the same frequencies as the currents. It turns out that these voltages can be sent out into space (along with varying amounts of magnetic phenomena—see the discussion of magnetism in Section 5-4) in the form of *electromagnetic waves.* There is no limit to the frequency of electromagnetic waves.

Such transmission through space is known as *radiation* and was discussed in the section on light (Section 4.5).

NOTE: Light is electromagnetic radiation too, but is at a much higher frequency than the most rapidly alternating electric waves, which occur at microwave frequencies (see following).

Electrical engineers have given names to the different ranges of the speed of electric alternations that are commonly met, either as electric current or as electromagnetic waves.

- *Audio frequencies* (referred to as AF) are oscillations at the speed which we can hear with our ears—between 15 and 20,000 hertz.
- *Radio frequencies* (referred to as RF) are those used by radio stations (Figure 5-8*a*)—3000 hertz to 300 gigahertz—to radiate electric signals that are picked up by the antennas of our radios. Included in this range are

  High frequencies (HF) from 3 to 30 megahertz.

  Very high frequencies (VHF), which oscillate in the 30 to 300 megahertz range (Figure 5-8*b*).
- *Ultrahigh frequencies* (referred to as UHF) oscillate in the billions of hertz range—the gigahertz range.
- *Microwave frequencies* are any oscillations faster than ultrahigh frequencies. Letter designations are used to designate these, and the ranges designated are known as *frequency bands.* These are *L band, S band, X band, K band,* and *Q band.*

NOTE: It is amusing to see the predicament electrical engineers have got themselves into. They went from "high" frequencies to "very high" to "ultrahigh," thinking in the early days of the twentieth century that no frequency would ever get higher than ultrahigh. But we now consider ultrahigh frequencies only "rather high." The engineers then had no recourse but to turn to letter designations. (What else could they have done? Designated a frequency band "very high frequency indeed"?)

c. The wide range of possible wavelengths All electromagnetic waves have a length dimension, which ranges from extremely short to extremely long.

This is the distance from peak to peak of the wave as it may be radiated into the air from, for example, an antenna of a radio or a television station (see Section 4.4.3 for a discussion of wavelength.)

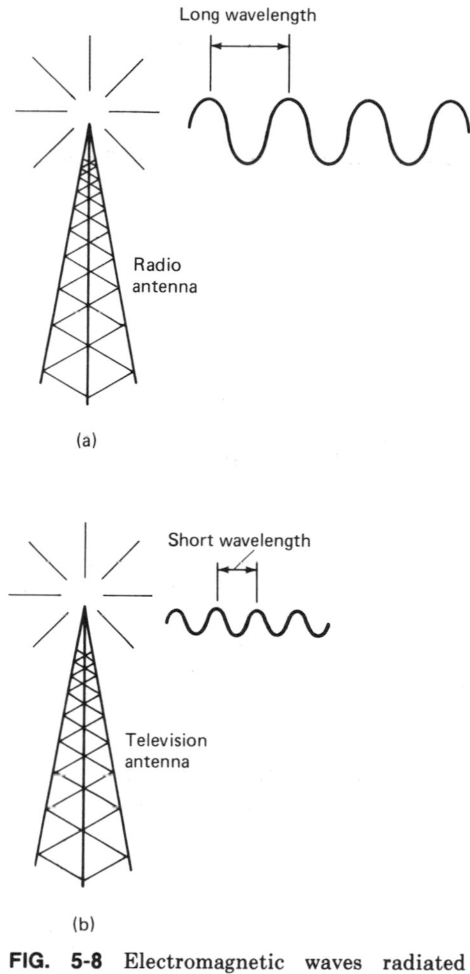

FIG. 5-8 Electromagnetic waves radiated from different antennas: (*a*) Radio frequency (RF), (*b*) very high frequency (VHF).

The higher the frequency, the shorter the wavelength.

Figure 5-8 illustrates the different waves that are radiated from two different antennas.

NOTE: The term *shortwave,* as applied to *shortwave radio,* simply means that "very high" (compared with other radio waves) frequencies are being radiated.

Thus we have wavelengths designated in ordinary length units:

- *Meter waves* (at radio frequencies)

- *Centimeter waves* (the wavelength of microwaves in microwave ovens)
- *Millimeter waves* (in the letter-band frequencies)
- *Micrometer waves* (the wavelength of light waves; also known as *micron waves*)

2. Two Kinds of Voltage—Direct and Alternating

Direct current is always associated with (driven by) *direct voltage.*

When the voltage impressed across any space or any part of a circuit is unchanging, with positive voltage at one spot and negative voltage at another spot (see "Positive and Negative Voltage" in Section 5.1.3), the electrons, being negatively charged, respond to that situation by always heading to the plus (positive) location.

Alternating current is always associated with (driven by) *alternating voltage.*

When the voltage impressed across any space or any part of a circuit is made to change direction (or *alternate*), the current which results also alternates.

This is because electrons, being negatively charged, always want to go in the direction of the positive voltage. If the charge at a given location rapidly changes from positive to negative to positive again, the electrons change their direction accordingly.

Batteries and dynamos are examples of devices that generate direct and alternating voltages (and currents) respectively.

a. Batteries Batteries generate voltages that are always direct and thus generate current flow that is always direct.

Because of the chemicals in the battery and the chemical makeup of the electrodes in the battery, one electrode must always be the positive one and the other must always be the negative one. Because of the way batteries operate, that situation can never be reversed (see "Batteries" in Section 6.3.4).

b. Dynamos Dynamos—more commonly called *generators*—generate voltages that are always alternating and thus generate current flow that is also always alternating (Section 5.5.2).

Generators are rotating machines that convert mechanical energy to electric energy (Section 5.5.2). They accomplish this by rotating their central structure (called the *armature*).

Every time the armature passes a given position as it turns, it takes on a positive charge; but as it passes the next position—still turning—it takes on a negative one.

It is the rotation process of the mechanical generation of electricity that gives us alternating voltages—whether we want them or not. The electric energy generated by the generators in our local power plant is therefore always alternating. And it is alternating electric energy that comes out the electric outlets in our homes, offices, and factories.

3. Two Kinds of Electric Circuits—Direct and Alternating

Since the available voltages and currents may be either direct or alternating, so must the electric circuits that are powered by these two forms of electricity be able to accommodate to one or the other—or sometimes both.

The essential part of your toaster is an electric circuit that does not care whether it is being powered by direct current or by alternating current. Either type of current will cause it to heat up and make toast.

Your transistor radio is probably run from a battery. If so, it runs by means of direct voltages and direct current.

The motor in an electric typewriter is run from the power out of an electric outlet. That is alternating voltage and alternating current.

4. Power Converters

Sometimes a circuit runs by means of one type of power and only the other type is available. That can be taken care of by the use of a device known as a *power converter.*

There are power converters that convert direct current to alternating current, and there are power converters that convert alternating current to direct current.

5.2 ELECTRIC POWER AND ELECTRIC ENERGY

The preceding sections of this chapter described how electrons move through materials, and as they move they may deliver energy at the same time. It is our capability to manipulate electric energy that results in the thousands of uses we make of electricity, from entertainment to home appliances to transportation.

5.2.1 Electric Power—The Watt

The unit of electric power is the *watt.* It is a measure of the amount of electric energy used in a second. In particular, the watt is the work done in 1 second by 1 volt in moving 1 coulomb of charge.

Electric power is the analogue of mechanical power (see "Force Energy, Power" in Section 4.2.2) in that it is the amount of energy used up in a period of time. Thus the electrical unit, the *watt,* is the analogue of the mechanical unit for power, the *horsepower.*

a. Symbol The symbol for the watt is the uppercase W.

NOTE: The watt is named after the nineteenth-century British scientist James Watt.

b. Definition Power P in watts is defined as the product of the current I in amperes times the voltage E in volts:

$$P = I \times E$$

For example:

where I = 3 amperes and E = 40 volts

$$P = 120 \text{ watts}$$

NOTE: The symbol P represents power, but the amount of that power is given in watts, symbol W. Some authors and some publications use the symbols for units such as volts (V) and watts (W) and others prefer to use the words. Either usage is acceptable; but usage should be consistent within one paper or book.

c. Range The number of watts may be *any size,* from infinitely large to infinitely small.

- A *megawatt* (MW) is a million watts.
- A *kilowatt* (kW) is a thousand watts.
- A *milliwatt* (mW) is a thousandth of a watt.
- A *microwatt* (μW) is a millionth of a watt.

NOTE: It is common usage to refer to the number of watts as the *wattage.*

5.2.2 Electric Energy—The Kilowatt-Hour

The unit of electric energy is a wattage term combined with a time term. The one most commonly used is the *kilowatt-hour*.

Example:

$$\text{kilowatt-hour} = \text{kilowatts} \times \text{hours}$$
$$\text{kWh} = \text{kW} \times \text{h}$$

a. Definition It is a measure of wattage maintained for a given period of time, that is, power exerted over time.

Electric energy is the analogue of mechanical energy (see "Force, Energy, Power" in Section 4.2.2). Mechanical energy used in doing work is expressed in *foot-pounds;* equivalent electric energy is expressed in such terms as watt-seconds.

Common units of electric energy are

• kilowatt-hour	kW·h	or	kWh
• watt-second	W·s	or	Ws
• watt-hour	W·h	or	Wh

b. Range The values of electric energies may be any size, from infinitely large to infinitely small.

The energy consumed by a watch battery will be in terms of milliwatt-years. The energy consumed by an electric drier in the laundry will be in terms of killowatt-hours.

5.3 ELECTRIC CIRCUITS

Given the fact that electrons move—with great ease—through certain materials (Section 5.1.1) and the fact that as they move they are involved in the useful transfer of energy (Section 5.2), we may ask, How do we control and harness this movement and this energy? The answer is, with the *electric circuit.*

An electric circuit is an *electrical system,* where *system* means an assemblage of many parts all of which function together to achieve a particular function. The circuit system is made up of devices that control the flow of electrons and shape the patterns of electric waves. (The manner of movement of electrons emerging from these devices is always different from what it was on entering them.) These devices—which modify the flow of electrons—are known as *circuit components.* Thus the electric circuit consists of these circuit components and the wires that connect them to each other.

5.3.1 Circuit Components

The following sections describe the major circuit components and some of the most important circuit formats and functions.

There are dozens of ways circuit components modify the flow of electrons. By the same token, there are dozens of different kinds of circuit components. All influence the flow of electrons. Electron flow may be constricted, as with the resistor (see "Resistors" in Section 5.1.1); or allowed to accumulate, "pile up," as with a capacitor; or slowed down, as with an inductor coil; or amplified into more voluminous flow, as with a vacuum tube or a transistor; or stopped altogether, as with an open switch. (These various components are described in detail later in this chapter.)

Each circuit component has its own symbol, and all these are described as each component is described. A summary of these and others may be found in Glossary J.

1. Discrete Components, Integrated Components

Components exist in two major physical forms—*discrete* and *integrated.*

a. Discrete components A discrete device has a separate physical identity, having the ability to be physically separated from other devices.

One can hold a discrete component in one's hand. It is a three-dimensional object. It was manufactured "one at a time." Even though many thousands could have poured from some mass-production machine, they poured out sequentially. Figure 5-1 shows discrete resistors.

b. Integrated components Integrated devices do not exist separately from other components.

They cannot be held separately in one's hand (without at the same time holding dozens and maybe hundreds or thousands of other components).

Integrated devices have been manufactured in such a way that they are physically linked to a support structure and to their wire connections and to other components so intimately that they cannot be separated physically (even though they still appear separate to the moving electrons). They have been made "all at once," born together like Siamese twins. They cannot be separated. They are *integrated,* which means the opposite of discrete.

NOTE: Do not confuse *discrete* with *discreet.*

- Discrete means separate.
- Discreet means judicious and modest.

2. Passive Components, Active Components

There are two major categories of circuit components—*passive* and *active.*

Although both modify the flow of electrons, the difference is that in a passive component the electron flow has the same amount of energy (or less) going out as going in, while an active component contributes some additional energy (which it extracts from some other source in the circuit).

NOTE: It is tempting to draw an analogy with people, some of whom are passive and some of whom are active. Passive people may just "sit there like a lump," but we can't ignore them. They modify the "flow" of people around them, if only by getting in the way. But active people do more than that; for example, they may attract more people.

3. Fixed Components, Variable Components

Most components are so made that they never change. Their qualities are manufactured into them. Thus they will always have the same effect on electrons. These are known as *fixed* components. But almost every kind of fixed component (described in the following sections) is also made in another form, one whose value can be varied at will. These components are known as *variable.*

For example, the knobs on your radio are attached to variable components. When you change stations, you are rotating the plates of a variable capacitor (see "Capacitors" in Section 5.3.2) and thus changing its value. When you change volume, you are moving a contact on a variable resistor (see "Resistors" in Section 5.3.2) and thus changing its value.

5.3.2 Passive Components

Passive components modify the flow of electrons. In the process, the amount of electric energy entering the component may emerge unchanged, or it may be lessened (having been converted to another form of energy such as heat), but it will not be increased.

Resistors, capacitors, coils, transformers, switches, meters, and transducers are passive components. In the following sections each of these will be briefly described and, where applicable, diagrammatic symbols for each will be given. Units and their abbreviations will also be detailed.

Although the diagrammatic symbols may appear in text within a typewritten line, they find their greatest application in circuit diagrams (see Section 5.3.5).

1. Resistors

Resistors are the most common circuit components. A circuit may use dozens or even hundreds of them.

NOTE: There are specially designed arrays of resistors, often very complex, called *resistor networks* (see Section 5.3.5).

a. Definition Resistors are devices which limit the amount of electron flow (the amount of electric current in amperes) because they have a limited and predetermined conductance (see "Resistors" in Section 5.1.1).

b. Symbol The symbol for the resistor is

c. Resistor size—resistance The amount that a resistor limits current is a measure of its *size.* This is called the *resistance* of the resistor, *R*, and is measured in ohms (see "Resistance and Resistivity" in Section 5.1.1).

The resistor may then be described, for example, as

"a 1000-ohm resistor"

or

"a 1000-Ω resistor"

depending on the author's preference.

NOTE: Resistance in ohms is a measure of electrical size in the same way that length in feet is a measure of physical size, or distance.

It is also possible to refer to the size of a resistor in terms of its *conductance,* expressed in mhos (see "Resistance and Ohms" in Section 5.1.1).

Conductance is merely a backward way of indicating resistance. Anything with a large resistance has a small conductance, and vice versa. As described in Section 5.1.1, the numerical value of conductance of any resistor is merely the reciprocal (1 divided by that number) of its resistance. Thus the same resistor might have

- A resistance of 100 ohms
- A conductance of 0.01 mhos

NOTE: Another term for a factor limiting the amount of electron flow is *impedance.* It can be used as a measure of the resistance of a resistor, but it also is a measure of the resistance (to alternating currents only) of other components such as capacitors (see "Capacitor Size—Capacitance" on p. 5.26) and inductors (see "Magnetic Effects in Coils—The Solenoid" on p. 5.30). It is also measured in ohms, but is symbolized by the letter *Z.*

d. The variable resistor—the potentiometer Some resistors are made with a moving part which makes it possible to utilize only a portion of the resistor at any time, rather than all of it. This is known as a *variable resistor.*

Its symbol is the same as that of a resistor, but with an arrow drawn through it:

Another name for the variable resistor is the *potentiometer.* (Or in conversation or informal text, it may be called the *pot.*) It is called this because changing the value of the resistance causes a corresponding change in the amount of voltage (potential) across that resistor. Section 5.1.3 describes why static voltages are called *potentials.*

2. Capacitors

a. Definition A *capacitor* is a circuit component which allows alternating current but not direct current. (See Section 5.1.5 for a description of these two kinds of current.)

Another name for the capacitor is *condenser.*

All capacitors are based on the same constructional elements, which are two metal plates which face each other and which are separated by an insulating region known as the *dielectric.*

Electrons may pile up on one of these metal plates, resulting in a negative charge there, and may at the same time be depleted at the other, resulting in a positive charge there (Figure 5-9*a*). Then, when the alternating current reverses itself in polarity (the positive becomes negative and the negative becomes positive), the same piling up and depletion takes place on the opposite plates (Figure 5-9*b*).

If the polarity does not reverse, the electron imbalance (the piling up) remains unchanged, or static (Section 5.1.3). The capacitor is then said to be *charged.* Capacitors are considered to be *energy-storage devices.*

When electrons find a new path which bypasses the dielectric insulator (*short-circuits* it), they may flow in a sudden rush. That is known as *discharging* the capacitor. The rush may be so intense that it appears as a *spark* (Figure 5.9*c*).

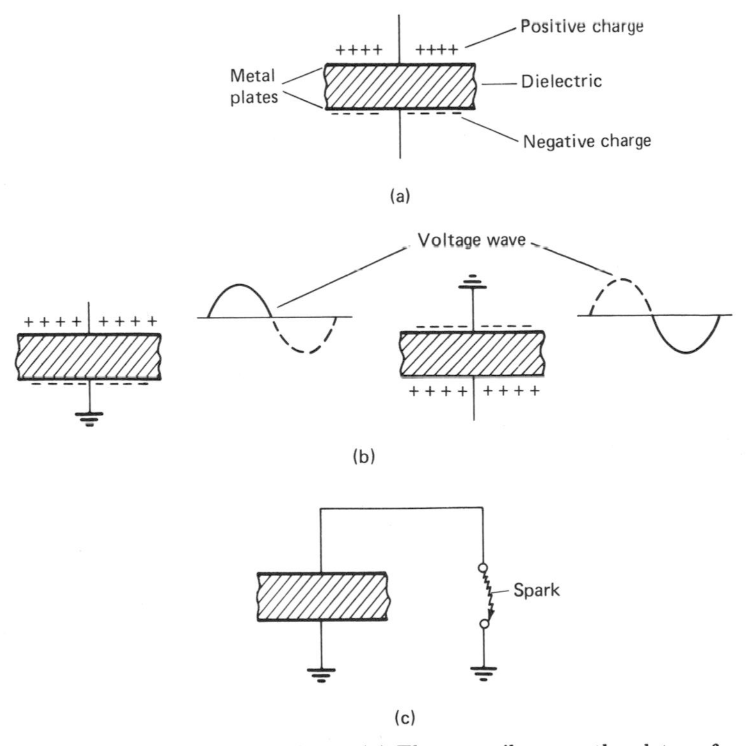

FIG. 5-9 Charges on capacitors. (*a*) Electron pileup on the plates of a capacitor. (*b*) Reversal of charge on capacitor with reversal of applied voltage (voltage is ac). (*c*) Discharging a capacitor with a shunt circuit.

FIG. 5-10 The cloud-earth capacitor—discharged by a flash of lightning.

NOTE: Lightning during a thunderstorm is such a spark. The clouds make up one plate of the capacitor, and the earth makes up the other plate. The air between is the dielectric. When there is enough moisture in the air, the moisture becomes a conductor and a short circuit results. The electrons then burst out of the cloud and plunge to earth in the spark we all recognize as lightning (Figure 5-10).

Any parallel path for electrons, like the *short circuit*, is known as a *shunt*.

b. Symbol The symbol for the capacitor is

The symbol is based on the primary constructional elements of the capacitor. The two horizontal lines represent the two facing plates, and the space between represents the dielectric.

c. Structure Discrete capacitors usually resemble discrete resistors. They commonly are small cylinders with a wire emerging from each end.

d. Capacitor size—capacitance Capacitors may be any size from very large to very small.

The electrical size of a capacitor (its *capacity* to store electrons) is known as its *capacitance*. It is symbolized in mathematical notation by C.

The larger the area of the capacitor plates, the larger the electrical size (capacitance) of the capacitor. The more insulating the dielectric which is between the plates, the larger the electrical size (capacitance) of the capacitor.

NOTE: The same symbol is used for conductance (see Section 5.1.1). The unit of capacitance is the *farad* (abbreviated F).

NOTE: The farad is named for Michael Faraday, a nineteenth-century British scientist.

The two more frequently used units are

- The *microfarad* (μF), one millionth of a farad.
- The *picofarad* (pF), one millionth of a microfarad.

e. Capacitive reactance Capacitors inhibit the flow of electrons in a manner analogous to that of resistors. But such inhibition by capacitors occurs only when alternating current is flowing. Resistance caused by capacitors is called *reactance,* in this case, *capacitive reactance.* It is measured in ohms, just as resistance by resistors is.

The term *capacitive reactance* is to distinguish it from *inductive reactance,* which is found in coils (see "Inductive Reactance" in Section 5.3.2).

Capacitive reactance is symbolized by X_C.

Inductive reactance is symbolized by X_L.

NOTE: *Impedance* was described in the discussion of resistance as a measure of resistance which is measured in ohms and symbolized by the letter *Z*. How does it differ from resistance *R* and the reactances described above? The answer is not much. It is still the effective resistance of a circuit to the flow of electrons. But it includes the resistance to alternating current flow contributed by the inductive and the capacitive reactances, if such reactances are present (in that case the circuit is known as an *RLC* circuit). The impedance *Z* of an *RLC* circuit is given by the expression

$$Z = \sqrt{R^2 + (X_L - X_C)^2}$$

f. The variable capacitor Capacitors may be made adjustable so that the user can select any desired value of capacitance. Such capacitors are known as *variable capacitors* or *variable condensers.*

Capacitance is varied by moving the capacitor plates so that they face each other along their entire areas, or do not face each other at all. or any amount of facing in between.

The symbol for the variable capacitor is the same as that for a capacitor, but with an arrow drawn through the symbol

3. Coils

a. Definition A coil is a circuit component which introduces a time *lag* between the oscillations of an alternating voltage and the alternating current that it causes.

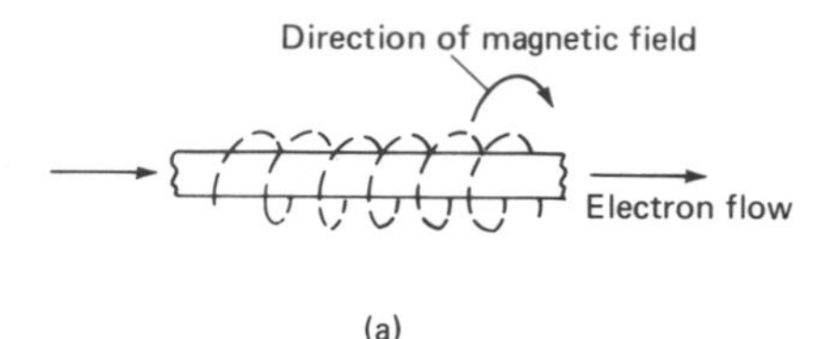

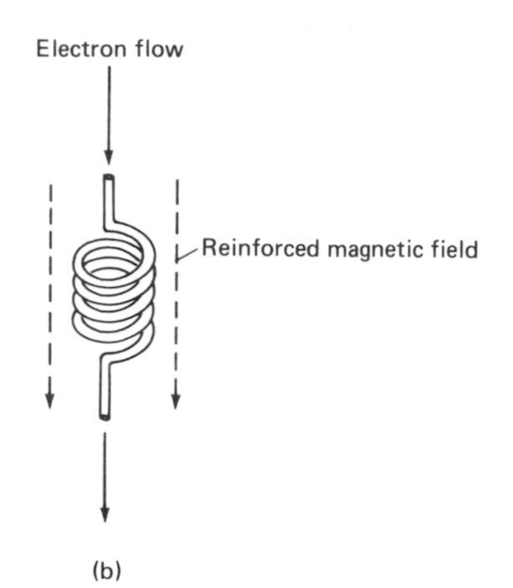

FIG. 5-11 Magnetic fields around current-carrying wires and coils. (*a*) The magnetic field rotating around the direction of electron flow in a wire. (*b*) A coiled wire. Magnetic fields reinforce each other. The electron flow lags the magnetic field flow.

These lags have very definite circuit effects, some of which are useful and some of which are harmful.

All coils are just that—turns of wire wound up on a spool like that used for thread.

What happens when electric current flows in a coil is that each winding influences the two windings on each side of it in a way that slows down the current in those windings, causing lag. If the wire were straight, without turns, there would be virtually no slowing down. This influence between windings is known as *induction,* and another name for a coil is *inductor.*

It is not necessary for our purposes to go into why this effect occurs, other than to note that the effect is due to magnetic fields that surround regions where current flows (Figure 5-11*a*), and the greater the amount of current flow in any winding, the greater the magnetic field that surrounds that winding. It is these fields that extend between windings that result in the lags we observe between current and voltage. (Section 5.4.1 discusses magnetic phenomena.)

There are several factors which control coil performance, all related to geometrical or material issues.

Induction is greatly influenced by the closeness of the turns to each other, and a tightly packed series of windings will be more inductive than a loosely packed or more open coil. Another factor is the form of the spool, which is to say the size and shape of the space the windings enclose. The material of which the spool is made is another important factor. It is known as the *core*. Most cores are made of iron alloys and are known as *magnetic cores*. When the spool is hollow, the core is kown as an *air-core*.

b. Symbol The symbol for the inductor coil is

Drawn that way, an air-core is understood. When there is an iron core, the coil is represented thus:

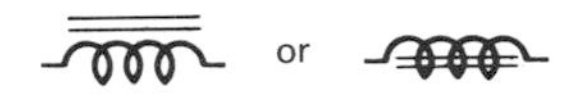

Like the symbol for the capacitor, this one makes sense in that it is a simplified picture of what it represents—a coil.

c. Coil size—inductance Coils, considered in terms of their electrical performance only (and not in terms of their physical dimensions), may be any size from very large to very small.

The electrical size of a coil (its capacity to induce lagging in adjacent turns) is known as its *inductance*.

The unit of inductance is the *henry* (abbreviated H).

The symbol for inductance in mathematical notation is L.

NOTE: The henry is named for Joseph Henry, a nineteenth-century American physicist.

The most frequently used unit sizes are

- The *henry* (H)
- The *millihenry* (mH), one thousandth of a henry
- The microhenry (μH), one millionth of a henry

d. Magnetic effects in coils—the solenoid When current flows in coils, the coils become magnets (see Figure 5.11*b* and Section 5.4). But the instant the current is shut off, they cease to be magnets.

The most common use of these magnets is in the circuit component known as the *solenoid.*

The solenoid is a coil attached to a switch. When the current is put on, the switch is closed (because it is pulled closed by the magnetic action of the coil).

NOTE: Solenoids are commonly found in the modern kitchen. One can hear the snap as the solenoid closes or opens to control hot water in a dishwasher or washing machine.

e. Inductive reactance Coils inhibit the flow of electrons in a manner analogous to that of resistors. But such inhibition by coils occurs only when alternating current is flowing. Resistance caused by coils is called *reactance, inductive reactance.* It is measured in ohms, just as resistance by resistors is.

The term *inductive reactance* is to distinguish it from *capacitive reactance,* which is found in capacitors (see "Capacitive Reactance" on p. 5.27).

Inductive reactance is symbolized by X_L.

Capacitive reactance is symbolized by X_C.

NOTE: The slowing down of electron reaction to voltage oscillation—expressed in reactance as described above—is called *choking.* Thus a coil is often called a *choke.*

f. The variable inductor Coils may be made adjustable so that the user can select any desired value of inductance. Such coils are known as *variable chokes* or *variable inductors.*

Inductance is varied by moving a metal core in and out of the central space of the spool around which the coil has been wound. If the core is entirely inside the core space, the inductance is the highest.

The symbol for the variable inductor is the same as that for the inductor, but with an arrow drawn through the symbol:

4. Transformers

A *transformer* is a circuit component consisting of two or more coils placed close to each other so that electric energy in one coil induces energy in the adjacent one.

The usefulness of the transformer lies in the fact that it can *transform* low voltages in one coil to high voltages in another coil. Thus if an engineer wants to achieve a high voltage, the transformer is one way to do it.

a. Turns ratio Transformer action depends on the *turns ratio.*

If the first coil has 100 turns, or loops, of wire and the adjacent one has 1000 turns, the turns ratio will be 10. The coil with the greater number of turns will have the higher voltage. Thus the voltage on the second coil will be 10 times greater than that on the first.

b. Primary and secondary coils in transformers The coil of the transformer where energy is put in is the *primary;* the coil where energy is taken out is the *secondary* (Figure 5-12).

c. Step-up and step-down transformers Since the most important use of transformers is to increase—or step up—voltages, they are often called *step-up transformers.* But they can be operated in reverse to decrease voltages, in which case they are known as *step-down transformers.*

d. Symbol The symbol for the transformer is two coils with the shared core represented by two or three lines between them:

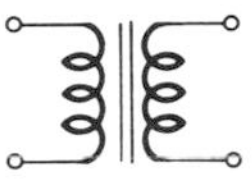

5. Switches

A *switch* is a component which opens an electric circuit so that electrons cannot flow or closes it so that they can flow.

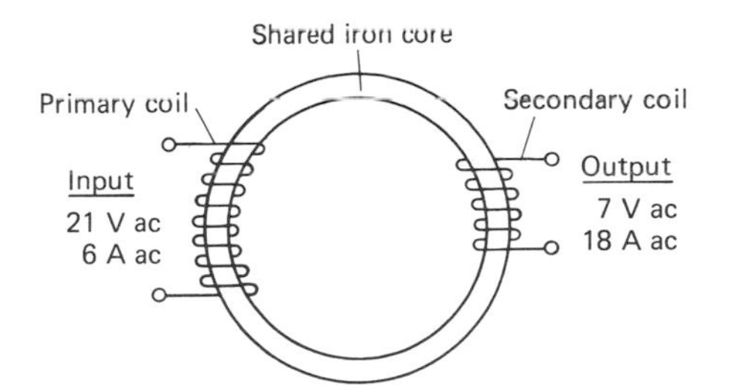

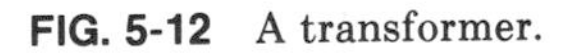

FIG. 5-12 A transformer.

In this transformer the core happens to be in the form of a ring (a common shape for transformer cores). The device is operating as a step-down transformer with a turns ratio of 3. Thus the voltage steps down by a factor of 3 (21 to 7) while at the same time the amperage steps up by a factor of 3 to compensate (18 to 6).

The opening and closing actions are mechanical, but the consequence is electrical.

a. Changing the direction of electron flow The opening and closing of a switch may have the effect of changing the direction of the electrons from that of one set of conductors to that of another set.

b. Poles The points which are opened and shut are extremities of conductors or of components. They are also called *terminals* (see "Terminals, Electrodes" in Section 5.1.3). When associated with switches, these points are more often called *poles.*

Opening and closing switches is also called *making and breaking.*

c. Multiple-pole switches Some switches perform many electrical functions (many openings and closings of circuits) with only a single mechanical action. They are then called *multiple-pole switches.*

The mechanical action of activating a switch is often called *throwing.* Thus the expression "to throw a switch."

d. Symbols The basic symbol of a switch, which will appear in text and in illustrations or circuit diagrams, is the hand-drawn

o—o⁄ o—o (open position) and o—o——o—o (closed position)

e. Many constructions of switches There are many ways to make and break contacts between poles by mechanical means. These give each switch type its own distinctive name.

Switch types include the following:

- *Toggle switch* (flipped by the finger)
- *Knife-blade switch* (the knifelike blades of the switch can be thrust into or out of narrow channels made to receive them)
- *Knob switch* (turned to open or close—and turned farther to bring in different levels of current, as for the burners of an electric range)
- *Button switch* (a button is merely pushed)
- *Membrane switch* (as one finds on some photocopy machines and kitchen appliances—a plastic "bump" is depressed)
- *Touch plate* (as one finds on new kitchen appliances such as microwave ovens—a finger is merely brought in contact with a glass plate and switching action is accomplished)
- *Relay* (usually a knifeblade switch, but may be massive in size to handle large quantities of current)

Figure 5-13 illustrates the many forms of switches.

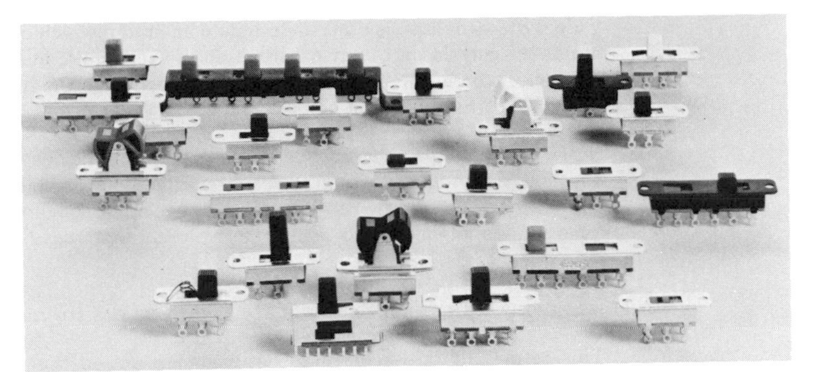

FIG. 5-13 Some switches. *(Photograph courtesy of Switchcraft, Inc., Chicago, Ill.)*

6. Meters

A meter is a device which measures and indicates the quantity of any phenomenon being observed.

Thus we have *ohmmeters, ammeters,* and *voltmeters,* which will tell what quantities of ohms, amps, and volts, respectively, to place with the symbols Ω, A, and V, respectively.

There are also meters specially constructed to deal with unusually large or unusually small quantities, for example, the *megohmmeter,* the *millivoltmeter, the microammeter,* and so on.

Many other kinds of meters with special uses are, for example, *conductivity meters, contact resistance meters, thermocouple meters, altimeters, capacitance meters, voltage breakdown meters,* and *pH meters.*

a. Meters as circuit components Meters may be inserted directly into a circuit, like any other circuit component.

Meters differ from all other circuit components in that they have no circuit function. They do not affect the currents, nor do they affect the voltages. They are there only to measure.

NOTE: The verb *to meter* means "to measure." Do not confuse it with the metric unit, which uses the same word to indicate length.

b. Constructions of meters Meters may be simple dial faces with a needle that moves clockwise as the quantity being measured increases, or they may present a set of numerals which increase or "count up" as the quantity being measured increases.

c. Analog and digital

- Meters having needle movements on a dial face are known as *analog* meters.

This term—analog—is used to designate an action which behaves in a manner correlated to any other action. For example, in an analog voltmeter, the needle moves exactly in proportion to the increase or decrease in the amount of voltage.

NOTE: *Analog* may also be spelled *analogue,* but engineers and scientists do not use that spelling. Thus *analogue* has come to be the literary form of the word, used when the general meaning—"something analogous to"—is intended, and *analog* the technical form, used when the meaning is as defined above.

- Meters having numerical readouts are known as *digital* meters.

This term—digital—is used to designate an action that provides abstract numbers only, without any behavior that is proportional to any other action.

NOTE: An "old-fashioned" wristwatch, with an hour hand, a minute hand, and a second hand, is an analog meter. We are so used to recognizing its analogy with time that we can tell the time even if the numbers are missing (and some wristwatches are made that way). A digital watch has no hands, and thus no action to be analogous with anything. We do not "tell" time with a digital watch; rather it tells us the time—with those regular flashings of sequential digits.

Figure 5-14 illustrates some of the more common meters.

7. Transducers and Sensors

The *transducer* is a special class of electric circuit component that converts one kind of energy into another kind of energy.

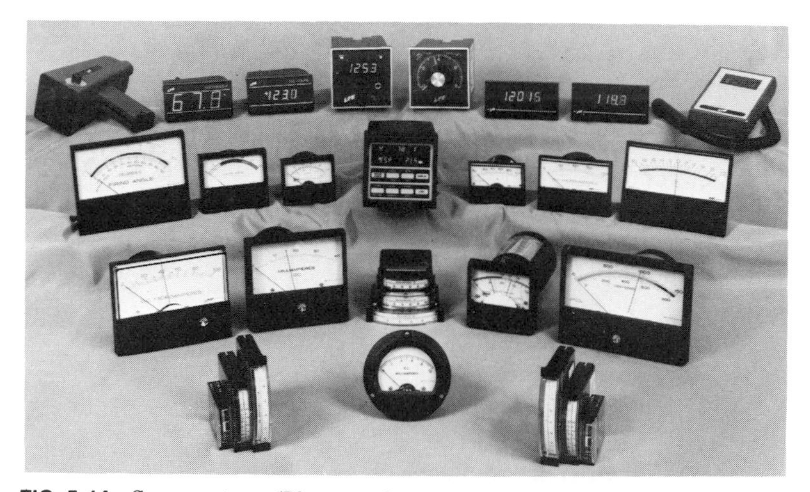

FIG. 5-14 Some meters. *(Photograph courtesy of LFE Corp., Clinton, Mass.)*

The conversion may be *to* electrical form, in which case voltages and currents result that will act in an electric circuit.

The conversion may be *from* electrical form, in which case the consequence may be mechanical, optical, acoustical, etc.

a. Transducers are analog All transducers behave in an analog manner. That is, the new form of energy is proportional in quantity to the previous form (see previous section).

b. Many forms of transducers There are many forms of tranducers.

- The *loudspeaker* is a transducer that transforms electric energy to acoustical energy.
- A *thermocouple* is a transducer that transforms heat into an electric signal used to measure temperature.
- A *photocell* is a transducer that transforms light into electric signals.
- A *motor* is a transducer that transforms electric energy into mechanical energy.

Table 5-1 lists the more common transducers and the phenomena associated with each.

c. Sensors It is evident that transducers do many things that our human sense organs do. A *photocell* "sees" light as our eyes do. So does the *camera tube* of a television transmission system. A *microphone* "hears." The *thermocouple* "feels" temperature, and so on. It is not surprising that since most transducers "sense" the world around them, they are called *sensors*.

8. Other Components

The components described in Sections 5.3.1 and 5.3.2 are the most common passive ones, and most of your work will probably concern them. But be prepared for others. There are many more components that are less frequently used, and others, which haven't been invented yet, will appear in future years.

5.3.3 Active Components

1. The Battery

A *battery* is a circuit component the function of which is to provide energy to the circuit. It does this by converting chemical energy to electric energy (Section 6.3.4). The energy provided is direct current (Section 5.1.5), so one electrode of the battery is always charged with a positive charge and the other with a negative charge.

TABLE 5-1
TRANSDUCERS

Name	*Type of Energy Change*	*Name of the Phenomenon*	*Application*
Acoustic transducer	Electric energy to sound waves	Magnetostriction	Sonar
Antenna	Electromagnetic radiation to electron current	Induction	Radar, radio, television
Humidity meter	Moisture to electric current	Electrochemical conduction	Humidity measurement
Loudspeaker	Electric power to acoustic power	Electromagnetism	Communication, entertainment
Magnetic pickup	Mechanical movement to electric current	Induction	Phonograph
Microphone	Sound to electricity	Electromagnetism	Communication, entertainment
pH meter	Acid level to voltage	Electrochemical conduction	Chemistry
Phonograph crystal	Pressure to electricity	Piezoelectricity	Phonographs
Photocell	Light to electricity	Photoelectricity	Switch
Photovoltaic cell	Light to electricity	Carrier generation	Energy conversion
Solar cell	Light to electricity	Carrier generation	Energy conversion
Strain gauge	Mechanical elongation to resistivity change	Change of resistivity	Strain measurement
Thermocouple	Temperature to electricity	The Seebeck effect	Temperature measurement
Thermoelectric cell	Electric power to heat	The Peltier effect	Energy conversion

The chemicals inside the battery chamber provide ions (Section 6.3.2). The battery electrodes, which are made of two different metals, provide a driving force such that the positive ions migrate toward and gather at one electrode and the negative ions migrate toward and gather at the other electrode. This is illustrated in Figure 5-5.

The electrode which attracts electrons is the positive one and is called the *anode*. The negative electrode is called the *cathode*.

When the two electrodes of a battery are connected with a wire, a circuit is formed, and the pileup of electrons at one electrode may be relieved by virtue of their flow in that circuit. But such a circuit is not useful. It is in fact a short circuit (Section 5.3.5), which will quickly drain the chemical energy from the battery. Rather, for the circuit to be useful, it must contain circuit components, as described in the preceding sections, which modify the electron flow as required to perform particular electric actions.

a. Batteries may be any size Batteries may be large, like the one in an automobile, or small, like the one in a hand calculator or watch, or in between, like the one in a flashlight.

In general, the larger the battery, the longer it will last. This is because the larger battery contains more chemicals, and they will provide more ions and electrons at the two electrodes before the chemicals are depleted.

b. Battery ratings The electrical size of batteries, as distinguished from the physical size, is called the *rating*. The rating states two facts:

- The *voltage difference* that will always be maintained between the electrodes
- The *life* of the battery, stated numerically as the amount of time it will function—delivering a given current in amperes—before it is depleted

Examples:

- A watch battery may be rated as providing 0.8 volts and 3.5 milliamps for 5000 hours.
- A flashlight battery may be rated as providing 1.5 volts and 100 milliamps for 1 hour.
- An automobile battery may be rated as providing 12 volts at 2 amps for 45 minutes (without recharging).

c. Battery cells Each compartment containing ions (Section 6.3.2) and a pair of electrodes is known as a *battery cell*.

Each cell contributes its own characteristic rating of voltage and delivery of electric current for a given time.

There may be only one cell in a battery, or there may be several.

If there is more than one cell in a battery, they are so arranged that the voltages add up.

Most automobile batteries consist of six 2-volt cells, so they add up to the 12 volts that is customary in the electrical systems of today's automobiles.

2. The Vacuum Tube

The *vacuum tube* is a type of circuit component that actively performs a series of electrical functions that are far more complex than those performed by passive components. Passive components merely *resist* or *store* or *induce* or *make or break*. The functions of vacuum tubes, which are active, include amplification, oscillation, rectification, modulation, adding signals, and subtracting signals.

The vacuum tube can perform these complex functions because it provides a vacuum chamber in which electrons can move more freely and unencumbered than in solid conductors or solid electric components.

Alternative names for the vacuum tube are *electron tube* and *radio tube*. The British call the tube a *valve*.

All vacuum tubes contain, at a minimum, a chamber or *envelope* to contain the vacuum (to keep the air out), a *cathode* (negative electrode) to provide electrons, and an *anode* (positive electrode, also called *plate*) to receive electrons (Figure 5-15).

a. The cathode The cathode is a small metal structure, usually a cylinder about the size of a pencil lead and less than an inch in length, which is located in the very center of the vacuum tube. When it is heated to a red heat it "boils off" electrons. These are then free, no longer part of the electron cloud of the metal of the cathode.

These are the electrons that can be manipulated in the vacuum space of the vacuum tube to perform active electronic functions.

The giving off of electrons at very high temperatures is known as *electron emission*.

The electrons of the electron cloud of a metal are always relatively free to move about within the confines of that metal. In fact, they move with a kind of vibration or oscillation. This vibration increases in intensity with increasing temperature until a temperature is reached at which they vibrate themselves out of the outer skin of the metal itself. It is at that point that electron emission occurs.

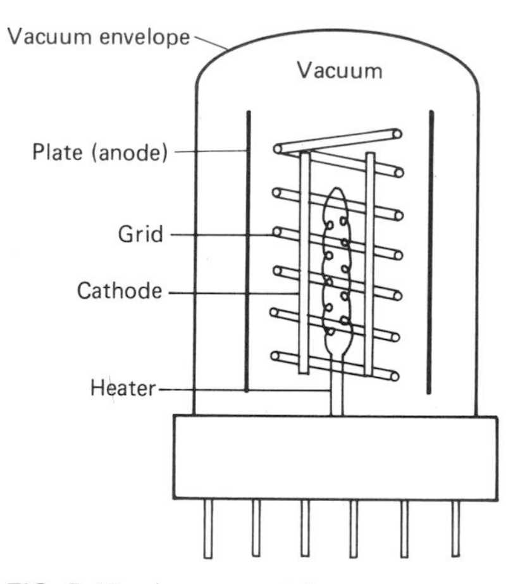

FIG. 5-15 A vacuum tube.

b. The heater Emission of electrons from the cathode proceeds at a higher rate the hotter the emitter. Thus cathodes normally are operated at red heat.

There is a little wire coil which electrically heats the cathode to enhance its electron emission. It is the *heater* (Figure 5-16). When you waited for your old television set to "warm up," you were waiting for the heaters in the tubes to become hot.

c. Grids A grid is a metal structure that is placed between the cathode and the anode of the tube.

The structure is made of wound wire or mesh so that it is "open" in the sense that the electrons making their way from the cathode to the anode can pass through it.

Some vacuum tubes *(diodes)* have no grids, some *(triodes)* have one, some *(tetrodes)* two, and some *(pentodes)* three.

d. The anode The anode (or *plate*) is a metal electrode, generally cylindrical, that surrounds all the elements of the vacuum tube.

Since the plate is kept at positive voltage, the electrons emitted from the cathode are attracted to it, and indeed are eventually gathered in by it, as the final act of active electronic function.

NOTE: The suffix "-ode" means "path" or "way." So, for example, in a triode, the electrons have three structural elements—cathode, grid, anode—into

(a)

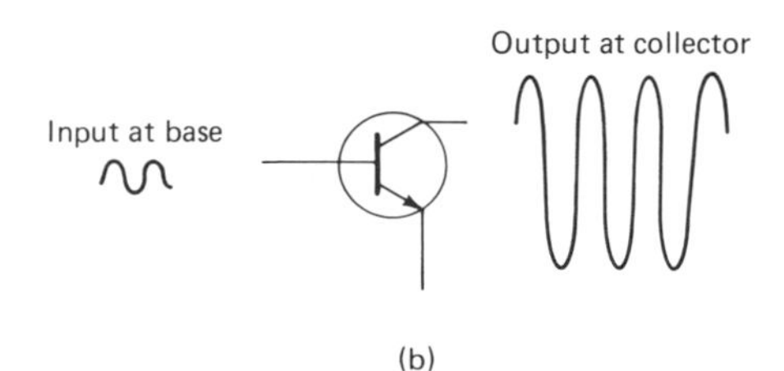

(b)

FIG. 5-16 Two examples of modulation in a vacuum tube: (*a*) Rectification in a semiconductor diode, (*b*) amplification in a transistor.

which they may enter and through which they may move. By the same token, an *electrode* is a "way for electrons."

3. Semiconductor Diodes and Transistors

Semiconductor diodes and transistors are the nonvacuum analogues—as active circuit components—of vacuum tubes. They perform the same type of circuit functions including rectification, amplification, and oscillation.

In vacuum tubes, electrons move freely in a vacuum which is created in the device during its manufacture. In semiconductors, electrons move just as freely in the vacuum that exists between the atoms of the semiconductor material.

NOTE: There can be no better vacuum than the one that separates atoms from each other. Between atoms there must be—truly—nothing (Section 4.8.1).

As with vacuum-tube diodes (see "The Vacuum Tube" in Section 5.3.3), semiconductor diodes consist of a source of electrons and a collector of electrons.

a. The emitter The source of electrons is known as the *emitter*.

The emitter is the equivalent in function of the cathode of the vacuum tube.

b. The collector The collector of electrons is known as the *collector*.

The collector is the equivalent in function of the anode (or plate) of the vacuum tube.

As with vacuum-tube triodes, in semiconductor triodes—which are known as *transistors*—there is an element which functions to modulate the flow of electrons from source to collector.

c. The base The modulation element in semiconductor devices is known as the *base*.

The base is the equivalent in function of the grid of the vacuum tube.

The equivalent to the vacuum-tube tetrode is the *silicon-controlled rectifier* (SCR).

Just as the vacuum-tube tetrode has two grids, the semiconductor tetrode (the silicon-controlled rectifier) has two bases.

4. The Active Functions of Vacuum Tubes and Semiconductor Devices

The major active functions of vacuum tubes and semiconductor devices are modulation, rectification, amplification, and oscillation.

a. Modulation Modulation of a flow of electrons is the control and changing of the amount of that flow.

- In a vacuum tube, the grid achieves modulation of the flow of electrons that would otherwise pass unchanged from cathode to anode.
- In a transistor, the base achieves modulation of the flow of electrons that would otherwise pass unchanged from emitter to collector.

The voltage on the grid or on the base behaves like a traffic police officer. It either "beckons" the electrons to move in greater quantity (if it is a positive voltage—to which the negative electrons are attracted) or it gestures to them to "hold up" (if it is a negative voltage—by which electrons are repulsed).

The important thing about the ability of the voltage on a grid (or a base) to *modify* the electron flow is that it takes only a very little change in the voltage on the grid (base) to cause a very large change in the amount of flow from the cathode (emitter) to the anode (collector).

Rectification and amplification are the two most important forms of modulation.

b. Rectification This is the process of converting alternating current (which oscillates between flowing in one direction and flowing in the opposite direction in each cycle, as described in Section 5.1.5) to unidirectional current (direct current).

Rectification occurs because only the cathode of a vacuum tube and only the emitter of a semiconductor device emit electrons. The tube anode and the semiconductor collector emit no electrons.

Thus electrons can go only *one way*. It is this one-way feature of these devices that achieves rectification. The process is illustrated in Figure 5-16*a*, where it can be seen that only half of the alternating current wave gets through the device.

NOTE: Electrical rectification has been compared to the flow of people going through a turnstile in a subway station. They can proceed in (once they pay their fare), because the turnstile turns once to let one person move through. But the turnstile moves only in one direction, so no one can go out through it. Thus the turnstile has "rectified" the flow of people, as the diode rectifies the flow of electrons.

Rectification is a form of modulation (see "Modulation" in Section 5.3.3).

c. Amplification This is the process of a small change having a large consequence. It is a form of modulation (see "Modulation" in Section 5.3.3).

Amplification is especially useful in electronics, where some electrical variations are so small that even the most sensitive meter cannot measure them, yet we wish to make use of them.

This occurs, for example, in a television set. The signal—which contains the program we wish to watch—is sent from the television station and is in the air all around us in the form of very-high-frequency or ultra-high-frequency waves. The antenna, which acts like a sensor (see "Transducers and Sensors" in Section 5.3.2), "feels" the presence of these waves (senses them) and conducts them down the antenna wire into the television set. There the circuit components amplify them millions of times to a power level so high that we now are easily able to perceive them—as pictures on the screen and as sound from the speaker.

In electronic amplification the small signal and the large signal are exact duplicates of each other. The only difference is size (Figure 5.16*b*).

Amplification of the flow of electrons may be compared to amplification of the flow of water from a reservoir. Consider that the reservoir is contained (held back) by a dam at one end (Figure 5-17). So long as the level of the water is lower than the top of the dam, no water will flow.

There is a gate in the dam.

NOTE: The grid in a vacuum tube and the base in a transistor are often called *gates*.

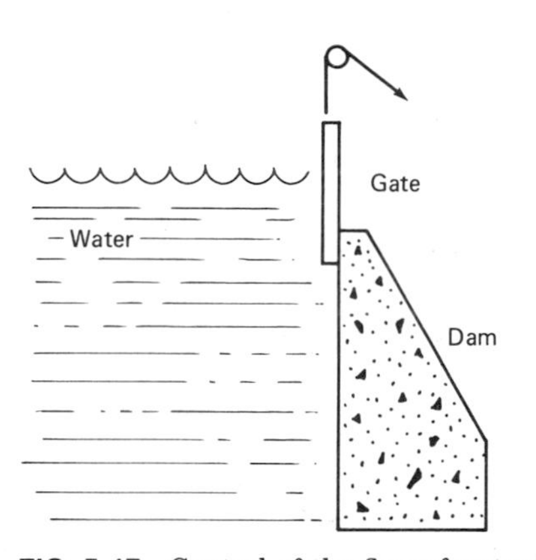

FIG. 5-17 Control of the flow of water in a dam.

The small amount of energy required to move the gate up and down releases a large amount of (potential) energy pent up in the water behind. In the same way, moving a small amount of voltage on the base of a transistor (or grid of a vacuum tube) releases a large amount of electrical current. This is—in all cases—modulation and amplification.

There are millions and millions of gallons of water in the reservoir (as there are millions and millions of electrons that can be emitted by the cathode of the tube or the emitter of the transistor).

These millions of gallons of water can be made to rush from the reservoir through the dam—if the gate is lowered. The important thing to notice is that it takes very little energy to lower the gate, but the energy that lowering releases in the rushing water is vast. In effect the energy invested in gate-lowering has been *amplified.*

Also, the gate may be easily moved up and down at some interval. The flow of water will be slower or faster as a result of this gate movement, and the rate of flow will exactly respond to and follow that movement. In effect, the movement of the gate *modulates* the flow of the water. And that modulation is exactly amplified as a result.

As the voltage on the vacuum tube grid or the transistor base goes up or down, it modulates the flow of electrons (just as the gate of the dam does), and that modulation too is amplified (just as the energy invested in gate movement is amplified in the energy change that occurs in the water flow).

d. Oscillation Another major function of vacuum tubes and semiconductor devices, in addition to rectification and amplification, is *oscillation.* This is the conversion of direct current to alternating current by nonmechanical means.

Such conversion from direct to alternating current takes place in the region between the emitter of electrons (cathode or emitter) and the gatherer in of electrons (plate or anode or collecter). The manner in which this occurs is complex and need not be dealt with here other than to note that as a result of various geometrical factors within the device combined with various voltage inputs to the electrodes, the direct current stream of emitted electrons takes on a rapid oscillation (alternating current) just as it arrives at the positive anode.

Many electronic systems require electric energy that will be radiated into space at very high frequencies as electrical waves (Section 5.1.5). An example is radar (see "Radar" in Section 4.5.1). Unless the beam of energy radiated from the radar antenna is oscillating at a frequency of many million cycles per second, the beam will not travel in a straight line, and it will not then be effective in indicating the exact location of an object at a distance, which is the purpose of radar. Vacuum tubes such as *magnetrons* work as generators of the required high-frequency radiation (oscillators).

Transistors can also perform as oscillators, but they differ from vacuum tube oscillators such as magnetrons in that they cannot deliver as much power because they easily overheat.

5. Semiconductor Materials and Processing

So much of the work of engineers and scientists working with semiconductors today is devoted to the manufacture of semiconductors—possibly as much as to their use—that the following sections on semiconductor materials principles and device fabrication are included.

a. The special electrical conductivity of semiconductor materials It is possible to design and build semiconductor devices to work as well as—and usually better than—vacuum tubes because of the special qualities of semiconductor materials as conductors of electrons. They are "halfway" between conductors and insulators (see "Conductors" and "Insulators" in Section 5.1.1) in their ability to conduct electrons—hence the name *semi*conductor.

But more important, the electrons are *available at room temperature.* They need no red-hot heater as the cathode of the vacuum tube does (see "The Heater" in Section 5.3.3).

This is because in the outer orbitals of the atoms of the semiconductor, there are free electrons ready to move at room temperature when propelled by voltage. Not every atom has a free electron—only about one atom in a million. But that is still many billions of electrons, and that is enough for the uses to which these devices are put.

NOTE: In conductors, every atom has a free electron, enough to make an electron cloud (Section 5.1.2). (If semiconductors have an electron cloud, it is very very thin.) Thus metal conductors are about a million times as conductive as semiconductors.

Insulators (see "Insulators" in Section 5.1.1) have so few free electrons that we call them nonconductors. But they have some free electrons—about a million times fewer than semiconductors. Thus metal conductors are about a million million times as conductive as insulators, and semiconductors are about a million times as conductive as insulators.

There are hundreds of semiconductor materials, but the most common is silicon. Next is the compound gallium arsenide and, after that, germanium.

b. The effect of impurities It is possible to control the conductivity of each semiconductor material more easily than that of any other kind of material and it is this factor that gives semiconductors their special usefulness. This is done by alloying, or "doping," the semiconductor material with atoms of particular metals (see "Alloys" in Section 4.7.2). Each added atom adds essentially one more free electron and so makes the semiconductor that much more conductive.

NOTE: Donors add one free electron per atom of dopant. Acceptors take away one free electron per atom with the result that an absence of an electron, called a *hole,* occurs. Holes behave like positive electrons. Thus in both cases, "essentially one more free electron" is added.

The metal alloy is very dilute. Thus the term *impurity addition* is more commonly used than *alloying*. The term *doping* is also used. Then the impurity added is called the *dopant*.

The essence of the engineering of making semiconductor materials is controlling the techniques of adding impurities.

The impurities added to semiconductors fall into two classes: *donors* and *acceptors*. When donors are predominatingly present, the semiconductor is N *type* (or negative type); when acceptors are predominatingly present, it is P *type* (positive type).

For the purposes of this book it is not necessary to define the differences beyond stating that donors contribute electrons and acceptors contribute holes. It is enough to understand that each impurity atom contributes essentially one additional free electron to the conductivity of the material. The common donors are phosphorus, arsenic, and antimony; the common acceptors are aluminum, boron, and gallium.

c. The special crystal form of semiconductor materials Semiconductor materials are manufactured for use in electric components as *single crystals*. This means that the atoms which make up the material are always arranged in orderly repeated spaces and patterns (Section 4.7.1).

Figure 4-12*b* illustrates a single silicon crystal such as is commonly used in the manufacture of semiconductor products. Thus semiconductor products differ from most manufactured products, which are made of many millions of tiny crystals, all jumbled together in a random array (Figure 4-21).

The reason for the single-crystal construction of semiconductor products is that the boundaries between crystals are barriers which interfere with the free movement of electrons. In metal conductors, such crystal boundaries do not perceptibly interfere with electron movement because there are so many more electrons present that any interference to their movement which the boundaries would contribute is insignificant.

Thus every semiconductor diode and transistor consists of only one crystal.

NOTE: The crystal form of semiconductors is that of the diamond lattice (see "Unit Cells" in Section 4.7.1 and Figure 4-17*c*). In fact, diamonds—the gems—could be made to function as (rather costly) transistors.

d. Slices, dice, chips In the manufacturing process, complex arrays of diodes and transistors—sometimes as many as thousands—are constructed on a single *slice* of semiconductor material.

That slice is still only a single crystal. It, in fact, is cut—as slices of salami are cut in a delicatessen—from a round rod of single crystal usually of salami size, known as the *ingot*. That ingot is *grown*, usually by the *Czochralski technique* (Section 4.7.1 and Figure 4-16*a*).

Such slices may in turn be cut into an array of smaller pieces, each of which is of a size to be a diode or transistor or other semiconductor device. This cutting process is known as *dicing*. Each piece is then called a *die*. The plural of die is *dice*.

Figure 5-18 shows several typical silicon slices already marked off into dice.

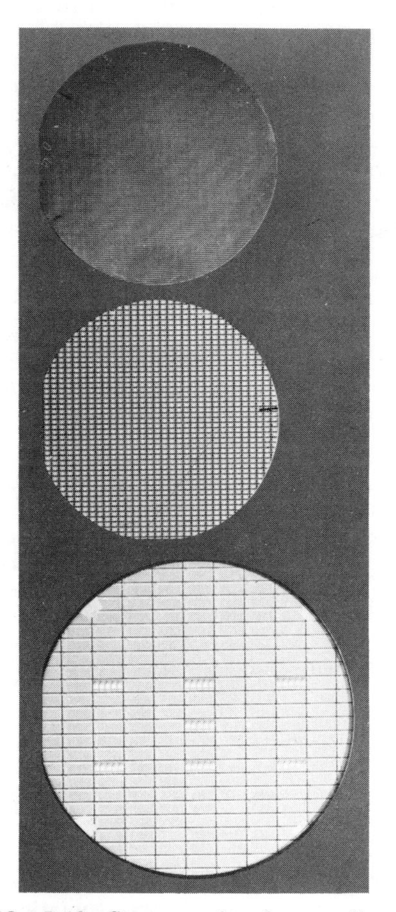

FIG. 5-18 Some semiconductor slices. *(Photograph courtesy of Raytheon, Mountain View, Calif.).*

Arrays of more than one semiconductor device (and other devices) on the same piece of semiconductor material are the basis of *integrated circuits* (see "The Integrated Circuit" in Section 5.3.3).

Another name for dice is *chips*.

NOTE: The term *chip* also refers to a finished component which in turn is made up of many components. Indeed, that finished component is itself an integrated circuit. Chips of this sort are described in "Chips" in Section 5.3.3. They are used most commonly in computer circuits (see "Chips" in Section 3.4.2).

e. Metallurgical fabrication techniques The different functioning parts of semiconducting diodes and transistors are nothing but *regions* in the device with impurity contents and types (N type or P type) different from those in adjacent regions (Figure 5-19).

Between all distinctive regions—which are either P type or N

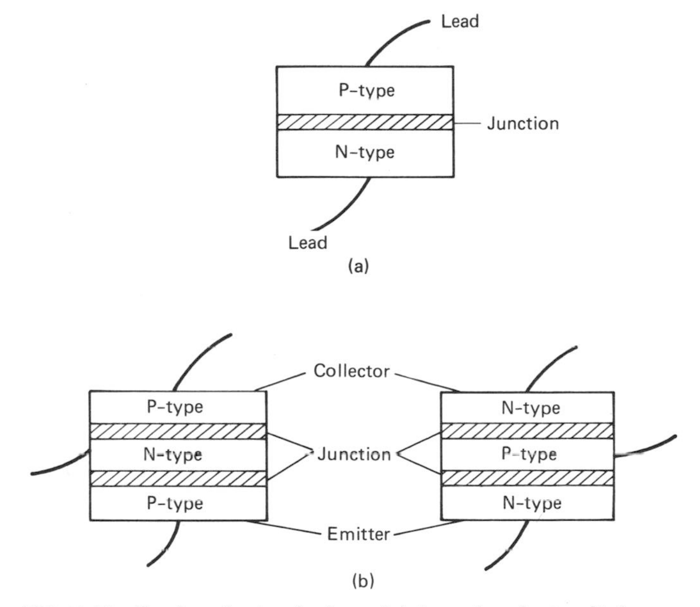

FIG. 5-19 Semiconductor devices. (*a*) A semiconductor diode.

Typically the N-type region functions as the emitter, giving off electrons, and the P-type region functions as the collector, gathering in the electrons. The junction region functions as the vacuum does in a vacuum tube.

(*b*) Transistors.

These may be either PNP or NPN. The region sandwiched in the middle is the base, which functions like the vacuum tube grid.

NOTE: The silicon controlled rectifier (SCR) has two bases.

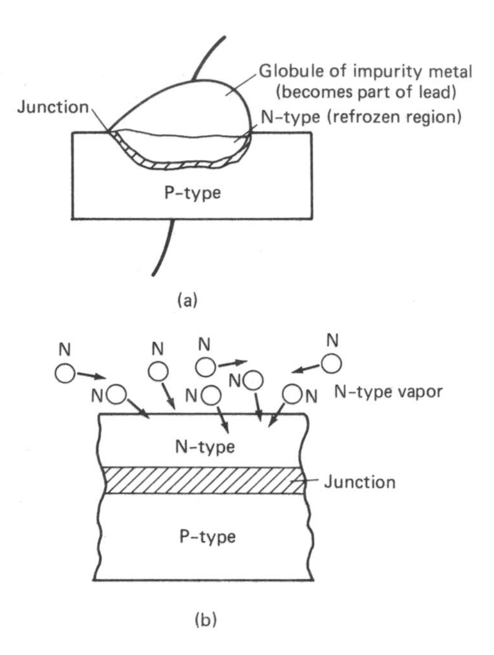

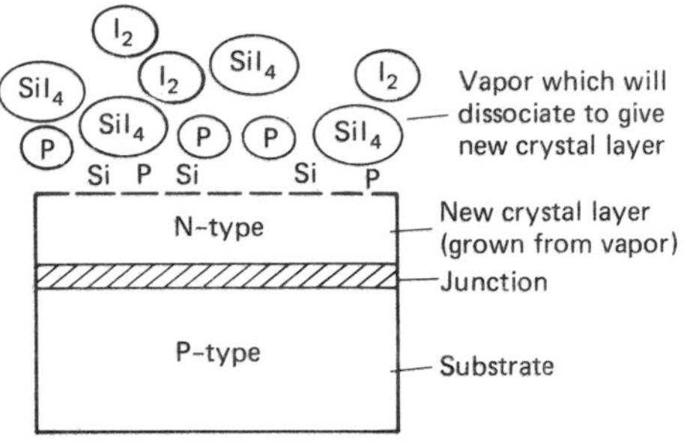

(c)

FIG. 5-20 Formation of a doped region by metallurgical methods. (*a*) Formation of a doped region by alloying. (*b*) Formation of a doped region by diffusion. (*c*) Formation of a doped region by epitaxy from vapor on substrate.

In (b), the entire slice has been subjected to diffusion before dicing (a fragment of the slice is shown). An N-type impurity vapor has diffused into what was originally a P-type slice. The concentration of N-type impurity atoms tapers off, the deeper they penetrate. Thus the junction region, where the N-type impurities become greater than the P-type impurities, is a gradual transition region—and the junction is thus wide.

type—there is always a transition region which is neither. This is the *junction* (Figures 5-19 and 5-20).

This class of circuit components—semiconductor products—differs from other circuit components, the basic parts of which consist of bits of wire and bits of metal ribbon or strip, welded or otherwise joined together. In semiconductors, there is no question of joining together, since the whole device is only one piece to begin with. Only the distribution of impurities within it changes.

Thus making semiconductor devices is a special kind of metallurgy involving ways of making very dilute alloys *(doping)* in semiconductor surfaces and confining the resulting doped regions to very close dimensional restrictions.

Three ways are most commonly used. These are *alloying, diffusion* of atoms, and *epitaxy* (or *epitaxial crystal growth*).

- *Alloying* is doping by means of melting an impurity on the surface of the semiconductor die.

  Part of the die melts and dissolves into the melted impurity globule. When the globule and the die portion refreeze, the portion of the die that has refrozen contains some of the impurity metal of the globule. That portion has been doped (Figure 5-20*a*).

- *Diffusion* is alloying without melting. A slice is heated in a furnace which contains a vapor made up of the dopant metal.

  The atoms *migrate in* from the surface and form substitutional solutions (see "Alloys" in Section 4.7.2), which are, in effect, doped regions. This migration is controlled by temperature and time in the furnace. The hotter and the longer the process goes on, the deeper the doping (Figure 5-20*b*).

- *Epitaxy* is crystal growth of a material from a gaseous chemical compound containing that material. The growth proceeds on a seed of the same material—which is already a single crystal—grown elsewhere, usually by another technique such as the Czochralski technique (see "The Growth of Crystals" in Section 4.7.1 and Figure 4-16).

  The slice is placed into a furnace containing a vapor such as the chemical compound silicon tetraiodide (SiI_4) (see Section 6.1.4). It may also contain another vapor of a chemical which can act as a dopant, for example, phosphorus.

  At the furnace temperature, the compound dissociates.

  This means that the iodine breaks away from the silicon and evaporates aways as iodine gas. What is left is solid silicon. The silicon was in gaseous form only so long as it was combined with the iodine. With the iodine gone, it comes out of the atmosphere of the furnace as a solid—just as ice crystals (snow) come out of the atmosphere of the air in the winter. The silicon then deposits on the slice (now known as a *substrate*) along with a tiny amount of the phosphorus. This process extends the thickness of the slice and forms a doped region (Figure 5-21*c*).

NOTE: All three methods of forming doped regions give the same result, but they do it very differently. Alloying requires melting in and refreezing. Diffusion requires no melting, only atom migration. And epitaxy is an adding-on process.

f. Lithographic techniques The fabrication methods that have been developed for semiconductor components are determined by the fact that the basic parts of the components are extremely tiny.

In fact a typical transistor is so small that hundreds can fit onto a semiconductor chip the size of a period typed by an ordinary typewriter.

This small size requires techniques of manufacture more related to photography than to mechanical manipulation of "parts." There are three main ways to fabricate semiconductor products. These are *photolithography, x-ray lithography,* and *electron beam scanning.*

- *Photolithography* is a process in which a thin layer (a film) of photographic chemicals is applied to the surface of a semiconductor slice. A pattern (of the semiconductor "parts") is then optically produced in the film and is thus transferred to the slice. This pattern is a photographic print of those parts—and in later processing, it becomes the parts themselves.

The film is exposed to an intricate light pattern consisting of microscopic dots, bars, lines, etc. It is then developed, and the development process removes only the film exposed to the light. After development, the whole slice is immersed in an acid which dissolves away just those regions where the light pattern had been, since the unexposed film layer protects the semiconductor region under it from the acid.

The film layer is known as *photoresist.*

The dissolving process is known as *etching.*

The acid which does the etching is known as the *etchant.*

The pattern of intricate dots, bars, and lines, which later become such parts of transistors as emitters, bases, and collectors, is achieved by shining light through a sheet of thin metal foil which is perforated with holes having shapes that correspond to those dots, bars, lines, etc. That thin metal foil is known as a *mask.*

The process of producing that light pattern is known as *exposure through a mask* or *masking* (Figure 5-21).

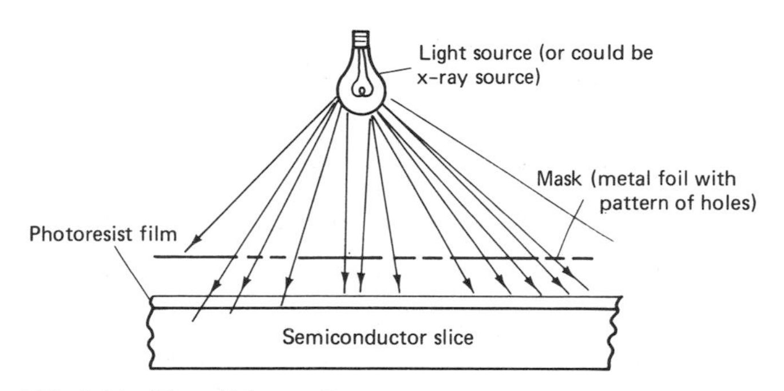

FIG. 5-21 Photolithography.

NOTE: This could be *x-ray lithography* if an *x-ray* source were substituted for the light source.

The exposed semiconductor regions can next be coated with metals which later, in a furnace, diffuse in and cause doping. Thus tiny regions have different impurity contents from those of adjacent regions and can function as separate semiconductor "parts."

- *X-ray lithography* is the same as photolithography except that x-rays, rather than light, are used to expose the film coating. This is done in order to create smaller patterns than are achievable with light.

The reason that smaller regions can be exposed with x-rays than with ordinary light is that the regions are of the same approximate dimensions as the wavelengths of light. And patterns cannot be accurately created in film when the waves used to expose the film are the same size as the patterns. X-rays have much smaller wavelengths than light waves and can thus create correspondingly smaller patterns.

- *Electron beam scanning* goes one step further than x-ray lithography in making patterns even smaller.

Here a mask is not required. An electron pattern is controlled electronically and it "writes" a picture on the film, just as electrons produce a picture on the inside face of a television tube. Electrons are even smaller than x-ray wavelengths, and thus there is practically no limit to the smallness—or the crowding of semiconductor components—that can be achieved on a single small chip of semiconductor material when electrons are used.

NOTE: A chip such as the one that controls electron flow in a product that most people own nowadays—a hand calculator—may be an eighth of an inch on a side. It contains as many as 5000 transistors as well as thousands of other components, all imprinted on its surface by one of the three ways: photolithography, x-ray lithography, or electron beam scanning. All these ways can make a complex product extremely small.

NOTE: The first computers, made in the 1940s, used vacuum tubes—since transistors hadn't been invented yet—and were the size of a three-story house. And they were far slower at performing a calculation than is the typical hand calculator that contains just one chip.

Also, the heaters that were in each vacuum tube to raise the cathodes to a temperature that is optimum for emitting electrons (see "The Heater" in Section 5.3.3) created so much heat that it was unbearable to approach the old-time computer. (One reason was that air-conditioning was not commonly used in those days.) No wonder the world had to wait for semiconductors before the computer industry became practical. This practicality is based on the fact that semiconductor products are small—and thus manageable—and that they work at ordinary temperatures. They require no heaters.

g. Ways to deposit thin films Once semiconductor regions have been exposed in required tiny patterns by the processes of photolithography, x-ray lithography, or electron beam scanning, it becomes necessary to lay thin films of metal into and onto those patterns for doping purposes and also to create conductive connections between certain of them. This is done by the techniques of *evaporation* or *sputtering.*

- *Evaporation* of metals utilizes a capsule containing the required metal placed in a chamber called an *evaporator bell jar* (Figure 5-22). The air is then evacuated from the bell jar. The capsule is heated, and the metal becomes a vapor that fills the space in the bell jar. When it encounters a solid surface, the vapor solidifies on that surface—as a metal film. This process is also known as *vapor deposition.*

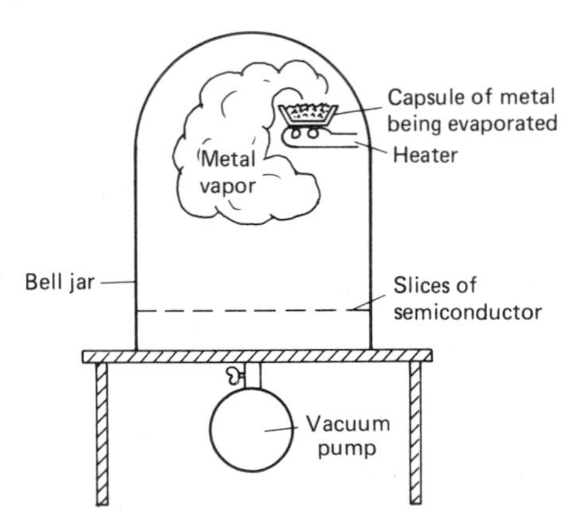

FIG. 5-22 Bell-jar technique for depositing films.

When slices of semiconductor material coated with an exposed layer of photographic material—so that there are photographically developed "holes" in that layer—are placed in the bell jar, they are coated with the new metal film. But the film forms on and attaches to the semiconductor surface only where the photographic coating does not protect it. The evaporated metal coats the top surface of the photographic film too, but later the film is dissolved away, taking the evaporated metal that is on it away with it. Thus only the exposed regions of the semiconductor surface become coated.

NOTE: The films of metal are so thin that it would take hundreds of layers to make the thickness of a human hair—but that thickness is quite enough.

- *Sputtering* is very similar to evaporation in that it too requires a bell jar and a heated capsule of the metal which will eventually coat semiconductor slices through holes in the photographic coating.

Sputtering is different from evaporation in that in the sputtering process, the atoms of evaporated metal are subjected to an electric field (a voltage), and as a result each atom acquires an electric charge—either positive or negative. In effect, each atom becomes an ion (Section 6.3.2). As the metal gas becomes ionized, it gives off light. Such a glowing collection of ions is known as a plasma (see "Plasma Emission of Light" in Section 4.5.2).

The semiconductor slice is now given an electric charge of the opposite sign to the one on the atoms, and as a result the atoms are drawn toward the slice (since opposite charges attract) (Section 6.3.1). The result is that the ions do not merely deposit softly on the slice—instead they "bang into" the surface of the slice, with great force. Indeed they penetrate a short distance (a phenomenon known as *ion implantation*). The result is that sputtered films adhere better than evaporated films.

h. Some semiconductor words and terms Semiconductor technology has developed a jargon of its own. Table 5-2 lists the most common of these terms.

6. The Integrated Circuit

An *integrated circuit* is a complete set of circuit elements including resistors, capacitors, semiconductor diodes, and transistors—and the wires or leads or conductors that connect them—all of which have been made at one time on a single semiconductor slice by the techniques described in the preceding sections.

This means that an integrated circuit contains no discrete components—where *discrete* means separate (see "Discrete Components, Integrated Components" in Section 5.3.1). Thus there are no separate resistors or capacitors or semiconductor diodes or transistors, each with its own separate wires or leads, waiting to be welded or soldered into place.

TABLE 5-2
SOME SEMICONDUCTOR WORDS AND TERMS*

abrupt junction	epitaxy
acceptor	epoxy bond
alloy junction	eutectic bond
alloyed region	evaporation
alloy transistor	field-effect transistor (FET)
amplification	gate
amplification factor (β, beta)	germanium transistor
avalanche	gold-bonded diode
avalanche diode	graded base
avalanche transistor	high-frequency transistor
avalanche voltage	hole
ball bond	hole conduction
base	hole-electron pair
base binistor chip	IMPATT diode
basewidth	impurity gradient
bias	inductance
binistor	input
built-in field	integrated circuit
capacitance	intrinsic region
chip	ion implantation
collector	ionization
common base circuit	junction
common emitter circuit	junction capacitance
controlled rectifier	junction transistor
cutoff current (I_{co})	large-scale integration (LSI)
dice	large-signal operation
die	lifetime (τ)
diffused diode	lithography
diffused region	mask
diffuse junction	masking
diffusion junction	mesa transistor
diffusion transistor	microwave diode
diode	mobility (μ)
dissipation	modulation
donor	monolithic
dopant	MOS transistor
doping	*n*-doped region
electron beam scanning	*npn* transistor
electron conduction	*n* type
emission efficiency (γ, gamma)	ohmic contact
emitter	ohmic region
encapsulation	output
epitaxial	package

passivated transistor
p-doped region
pentode
photodiode
photolithography
photoresist
phototransistor
PIN diode
planar transistor
pnp transistor
point-contact transistor
post-alloy-diffused transistor (PADT)
power rectifier
power transistor
preform
p type
rectification
rectifier
reference
reference diode
reference voltage
resistance
resistivity (ρ, rho)
saturation current (I_s or I_{sat})
saturation voltage (V_{sat})
silicon-controlled rectifier (SCR)
silicon transistor
skin effect
slice
small-signal operation
solder
space charge
space-charge region
sputtering
step-recovery diode
surface-alloy-diffused transistor (SADT)
surface-barrier transistor (SBT)
switching
tetrode
thick film
thin film
time constant
transistor
transition region
trinistor
triode
tunnel diode
varactor
variable reactance diode
varistor
very large scale integration (VLSI)
via
voltage gradient
wedge bond
wire bond
x-ray lithography
zener diode
zener voltage

---

*Refer to Dictionary in Part Two for definitions.

NOTE: All semiconductor devices are designated by commercial device numbers. These are usually a combination of numbers and capital letters.

Diodes start with "IN"
IN21 IN341A IN82
IN789 IN1734

Triodes start with "2N"
2N43 2N716 2N89
2N722 2N4521

Tetrodes start with "3N"
3N79 3N2017
3N1891

But there are exceptions for which there seems to be no logic:

CK716 WX3347
T18A

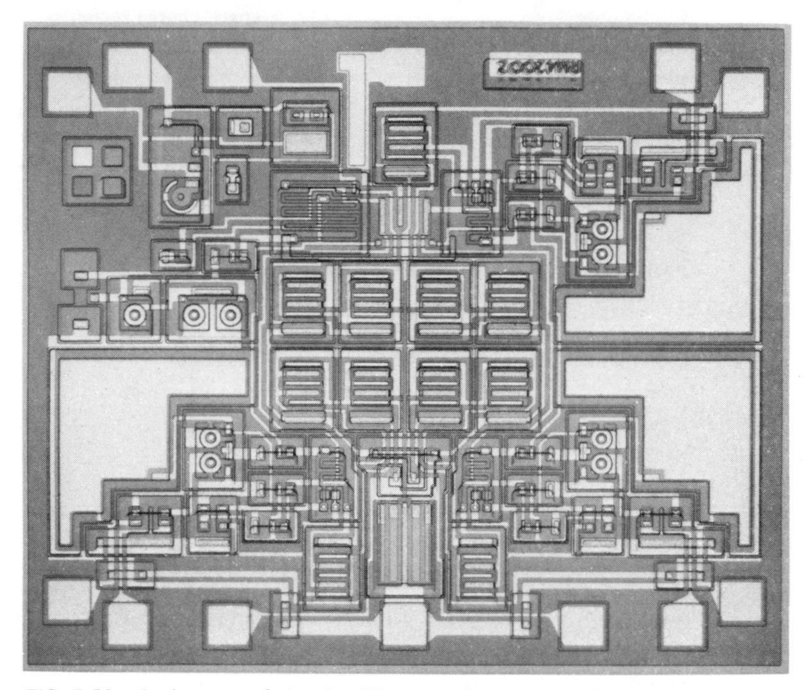

FIG. 5-23 An integrated circuit. *(Photograph courtesy of Raytheon, Mountain View, Calif.)*

This also means that integrated circuits are extraordinarily small, like the one in the hand calculator described in the preceding section. Figure 5-23 shows a typical integrated circuit. It contains many dozens of transistors and diodes and other components as well as an intricate system of interconnections. Yet is only a single die—out of hundreds of dice on a single slice—and it measures no more than a sixteenth of an inch across in its largest dimension.

a. LSI *LSI* is an acronym that stands for *large-scale integration.* It applies to cases in which "many" circuit components are present on a single chip—more than on "ordinary" integrated circuit chips.

LSI means that a single chip may contain more than 100 individual interconnected devices which have been formed at the same time in one manufacturing process, and all function together electrically in a complex circuit function.

LSI is a relative term. Before it came into use, ordinary integrated circuits were considered to have many circuit elements on a single chip.

b. VLSI *VLSI* is an acronym that stands for *very large scale integration.* It applies to cases in which "many more" circuit components are present on a single chip—more than on "ordinary" LSI integrated circuit chips, and certainly more than on ordinary integrated circuit chips.

VLSI means that a single chip may contain more than 1000 individual interconnected devices which have been formed at the same time in one manufacturing process, and all function together electrically in an (extremely) complex circuit function.

VLSI is a relative term. It simply means "more than LSI" by about a factor of 10.

c. Chips The integrated circuit—whether it is a "simple" integrated one or an LSI or a VLSI—consists of a single fragment of semiconductor material with all required circuit elements on or in it. When it is encased in a plastic or metal container (this process is known as *packaging*, and the container is known as a *package*), it is then known as a *chip*. A packaged chip in this sense is an active circuit component.

Chips, the circuit components just described, are always little black rectangles with numerous metal connections emerging at both long edges so that they look like centipedes as they might be depicted by a modern artist, which is to say, with straight edges and square corners rather than rounded ones (Figure 5-24). The term *chip* has two meanings, and this can be confusing because these meanings overlap in semiconductor technology.

FIG. 5-24 Chips as circuit components—with a silicon slice such as the one that gave them birth. *(Photograph courtesy of Raytheon, Mountain View, Calif.)*

- A *chip* as just described is a packaged integrated circuit (which may be "simply" an integrated circuit or an LSI or VLSI circuit).
- A *chip* is also a die (one of many dice cut from a semiconductor slice (see "Slices, Dices, Chips" in Section 5.3.3). Such a die may contain one semiconductor element such as a diode or a transistor, or it may contain many integrated circuit elements. In any case, considered in this sense, it has not been packaged yet, which means that the leads have not yet been attached to it, and the plastic or metal enclosure has not yet been applied to it.

NOTE: It is ironic that once this most integrated circuit has been packaged—and all of what may be thousands of nondiscrete circuit components enclosed in one tiny container (and the whole is then called a chip)—then that chip is considered to be just another discrete circuit element. It is considered discrete because it *looks* discrete, that is, it is a separate little part, used electrically in an even more complex circuit system, along with dozens of other chips and other discrete circuit components.

5.3.4 Circuit Formats

So far in this chapter we have described how electrons move in conductors and in circuit components, both active and passive. But several questions remain. How are these elements put together? What does the final assembled system look like? And are there different classes of assembled circuit systems, in different formats?

The most common electronic circuit formats are the *chassis, printed circuits* (in many subformats such as *circuit cards*), and the *hybrid circuit*. All these circuits are known as *hardware*. Circuit is often abbreviated *cct*.

Hardware is the tangible structure which makes up the circuit. It is what one sees when one looks at a circuit, what one holds in one's hand, what has been put together. It is usually made up of "hard"—though little—things; therefore its name. (And the way they are put together is, of course, the *circuit format*.)

NOTE: Computer people differentiate between *hardware* and *software;* by the latter they mean the computer programs which are applied to computer circuits (Section 3.3.3). The term *software* presumably was invented to illustrate that there is nothing "hard" about it—nothing one can see or hold. Software is no more hard than a set of words and numbers printed on paper or recorded on tape is hard.

1. Chassis

A *chassis* is the metal frame on which circuit components and connections are mounted.

It is usually in the form of a metal box and is made of sheet aluminum or galvanized iron, although copper chassis have been used.

NOTE: *Chassis* comes from the Latin word for "box." The plural of chassis is *chassis.*

Circuit components and connections and wires are mounted to the chassis by either soldering metal leads to the chassis surface, or screwing them down with screws in threaded holes in the chassis, or clipping, or bolting. Any way to attach that makes good mechanical contact is acceptable.

- Soldering, screwing down, clipping, or bolting metal leads to metal chassis achieves good electric as well as good mechanical contact. Then the chassis becomes part of the circuit. It is usually then the *ground* connection (see "Ground" in Section 5.1.3).
- An electrically insulating pad can be inserted during the mechanical attachment so that while mechanical attachment is achieved, electrical isolation is maintained. Such connections are electrically either *above ground* or *below ground.*

Chassis construction formats are most frequently encountered in consumer products such as television sets, hi-fi systems, etc. Figure 5-25 illustrates a typical chassis with an array of discrete components mechanically attached.

2. Printed Circuits

The *printed circuit* is a format in which a flat insulating board (usually made of a plastic–glass fiber composition such as Fiberglas) is imprinted with a pattern of metal lines which become the connections, the conductors, between circuit components, which are added later in a separate soldering operation.

FIG. 5-25 A circuit in chassis format.

a. Photographic printing process The printing process is photographic in nature.

The insulating board is made with a continuous metal slab glued (laminated) to it. Then a liquid photographic film is flowed over it, and this hardens. Then when the film is exposed to the negative of the desired wiring pattern, using light through a mask (exactly as in making integrated circuits—see "Lithographic Techniques" in Section 5.3.3), and developed, some of the film remains (in the desired wiring pattern) and the rest of the film is gone—developed away. Chemical etching can now be done to dissolve away the unwanted metal (it is the portion no longer protected by photographic film after development). What is left is the desired pattern—in metal.

b. "All-at-once" assembly process The advantage of printed circuits over chassis circuits is that an assembler does not have to handle and attach dozens or hundreds of wires to designated positions. Furthermore, because of the nature of the photographic process, there is almost no limit to the fineness of the wire pattern that may be imprinted.

The wires all appear at once, printed like a photographic print. Also printed circuits can be made smaller than chassis circuits—a lot more electronic capability can be crowded into a smaller space.

Another advantage of the printed-circuit format is that once discrete components are positioned over the wires that pertain to them, the whole assembly may be dipped into molten solder (in processes known as *solder flow* and *wave soldering*) so that all connections are made at one time, automatically, without the need for assembly workers' handling each piece separately. Thus printed circuits tend to be less expensive than chassis circuits.

Printed-circuit formats are most frequently encountered in small consumer products like digital watches and transistor radios. But they also are found in the most sophisticated electronic systems like radar or computers. Figure 5-26 illustrates a typical printed circuit with an array of discrete components attached by flow soldering.

c. Strip line A special form of printed circuit is made of a thin plastic strip, usually glued or otherwise attached to a chassis surface. It is called a *strip line.*

Its use is confined to microwave frequencies.

d. Printed circuit cards A variation of the printed circuit is the printed circuit *card.* This is constructed to have a set of electric contacts or terminations at one edge.

All input and output signals are located at these terminations as well as all required power connections. Once every contact is made, the board will function.

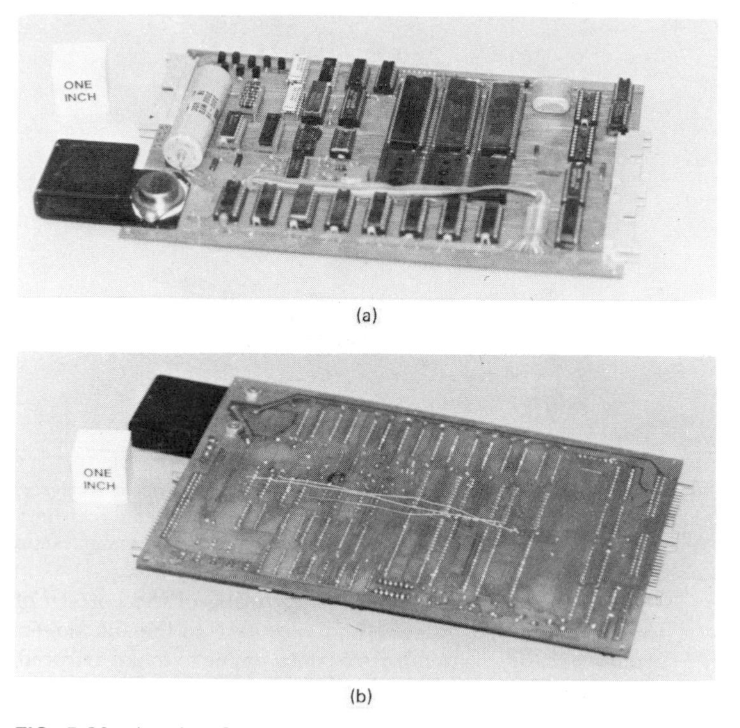

FIG. 5-26 A printed circuit: (*a*) Top side, showing many chips and other components, (*b*) reverse side, showing solder flow connections.

Printed circuit cards are indeed made like cards—very often a set of printed circuit cards resembles a deck of playing cards.

The advantage of the printed circuit card is that in the event of a malfunction, the repair technician can fix the problem without laboriously tracing which component failed so as to replace it. Rather, the whole card can be replaced, a much simpler and faster procedure.

Although this may appear wasteful—since so many good components are replaced too—it is much more economical. When components are integrated and wires are printed, it is cheaper to replace a whole card than to pay for hours of the technician's time.

Figure 5-27 illustrates a typical printed circuit card. Note the array of terminations at the bottom edge.

3. Hybrid Circuits

Hybrid circuits are very tiny printed circuits.

The method of imprinting is different from the photographic one used in printed circuits and the substrate is different; it is a postage-stamp-size slab of thin ceramic rather than a slice of plastic–glass fiber composition;

FIG. 5-27 Printed circuit card.

also, there is no slab of metal bonded to it to be selectively etched away. But the final structure is the same as a printed circuit in that an entire array of connection wires is produced in a single manufacturing step.

a. Silk screen printing Imprinting of the pattern of connection wires is achieved by a method very akin to the silk screening of decorative patterns onto fabrics or onto paper to make art products such as silk screen prints.

A *paste* or *ink* is squeegeed through a porous pattern onto the *substrate* so that the resulting deposited pattern matches that of the porous pattern. When the entire structure is placed into a furnace and heated, the paste or ink melts and bonds to the substrate. Since the paste is made of gold or copper powders, what results is a network of "wires"—the desired metal connections.

NOTE: To distinguish this silk-screening process from that in which metal patterns are deposited by evaporation, it is given the name *thick film* processing. The substrate is usually made of ceramic (rather than of Fiberglas as in printed circuits) so that it may survive the heating.

b. Die attach As a second step, tiny electronic components are glued or soldered, one by one, to the correct locations on the wire pattern. These locations are a little wider than the connection lines and are known as *pads*. Wave soldering cannot be used as it is in printed circuits because the components are so tiny that the wave of molten solder would wash them away. The process is known as *die attach*.

The components may be any of those discussed in this chapter, such as tiny resistors or capacitors, but most commonly they are semiconductor elements like diodes, transistors, or integrated circuits of the ordinary type or of the LSI or VLSI types (section 5.3.3). The fact that both kinds of components—semiconductors and nonsemiconductors—are used

FIG. 5-28 Hybrid circuit chip. *(Photograph courtesy of Raytheon, Lexington, Mass.)*

together in a hybrid circuit is the reason for the name *hybrid.* It distinguishes such circuits from integrated circuits, which are entirely made of semiconductor material.

c. Packaging As a final step, hybrid circuits are enclosed in a hermetically sealed package.

The circuit is then about the same size and appearance as a semiconductor integrated circuit package, sometimes known as a *chip,* and is therefore often called a *hybrid circuit chip.*

NOTE: Systems using hybrid chips are called *microelectronic systems,* and the chips are often called *microelectronic chips.*

Figure 5-28 illustrates a hybrid circuit against a background of the same circuit analyzed.

4. Cables and Interconnects

a. Cables These are the metal connection wires that lead from one circuit to another in a larger electronic system, one that has many circuit subsystems.

The only thing distinctive about them is that they come in different formats to accommodate the different circuit formats.

b. Plugs, jacks, interconnects, terminals At the "ends" of cables are *plugs* (which fit into *sockets*). Such plugs are also called *jacks* (see Figure 5-29).

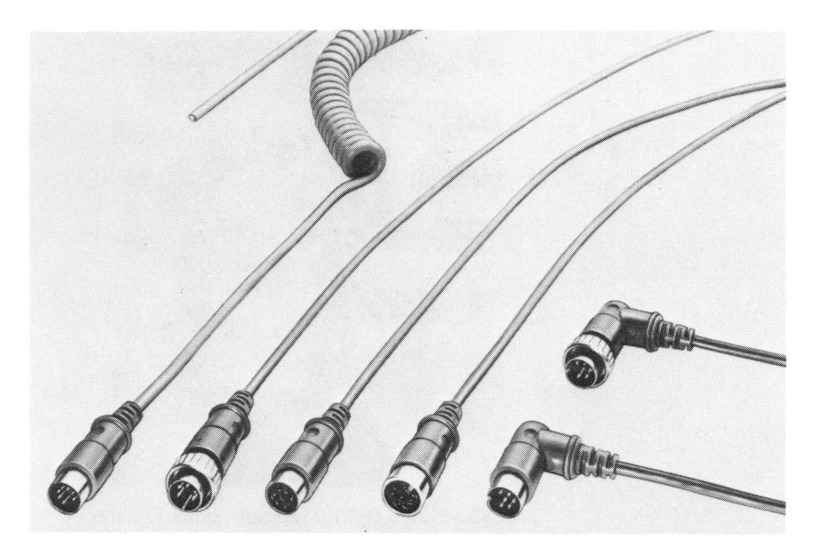

FIG. 5-29 Jacks. *(Photograph courtesy of Switchcraft, Inc., Chicago, Ill.)*

Plugs or jacks terminate the wire portions of the cables and are designed to ease the operation of opening and closing (electrically) a large set of contacts with one motion, rather than each contact pair, one at a time. These pairs of connection elements (plugs and sockets) are also known as *interconnects,* as *disconnects,* and as *circuit terminals.* They come in different shapes to accommodate different circuit formats.

NOTE: The array of terminals imprinted at one edge of a printed circuit card and the matching set in the main frame of the electronic equipment into which the cards may be inserted are an example of such a pair of interconnects or terminals with a capability for easily opening and closing a large set of connections with a single motion (as when one card is replaced by another).

5. Fiber Optic Circuits

A technology that may be viewed as a variation and a complication of what has been described in this chapter, namely, different circuit components and circuit formats, is *fiber optics.* It shares so much with ordinary electronic circuitry in terms of using many of the same elements and formats that it is necessary for our purposes only to show where it differs.

a. Light instead of electricity The primary difference in principle is that variations in light intensity rather than variations in electric current intensity constitute the signal.

b. Glass fibers instead of metal wire This leads to the primary difference in physical structure. Strands of fine glass fibers, finer than a human hair (not metal wire carrying electric signals), carry the light signals from a faraway place to the circuit.

It has been shown (Section 4.5.1) that light will travel in transparent media like glass. In fiber optics, glass is drawn into very fine filaments, and those are used in exactly the same way that fine filaments of wire are used to conduct electrons in electronic circuits. The fine filaments of glass conduct the light. The glass is of such a sort that light can come in and emerge only at the ends. None can escape from the side skin of the fiber.

Since glass fibers are light conductors, we can think of them as though they were glass "wires," as an analogue of metal wires which conduct electrons.

c. Final circuit functions are electrical Once the light enters the circuit (through the glass fibers) it is converted to electric currents through transducers (see "Transducers and Sensors" in Section 5.3.2), and from that point on the circuit is an electric circuit, and all that has been said about electric circuits pertains.

NOTE: Fiber optics is finding its largest application in telephone lines, where lightweight glass fibers substitute for heavy copper cables.

d. Optical cables The array of optical fibers that conduct optical signals between circuits is known as a *fiber optic cable.*

e. Optical interconnects There are plugs and sockets for optical cables and the circuits they serve that serve the same function as electric plugs and sockets. They are known as *fiber optic interconnects* or *fiber optic terminals.*

They function by matching the ends of one set of fibers with those of another set, so light signals may come out the ends of one set and be picked up by the ends of the other set.

5.3.5 Circuit Functions

The preceeding sections of this chapter describe how electrons move through materials, how electric energy is controlled, the individual roles of individual components, and how these components are placed into different circuit constructions or formats. We should think of these concepts as the bases of building blocks.

We are now in a position to describe how these building blocks can be put together in different combinations to perform a variety of useful tasks—*circuit functions*—which result from these combinations.

1. Networks

It has been pointed out that circuits are paths, often complex, made mostly of metal wires (or imprinted metal wires), down which electrons

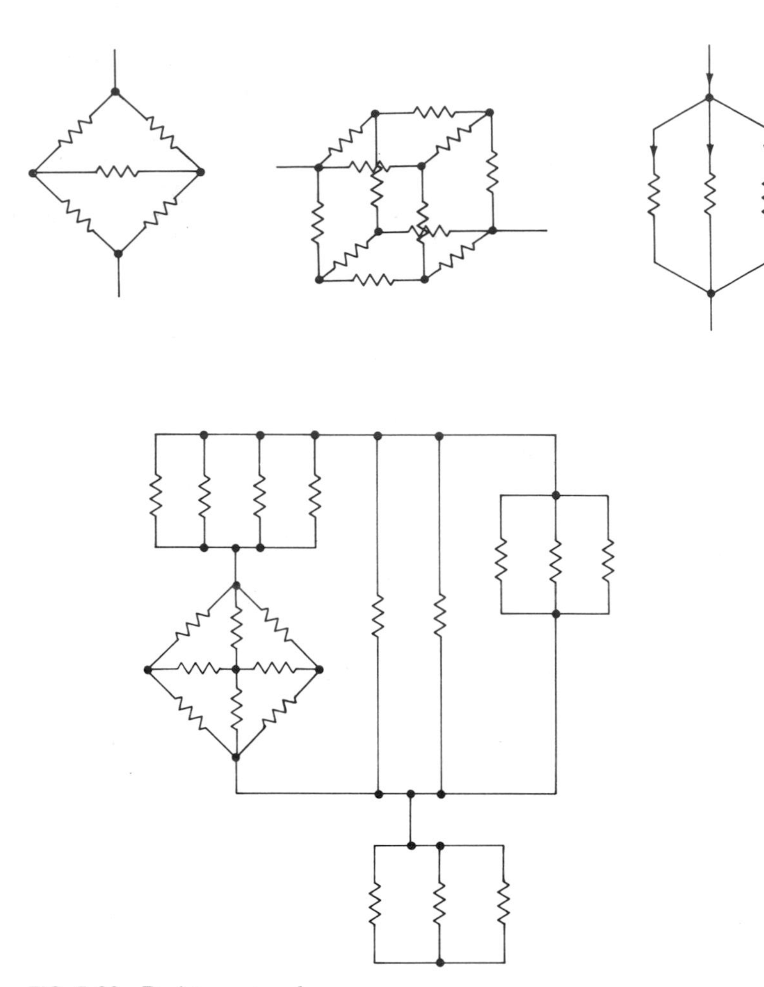

FIG. 5-30 Resistor networks.

are sent. These paths are called *networks*. Figure 5-30 shows some resistor networks.

> A complex network consisting of several subnetworks is called a *mesh*.
>
> There is a separate field of study for electrical engineers called *network theory*.

a. Circuit diagrams Electric paths are like any paths. They are merely channels, roads, or tracks down which it is easier to move than through regions that are not laid out in such paths. A road map is a guide to the location of such a network of paths. The electrical equiva-

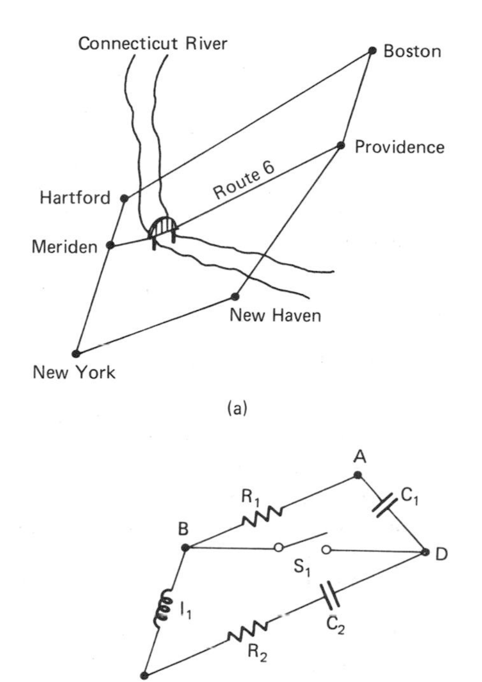

FIG. 5-31 Series and parallel: (*a*) Boston to New York, (*b*) A to C.

lent of the road map is the *circuit diagram.* It includes paths for electrons and the electric components they must pass through.

Figure 5-31*b* illustrates a typical circuit diagram. Drawing such diagrams is discussed in Section 5.8.

b. Series and parallel Just as a road map shows different optional ways to travel from one point to another, so a circuit diagram shows different optional ways for an electric current to go from one point to another.

A road map will provide the information that we can drive from Boston to New York through Hartford by going in two steps: first from Boston to Hartford, then from Hartford to New York (Figure 5-31*a*).

In the same way, the circuit diagram of Figure 5-31*b* shows that electrons can get from point A to point C by way of point B in two steps: first from A to B through the resistor R_1, and then from B to C through the inductor I_1.

- These are *series* paths. A series path is one in which the traveling objects move from one position to another through segments that are positioned end-to-end. In the circuit diagram, *A* to *B* to *C* is a *series circuit* (as Boston to Hartford to New York is a series route).

We could decide to go by another route. We could use the series route from Boston to Providence and then from Providence to New York (passing through the rush-hour slowdown in New Haven). In the same way, the electrons in the circuit shown in Figure 5-31*b* could proceed in a series circuit getting from *A* to *C* by way of *D*, passing through the capacitors C_1 and C_2 and resistor R_2 on the way.

NOTE: The analogy with the auto trip applies, since in a way, the capacitor C_2 is an impediment to flow as is New Haven traffic.

- These are called *parallel* paths. Parallel paths are optional paths: either can be taken. Or if there are many vehicles or many electrons traveling the same route, the traffic or current can divide. Some vehicles or electrons can go down one path and some down the other. But in the end, since both paths come out at the same place, both groups arrive at the same end point (destination). In the circuit diagram, *A* to *D* to *C* is a parallel circuit to the optional circuit (or pathway) *A* to *B* to *C*. Both parallel paths connect *A* with *C*. In any circuit diagram there can be any number of parallel circuits between any two points.

c. Loops and bypasses Once in New York—and having completed our sojourn there—we return to Boston. We might have gone by way of Hartford and returned by way of Providence. In the same way, the electrons might have gone from *A* to *C* by way of *B* and returned by way of *D*.

- This is called a *loop* path. A loop path is one that goes one way and returns another way—without retracing its steps. It makes a "circle" (or circuit). In the circuit diagram of Figure 5-31*b*, *A* to *B* to *C* to *D* to *A* is a *loop circuit*.

We could be heading toward New York on the Providence road and realize that would be a longer drive than the Hartford road. We might decide to change routes. The road map shows that we could turn off at Route 6 and rejoin the other road at Meriden. That is a shortcut. In the circuit diagram there is a similar shortcut—a *short circuit*—that may be taken by electrons between *D* and *B*. This is also known as a *circuit bypass*. (Indeed Route 6 is a road bypass.)

NOTE: We might regret having taken Route 6. A bridge over the Connecticut River just east of Meriden might have been washed out in last night's heavy rain. We would be stuck. We could not go on. We would have to either stop right there for the night or turn back. In the electric circuit there is a switch S_1. If it is open, the electrons trying to get from *D* to *B* face the same dilemma. They confront an *open circuit* (which is the opposite of a short circuit).

2. Stages

Electronic circuits are divided into *stages*.

There can be any number of stages, from a single stage in a simple circuit to dozens or hundreds of stages in complex circuits.

a. Definition A circuit *stage* is a portion of a circuit that performs an active function as distinguished from a passive one (see "Passive Components, Active Components" in Section 5.3.1).

That is, it contributes energy from another source in the circuit rather than merely modifying the flow of electrons as a resistor or a capacitor alone would do.

b. One active component per stage Every circuit stage contains one active component such as a vacuum tube or a semiconductor diode or transistor.

c. Associated passive components Every circuit stage contains all the passive components that "go with" the active component.

These passive components are required in order for the active component to perform its unique function. Their usefulness is related primarily to that stage and only secondarily if at all to other stages.

d. One function per stage Every stage performs a unique function.

The functions of rectification, amplification, and oscillation were described in Section 5.3.3. There are many dozens of other such functions, any one of which might be performed by a stage.

e. Each stage an ingredient of a larger system No function (no stage) has value by itself (except in a simple single-stage circuit). It is valuable only insofar as it forms part of an *electronic system.*

In today's world we are surrounded by electronic systems. We depend on them to entertain us with radios, television sets, and stereo systems, to reproduce what we write with word processors, to stimulate faulty hearts with pacemakers, to perform complex calculations with computers, and so on.

The important thing to keep in mind about electronic systems is that each system is but the sum of the stages that constitute it.

The stages of an electronic system may be compared to the parts of an automobile engine. There are the engine block, the pistons, the battery, the ignition, the radiator, and so on. These are in effect the "stages" of an automobile system. The automobile is but the sum of the stages that constitute it.

Every one of these parts is useless by itself. It is only when they are combined in the automobile system that the engine will run and the vehicle will move.

3. The Depiction of Circuits

Circuits are depicted in circuit diagrams and in block diagrams.

a. Circuit diagrams These diagrams are arrays of symbols which stand for circuit components—both active and passive. The symbols are connected to each other by straight lines which represent wire or metal imprinted connections (Figure 5-32*a*).

All the most common symbols have been described and are summarized in Section 5.8. For a more complete list, see Glossary J.

- Circuit diagrams are linear representations of the course of electric currents and in that sense tend to proceed across the printed page like text—from left to right.

There are exceptions in that some circuits have paths that double back in loops (see "Loops and Bypasses" in Section 5.3.5).

- Stages may be indicated by boldface numerals (Figure 5-32*b*).
- Individual components may be labeled with arbitrary numerical or letter designations, or with their manufacturers' type numbers, or with their values in pertinent units, or some combination of type numbers and values.

NOTE: Stages may not be marked at all, with the supposition that the engineer simply knows where each stage begins and ends.

b. Block diagrams These diagrams are shorthand ways of showing stages. A stage is represented by a box with the name of the stage function printed in it, or only a numeral may be employed.

Block diagrams are made with the assumption that the reader understands so well what circuit components make up that stage that it is unnecessary to detail them with individual symbols.

Sometimes a large number of circuit functions are represented by a single block. This is common when integrated circuit chips are represented.

Figure 5-32 shows a single electronic system, for an audio amplifier, both ways—with a detailed circuit diagram with every component depicted, and with a block diagram.

5.4 MAGNETISM

Magnets are part of our everyday experience. They are materials that exert a force which emanates from the material itself.

The horseshoe magnet of Figure 5-33 attracts bits of steel. The magnet in the form of a small slug holds notes and reminders on our refrigerator door—because it is attracted to the steel of which that door is made. In fact, the magnet itself may be made of iron.

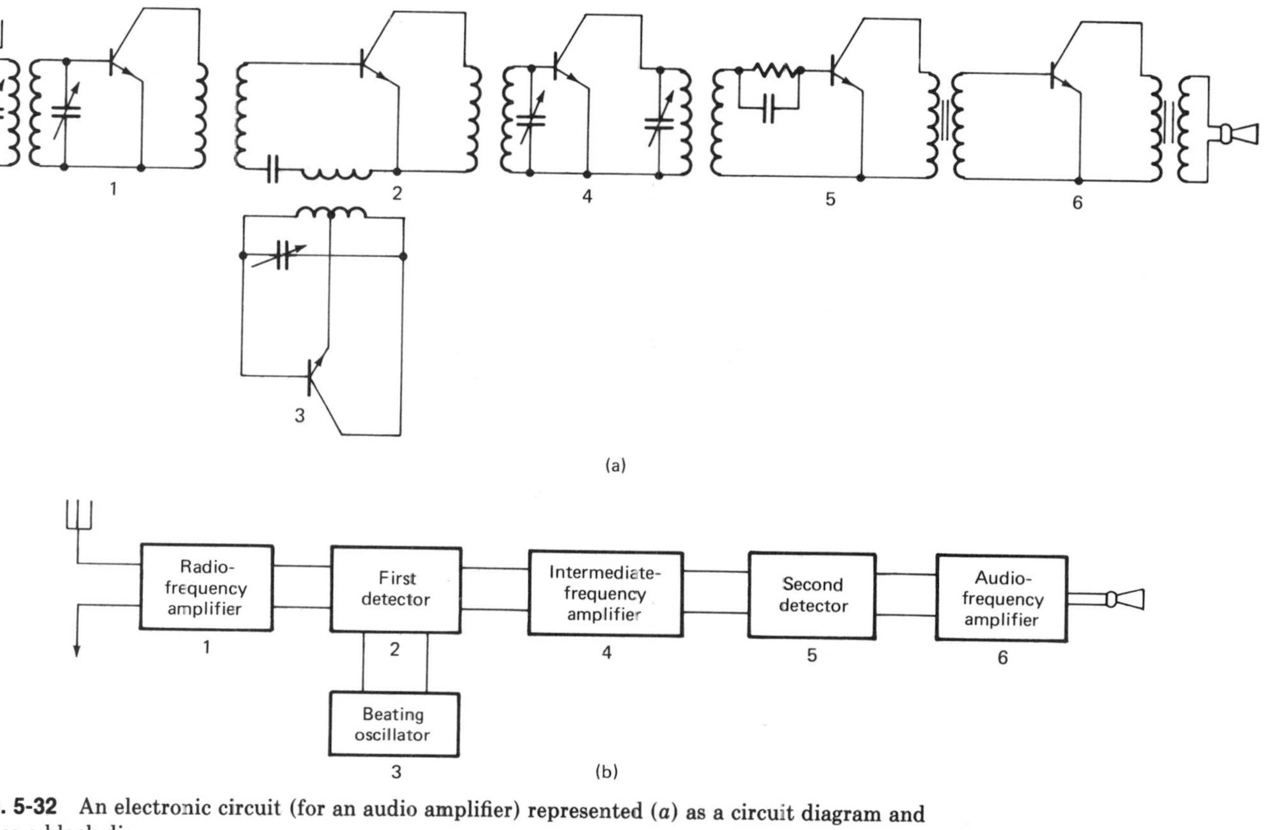

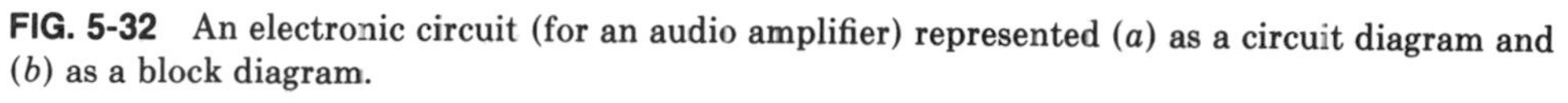

FIG. 5-32 An electronic circuit (for an audio amplifier) represented (*a*) as a circuit diagram and (*b*) as a block diagram.

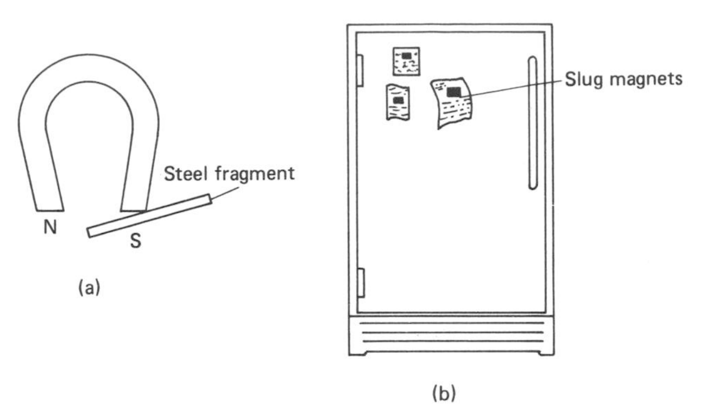

FIG. 5-33 Magnets: (*a*) Horseshoe, (*b*) slug.

Many electric motors contain magnets.

We may ask two questions. What is there about the metal iron (of which steel is made) that makes it capable of becoming magnetic? What is there about magnetism that relates it to electricity?

This section will show that it is the electrical properties of the iron atom (and certain others like it) that make it magnetic. It will also show that magnetism is as much an electrical phenomenon as the flow of electrons down a wire.

5.4.1 Fields and Forces Caused by Electric Currents

When a current of electrons flows down a wire, it carries a field of magnetic force along with it.

- A *field* is a region in space in which a certain kind of force is present and in which that force is pointing in a certain direction. It is also called a *flux*.
- The *force*, as it applies here, is mechanical—a push or a pull.

1. The Left-Hand Rule

The magnetic field associated with an electric current has a particular geometrical relation to the direction of flow of that current. The magnetic field forms a sheath around the flowing electrons, spiraling around them like a thread on a screw. But it spirals in just one direction—clockwise. This is described by the *left-hand rule* (see also Section 5.5.1).

If the current is considered to be flowing in the direction of an extended left thumb, then the magnetic field moves in a spiral in the direction of the fingers of the left hand and at a right angle to the direction of the current. (Figure 5-34).

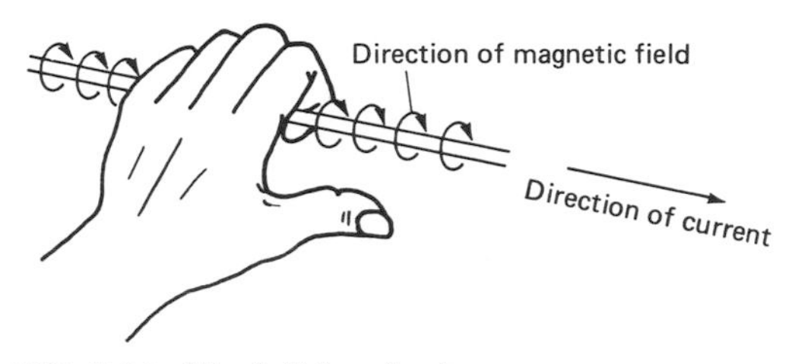

FIG. 5-34 The left-hand rule.

2. Coils and Electromagnets

Arranging a wire into a *coil* shape is a way to increase the amount of magnetic force in the direction of the line at the center of the coil. Such a coil is an *electromagnet.*

When a wire that carries an electric current is formed into the shape of a coil, the magnetic forces around each turn of the coil find themselves all pointing in the same direction. These forces add up; consequently they reinforce each other and cause a strong magnetic field to appear in that direction (see Figure 5-11*b*).

a. The solenoid This is the basis of the circuit component the *solenoid,* which can become a magnet—or not a magnet—on the basis of whether the current in the wire of its coil is turned on or off.

Solenoids are used to activate switches such as the ones found in washing machines to turn water on and off. The magnetic field is high enough to "pull the plug" or shut off the water in the pipe as required by the washing cycle.

b. Motors Electromagnets are also the basis of motors (Section 5.5.4). The coils in a motor are arranged so that the magnetic forces cause the central portion of the motor structure to rotate.

3. Electric Currents in Atoms Which Cause Magnetic Fields

The electrons that move in circular orbitals (see "Atomic Structure" in Section 4.8.1) around the nuclei of atoms are a form of electron current—anytime electrons move, that movement is defined as an electric current (Section 5.1.4).

a. Electron spin The rotation of electrons in an atom is called *electron spin.* This spin generates a tiny magnetic field around each electron. In elements that have an even number of electrons, the fields around those spinning in one direction are canceled by the fields around those spinning in the opposite direction. In elements with an odd number of electrons, more electrons are spinning in one direction than in another, and the atom as a whole generates a magnetic field. When the polar directions of these atomic magnets are distributed in random fashion throughout a material, the material as a whole will not display magnetic properties. It is in the so-called *ferromagnetic metals* (iron, cobalt, and nickel) that the fields around the individual atoms are well enough lined up to reinforce one another, and the substance as a whole generates a magnetic field. This is somewhat similar to the reinforcement that occurs for neighboring turns of a coil carrying electrons (see "Coils" in Section 5.3.2).

b. Crystal effects Crystals are arrangements of atoms in which the same geometric spatial relationship (such as the center and corners of a cube) is repeated continuously (Section 4.7.1). Within metal crystals there are sometimes regions called *magnetic domains* in which all the atoms have magnetic fields that are aligned in the same direction. The individual fields of the billions of tiny atomic magnets reinforce one another to create an overall magnetic field in one domain of the crystal. The extent to which the poles of the various magnetic domains in a crystal are aligned with each other determines how strong a magnet the entire crystal is. Some crystals, whose magnetic domains are only partially aligned, exhibit *partial magnetism.*

One end of the magnetic domain is always a north pole and the other is always a south pole (Section 5.4.3). This means that each domain is a tiny magnet. A magnet behaves as a summation of all the little magnets that constitute it (Figure 5-35).

If the domains are all pointing in the same direction, the magnet has greater magnetic strength (greater magnetic pull and push forces—see "Forces of Attraction and Repulsion" in Section 5.4.3) than if the domains are arranged in a random jumble of directions.

One way to line up domains is to stretch the metal by pulling it. That action lines up the domains mechanically and increases the magnetic forces in the material. Mechanical "working" is often done to certain materials to enhance their magnetic properties.

c. Electron spin added to crystal effects When most of the outer-orbital electrons spin in the same direction and the domains (which contain the atoms that have those spinning electrons) are also all lined up in the same direction, a state is created that favors magnetization.

This condition occurs in iron to a higher degree than in other metals, which is why iron tends to be a better magnetic material than other metals.

Nonmagnetic materials are nonmagnetic because they lack *both these qualities* (lacking one alone is not enough).

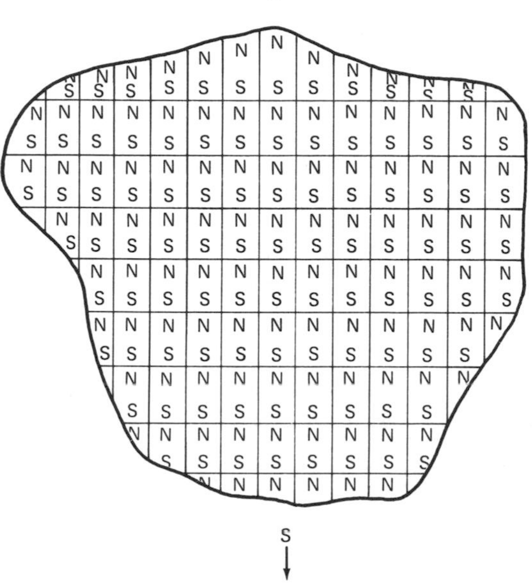

FIG. 5-35 Domains in a magnetic crystal.

A crystal is shown with the random shape that would occur in a material after casting (the shape being determined by the fact that adjacent crystals are pressing on it), but the atoms within the crystal are arranged regularly. Since the crystal is magnetic, it is subdivided into domains, all lined up in the same direction (with their north and south poles pointing the same way).

- The spins of electrons in their outer orbitals are not lined up with each other and are thus not reinforcing. The spins are random.
- The materials are not crystals. Thus they contain no magnetic domains that can line up to reinforce each other.

5.4.2 Magnetic Materials

There are only a few materials whose atoms have exactly the right combination of electron spins and crystal arrangements (in domains) to result in strong magnetism. They are

- Iron (and most steel alloys—made from iron)
- Nickel (and many of its alloys)
- Cobalt (and many of its alloys)
- Alnico (the trade name of an alloy of iron, nickel, and cobalt)

- Manganese
- Magnetite (a mineral or iron oxide—also called *lodestone*)
- Ferrites (a class of materials consisting of magnetic iron oxides containing other metals, such as nickel and manganese, in their crystal structure.

Materials have been categorized in accordance with their relative magnetic strengths and have been given class names: *ferromagnetic, paramagnetic, diamagnetic,* and *nonmagnetic.*

1. Ferromagnetic Materials

The alloys of iron are the most magnetic of materials. Since the prefix "ferro-" means "iron," these alloys are called *ferromagnetic.*

Ferromagnetic materials are called *permanent magnets* to distinguish them from electromagnets, which behave like magnets only while electric current flows in their coils.

2. Paramagnetic Materials

Materials that are weaker magnets than ferromagnetic materials—such as those based on manganese—are called *paramagnetic.*

3. Nonmagnetic Materials

All other materials in the world—neither ferromagnetic nor paramagnetic—have no magnetic qualities and are *nonmagnetic.*

NOTE: Diamagnetic materials are materials that are weakly magnetic but behave in a characteristic manner: rather than align themselve with the north-south direction of another magnet, as most magnetic materials do, they point themselves at right angles to that direction. Such materials are called *diamagnetic.* Almost all materials, even those considered nonmagnetic, show some diamagnetic behavior; but the effect is so small for most materials that it is not significant.

5.4.3 Magnetic Forces

Magnetic fields always exist together with mechanical force fields. It is this fact that distinguishes them from other (nonmagnetic) materials.

1. Magnetic Poles

All magnets have *poles.* These are merely the physical ends or extremities of the magnet. But it is there that magnetic forces concentrate, going in opposite directions at each end.

For every magnet there is always a *north pole* and a *south pole.*

These are just arbitrary names. They emphasize the fact that magnets are always double-ended.

NOTE: The earth is a magnet (which means that there is some sort of electron spin lineup and lineup of magnetic domains along one diameter). Thus its north and south poles are *magnetic poles,* just like those of any other magnet.

This resulted in the invention of the *compass* about 900 years ago. (When Columbus ventured across the sea, it made him more confident about so venturing, because compasses had come into general use in his day.) It consisted of a needle of the mineral lodestone. It "wanted"—as all magnets do—to line up with the magnetic field of any adjacent magnet. In those days there was only one magnet around which had perceptible magnetic behavior that wasn't another piece of lodestone. It was the earth. So navigators could easily determine which was north.

2. Forces of Attraction and Repulsion

Magnetic forces are lined up so that they "push" at one end of the magnetic structure and "pull" at the other end.

It is these ends that are called *poles* and assigned the names *north* and *south.*

NOTE: The tiny magnetic domains in magnets are each complete magnets with a north and a south pole for each (Figure 5-35).

- **If a pulling pole and pushing pole come close to each other, the pull of one will reinforce the push of the other, and they will move toward each other. They will show *attraction.***
- **If two pulling poles come close to each other, these forces will not reinforce. In effect, they will pull apart. They will show *repulsion.***
- **Repulsion also occurs when two pushing poles come close to each other. They will push apart.**

In magnetism, similar poles repel each other and opposite poles attract each other.

NOTE: If a magnet were to be cut into many pieces, each piece would then become a magnet in its own right with its own poles (Figure 5-36). Then oppo-

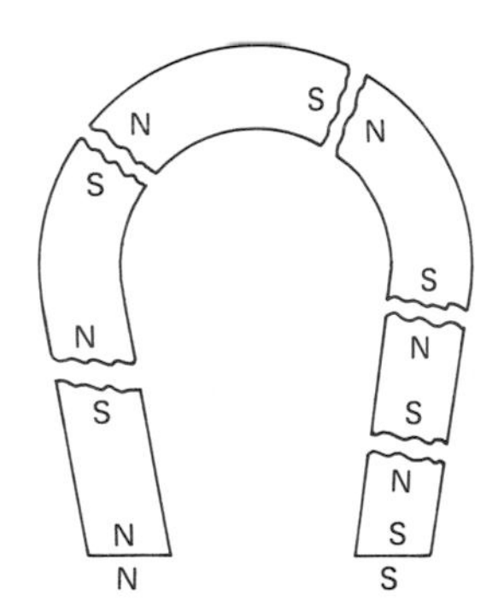

FIG. 5-36 Pieces of magnets are magnets.

site poles would face each other, and the fragments would try to rejoin with each other as they had been before cutting.

3. Magnetization and Demagnetization

Magnetic materials are not always as strong magnets as they might be. We can make them stronger by *magnetizing* them.

The process of magnetizing is the placing of the material into a strong magnetic field (created by another magnet). What happens is that those magnetic domains inside the material which are lined up randomly rotate so that most of them line up with the imposed magnetic field. Once they are so lined up, they reinforce each other's magnetism (they add up their magnetism) so that the resulting push and pull exerted at the north and south poles of the material is increased over what it had been before the domains lined up. We then say that the material has been *magnetized.*

If the magnetized material is subjected to a randomly moving magnetic field—one that keeps changing direction—it is possible to randomize the lineup of the domains again. Then the magnetic force at its poles is decreased. The material has been *demagnetized.*

There are a number of terms that relate to the way different materials react to being magnetized and demagnetized.

a. Susceptibility The more easily a material can be magnetized or demagnetized the more *susceptible* it is said to be. This is designated as the magnetic *susceptibility.*

b. Reluctance The difficulty of demagnetizing an already magnetized material is measured by its magnetic *reluctance.*

c. Coercive force The amount of opposing magnetic field required to demagnetize a magnet is called the *coercive force.*

d. Retentivity The amount of magnetic flux remaining in a magnet after it has been subjected to demagnetization is called its *retentivity.*

e. Hysteresis When magnets are subjected to alternating magnetizing and demagnetizing, some of the electric energy used goes into heat rather than into making or unmaking domain alignment. That heat loss is the result of an effect known as *hysteresis.*

Hysteresis is the lack of instant response of a domain to a reversal of direction of a magnetic field. The domain wants to turn instantly as the field turns—responding to it—but it is sluggish. It turns, but only part way. The energy that would have been used up in turning the rest of the way remains in the domain, and heats it up instead.

TABLE 5-3
MAGNETIC UNITS

Parameter	*Symbol Used in Equations*	*Unit*
Coercivity	H_c	oersted
Magnetic field intensity	H	oersted
Magnetic flux	Φ	maxwell; weber
Magnetic flux density	B	gauss
Magnetomotive force	F	gilbert; amp-turn
Permeability (magnetic)	μ	gauss per oersted; henry per meter
Permeance	P	maxwell per gilbert; henry
Reluctance	R	gilbert per maxwell; amp-turn per weber
Remanence	B_r	gauss
Susceptibility (magnetic)	χ_m	dimensionless

5.4.4 Magnetic Units, Other Terms of Magnetism

Table 5-3 lists the more common magnetic units.

5.5 GENERATORS AND MOTORS

Electric currents have the effect of causing magnetic fields and associated mechanical forces (Sections 5.4.1 and 5.4.3). The fact is, these three phenomena are interrelated, and—given the right conditions—the cause and effect reactions can all be reversed as well. That means that mechanical forces can cause electric currents or can cause magnetic fields. We can then make use of this three-way relation to generate electric power mechanically, as in power stations, and mechanical power electrically, as in motors.

For example, the mechanical movement of water over Niagara Falls supplies much of New York State and much of the adjoining region of Canada with electric power. (If you have ever been to Niagara Falls, you must remember that the roads that approach it are lined with the most massive array of towers and power lines in the American northeast.) The falling water turns enormous waterwheels (*turbines*) which, as they turn, pass magnets and coils near each other, so that the magnetic fields cause currents in the coils. It is these currents that are led through these power lines from the falls, supplying the region with electric power.

The power proceeds to factories and homes (by way of those power lines—which of course are just long wire conductors; see "Conductors" in Section 5.1.1). It then enters mechanisms which look like miniatures of the turbines in Niagara Falls. They are just that. They differ in that they are run backward. Electric current goes in, and out comes mechanical movement. They are *motors*.

5.5.1 The Left-Hand Rule (Three-Digit Version)

The three-way relation between the directions of electric current, magnetic field, and mechanical force (or *thrust*) may be shown by using the left hand.

It is important to electrical engineers to know the directions of all these phenomena; for example, the engineer wants to control whether a motor will rotate in a clockwise or a counterclockwise direction. So don't be surprised to see such a person looking reflectively at the extended fingers of his or her left hand.

If the thumb, first, and middle fingers of the left hand are held such in a way that each digit is at right angles to the other two, the thumb will point in the direction of the mechanical force, the first finger will point in the direction of the magnetic field and the middle finger will point in the direction of the current flow (Figure 5-37).

NOTE: The hand has been used already, in another left-hand rule, to show the spiral form of a magnetic field around a current in a wire (see "The Left-Hand Rule" in Section 5.4.1.) (And yes, there is a right-hand rule too, which shows similar relations, but which will not be described here.)

5.5.2 The Electric Generator (the Dynamo)

A *dynamo* is a device which converts mechanical motion into electric energy. It produces alternating current (ac). The turbine at Niagara Falls is a dynamo.

NOTE: The word *dynamo* is considered somewhat archaic and is not frequently used today. The term *electric generator* is more common.

5.5.3 The Parts of an Electric Generator

All electric generators consist of two major elements: a stationary structure, the stator, which supplies a magnetic field and a rotating structure, the armature, which contains metal wire wound into a circular form of a coil. This functions like the ordinary coils already described in that it conducts current in such a way that there is an interrelation between that current and associated magnetic fields and mechanical force. Additionally, there are elements which conduct current into or out of the armature. These are the brushes and associated cables.

1. The Stationary Structure—The Stator

The stationary structure of a generator is a magnet. It is known as the *stator*. For large generators it may be massive and may weigh tons. Like

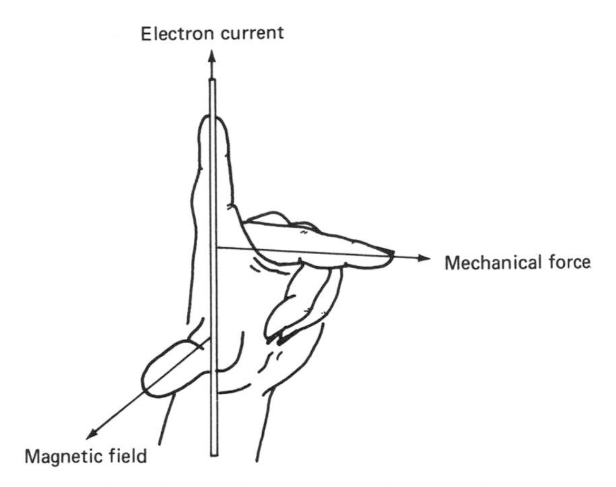

FIG. 5-37 The left-hand rule for the directions of current, magnetic field, and mechanical force.

all magnets, it has north and south poles. Parts of this structure are shown in Figure 5-38. They include

- Field poles (north and south)

  NOTE: Although Figure 5-38 shows only one pair of poles, there may be more.

- Field yoke (the portion of the magnet between the poles; it also serves as a mechanical structure)

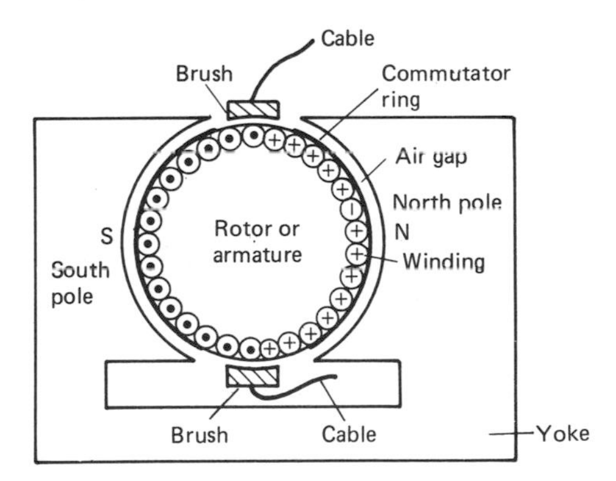

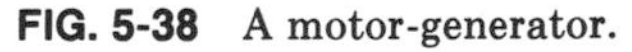

FIG. 5-38 A motor-generator.

NOTE: The coil winding is shown according to the convention that the symbol ⊕ means that the direction of current is into the page and the symbol ⊙ means that the direction of current is out of the page.

2. The Rotating Structure—the Armature

The part of the generator which moves is placed centrally within the stationary magnetic structure. It moves on its own axis. It is called the *armature*. It is essentially an array of wire coils arranged in a circular form around the armature axis. The parts of the armature include

- Coils
- Axle (supported on mechanical bearings)
- Commutator rings

The armature is also known as the *rotor*.

3. Brushes

As it is generated, current must be conducted out of the coils of the armature. It is impossible to have a permanent solid connection to something that is rotating. The best that can be done is a "touching" or "wiping" (or *brushing*) contact. The parts that perform this touching are called *brushes*.

Brushes touch a "slippery" set of rings on the armature known as *commutator rings*. The brushes are kept in contact with the commutator rings by mechanical springs.

Brushes are connected to *cables* through which the generated current is conducted.

4. Air Gaps

We do not normally consider empty space as a part. However, between the magnet structure and the armature of generators, there must be air spaces in which magnetic fields operate. These air spaces truly influence the way in which the generator operates and are part of the design of the machine. They are known as *air gaps*, or simply as *gaps*.

5.5.4 Motors

A *motor* is nothing but an electric generator (a dynamo) run in reverse. Thus all the descriptive material in the previous section pertains, including the names and forms of the parts of the generator. They are the same for motors. Therefore, Figure 5-38 pertains equally to motors and generators.

Running a generator in reverse, so that it is a motor, means connecting a voltage to the brushes (as we do when we plug in a motor). That voltage generates a current within the coils of the armature, and magnetic fields result around those coils (Section 5.4.1). When these magnetic fields interact with the fields from the stator, mechanical forces result that act on the armature in such a way as to cause it to rotate. It is that rotation which comes out of the axle of the motor. We use it for thousands of domestic and industrial applications varying in magnitude from turning the fan blades of a hair drier to turning the wheels of a subway locomotive.

1. Motors as Electromechanical Devices

Motors are *electromechanical devices*. This means that they interconvert electric and mechanical energy. Electric energy is invested; mechanical energy is derived.

2. Ratings

Motors are described with regard to their effective size by numerical designations known as *ratings*. Ratings are expressed in either electrical terms or mechanical terms or both.

a. Electrical ratings A small motor may be a 10-watt motor. A large motor may be a 10-kilowatt motor.

b. Mechanical ratings The mechanical rating of a motor indicates the mechanical power it delivers. The unit used is usually the horsepower (see "Force Energy, Power" in Section 4.2.2).

A small motor may deliver 0.1 horsepower (then the motor is often referred to as a "tenth-horse motor"). A large motor will deliver 100 horsepower or more (a "100-horse motor").

NOTE: Mechanical force is described as a combination of quantity of force and direction of force. Since motors always deliver power by rotating, the direction of the force is curved. Curved forces are called *torques* (see "Applying Force to an Object" in Section 4.2.3). Motors may also be rated in terms of the torque they deliver.

5.6 ELECTRICAL SYSTEMS

As we most commonly make use of them, electrons move in solid materials, and electromagnetic waves radiate in space. Both are controlled in electric and electronic circuits by means of a multiplicity of circuit components and interconnections. These circuits, along with electric machinery such as motors and generators, can be considered as tools and building blocks for electronics and electrical engineers. With these tools and building blocks they can create *electrical systems*.

1. Thousands of Electrical Systems

There is no limit to the number of electrical systems that scientists and engineers can devise. Or if they are not devising new ones, they are improving and revising old ones.

There is also no limit to the size and the power of electrical systems. A power network consisting of power lines that stretch hundreds of miles delivering millions of megawatts is no more and no less an *electrical system* than the microcircuit in a digital wristwatch that can be examined only with a microscope and which delivers only microwatts. Everything that uses or generates electricity is an *electrical system*. We can classify electrical systems into the following main categories.

a. Power This means the use of electrical systems to heat or physically move objects. This includes the heating or cooling of space, providing illumination, transportation as in electric trains or subways, the industrial processing of raw materials and the fabrication of these into products—and many others.

b. Communication This means the use of electrical systems to provide means for the transmission of information as in telephones, telegraph, microwave transmission, etc. A special case of communication is entertainment including radio, television, video and audio records—and many more.

2. Computers

This is a special case of communication. It refers to the processing of information as in business accounting systems, reservation terminals, library retrieval systems, hand calculators, computers of many sorts—and many others. This family of electrical systems has leaped into our lives in the last two decades with an explosive effect. It is so significant and so diverse that a separate chapter—Chapter 3—is devoted to it.

Basic Practice—Techniques and References

5.7 ELECTRICITY AND ELECTRONICS NOTATION

Many of the devices and concepts described in the first part of this chapter, "Basic Principles," are commonly referred to by universally accepted conventions, symbols, and abbreviations. The most important of these are described here, together with techniques for reproducing them both in text and in diagram form.

If difficulties or questions arise about any notation or usage, the reader is advised to refer to the descriptive material about that particular item in the preceding section, where clarification will most likely be available. Reference to accompanying illustrations and tables in that section will probably also be useful.

Most of the text that will be dealt with in the fields of electricity and electronics will be in the form of the scientific notation already covered in the chapter on mathematics. Refer to the "Basic Practice" part of that chapter as required.

5.7.1 Symbols, Terms, and Units of Conductance and Resistance

The symbol for conductivity is the lowercase Greek sigma.

$$\sigma$$

The symbol for conductance is the uppercase C, and its unit is the *mho*.

The symbol for resistivity is the lowercase Greek rho.

$$\rho$$

The symbol for resistance is the uppercase R, and its unit is the *ohm*.

The symbol for the ohm is the uppercase Greek omega.

$$\Omega$$

5.7.2 Symbols for Voltage and Volts

The symbol for the volt is the uppercase V. In equations, the uppercase E is used to represent voltage, or electric potential (but not the amount of voltage; V or the word *volts* is used for that).

For example, using Ohm's law (see Section 5.1.1):

$$E = I \times R = 10 \text{ volts} = 10 \text{ V}$$

Be sure to use the uppercase V and not the lowercase v to designate the volt.

CORRECT: V
INCORRECT: v

There is another kind of symbol, drawn freehand, to indicate the presence of voltage. It is

This symbol generally appears in circuit diagrams. It refers to voltage derived from a direct current (dc) (see Section 5.1.5) voltage source—or to that voltage source itself.

NOTE: It is also the symbol for a battery (which also provides dc voltage) (see "Batteries" in Section 5.1.5).

For an alternating current (ac) (see Section 5.1.5) voltage source (a single-phase generator or one phase of a three-phase generator), the symbol

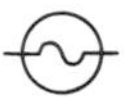

is used in circuit diagrams.

5.7.3 Electrical Ground

The symbol for electrical ground in text, notation, or illustration is the

drawn freehand or with the aid of a template. Other (less used) forms are

and

5.7.4 Symbols and Abbreviations for Current and Amperes

The symbol for the ampere is the uppercase A. In equations the uppercase I is used to represent current (but not the amount of current; A or the word *amperes* is used for that).

For example, in Ohm's law:

$$E = I \times R$$

and where $E = 10$ V and $R = 2\ \Omega$,

$$I = 5 \text{ amperes} = 5 \text{ A}$$

Be sure to use the uppercase A and not the lowercase a to designate amperes.

CORRECT: A
INCORRECT: a

The abbreviation for ampere is *amp*. The abbreviation may also be used in text as a word in its own right, in which case it means "ampere."

NOTE: The symbol A as a unit of measure is preferred to amps.

5.7.5 Direct and Alternating Current

The symbol for direct current is *dc* and for alternating current, *ac*.

INCORRECT: D.C. and A.C.
INCORRECT: AC and DC
CORRECT: dc and ac

NOTE: An exception is made when these symbols are used to start a sentence or in an all uppercase setting. Then AC and DC are acceptable.

5.7.6 Symbols and Abbreviations for Power and Watts

The symbol for the watt is the uppercase W. In equations, the uppercase P is used to represent power (but not the amount of power; W or the word *watts* is used for that).

For example: power is defined as the product of the current in amperes times the voltage in volts

$$P = I \times E$$

where $I = 3$ A and $E = 40$ V

$$P = 120 \text{ watts} = 120 \text{ W}$$

Be sure to use the uppercase W and not the lowercase w to designate the watt.

CORRECT: W
INCORRECT: w

There are two ways of abbreviating units such as kilowatt-hour or watt-second. They are used about equally without particular preference of one over the other—but in any text, remember to be consistent.

kilowatt-hour	kW·h	or	kWh
watt-second	W·s	or	Ws
watt-hour	W·h	or	Wh

NOTE: Like the symbol for multiplication, the dot is placed one half index up.

5.8 CIRCUIT DIAGRAMS

There are two kinds of circuit diagrams. There is the detailed depiction (known as a *circuit diagram*) in which every electric component is represented and in which every connection between components is shown. There is also the *block diagram,* which leaves out all representation of components and their interconnections and instead is made up of a series of boxes (blocks) in which groups of components that work together functionally are named, but not shown.

Groups of components that work together functionally are known as *stages.* Stages are depicted in detail in circuit diagrams and in "shorthand" boxes in block diagrams.

The Role of the Technical Secretary

It is common for the technical secretary to be required to draw simple circuit diagrams with one to four stages and block diagrams with a half-dozen stages. But when the complexity exceeds this level, aid should probably be sought from the graphics department.

5.8.1 Drawing the Circuit Diagram and the Block Diagram

A typical circuit diagram is shown in Figure 5-32*a*. The block diagram for the same circuit is shown in Figure 5-32*b*.

In *block diagrams,*

- Note that each stage (each block) is designated by a boldface numeral of the sort that is drawn with the aid of a template.
- Note that stages are interconnected by one or two straight lines.

In drawing block diagrams, always use a fine black felt-tipped pen—and be neat.

In *circuit diagrams,*

- Each stage is designated by a boldface numeral, as is the case for block diagrams.

  Some authors surround each stage (each group of components) with a dashed-line box. Or, as in Figure 5-32*a*, such boxes may be absent.

- Individual components may be labeled with manufacturers' type numbers or with their values in pertinent units, or labels may be omitted, or any combination of these.

Essentially, drawing circuit diagrams comes down to drawing symbols for circuit components. (Such symbols may also be drawn in text, individually, separate from circuit diagrams.)

What follows is a selection of the most commonly used symbols for circuit components and instructions for how to produce them. A more extensive tabulation may be found in Glossary J.

5.8.2 Circuit Components

The symbol for the resistor is

which may be hand-drawn either within text or in a sketch (See Figure 5 9).

The symbol for the variable resistor is the symbol for the resistor with an arrow drawn through it:

The symbol for the capacitor is

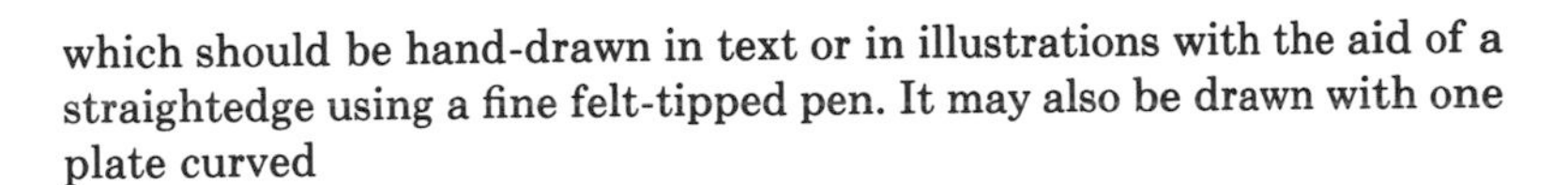

which should be hand-drawn in text or in illustrations with the aid of a straightedge using a fine felt-tipped pen. It may also be drawn with one plate curved

or rotated 90 degrees

The symbol for the variable capacitor is the symbol for the capacitor with an arrow drawn through it:

The symbol for the inductor coil is

which should be drawn freehand in text or in illustrations. When the symbol is drawn that way, an air core is understood. When there is an iron core, the coil is represented thus:

or

The symbol for the variable inductor is the symbol for the inductor with an arrow drawn through it:

The symbol for the transformer is two coils with the shared core represented by two or three lines between them

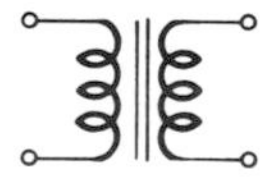

The basic symbol for the switch in text and in illustrations is the hand-drawn

(open position) and (closed position)

Multiple switches may also be similarly depicted, and the secretary will be expected to be capable of drawing them. Always use a straightedge.

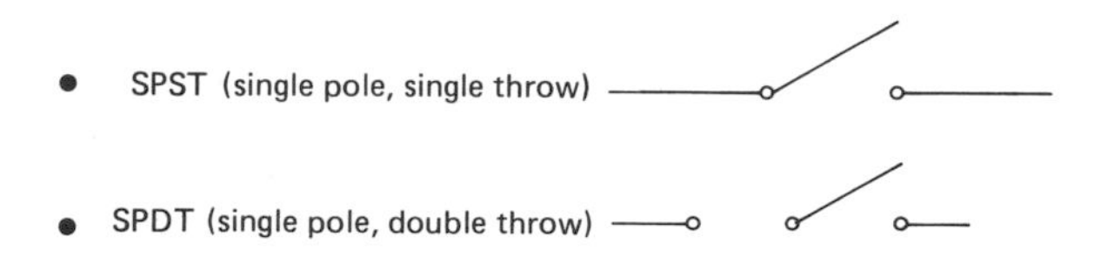

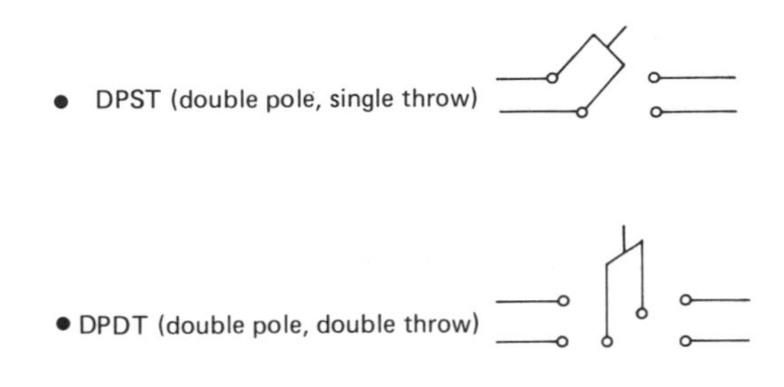

NOTE: The term *pole* stands for the termination of a wire at the location where it joins the switch. Thus *single-pole* indicates that a single wire is being opened and closed by the switch action.

The term *throw* stands for the possible options of switch action. *Single-throw* means the switch can open or close.They contact with only one or one set of poles. *Double-throw* means it can open or close contact with two or two sets of poles, and so on.

The symbol for the meter in text and in circuit diagrams is

or

The symbol may be made specific as

for voltmeter

for ammeter

Transducer symbols are sometimes used. Some transducers have symbols used in circuit diagrams and in text, such as the loudspeaker

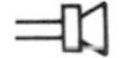

and the photocell

But in most cases there is no formal symbol.

Made-up symbols are sometimes encountered. It is not uncommon for the author of a text to make up a symbol, just for that one document. For example, the *thermocouple,* which converts heat to an electric voltage, is made of two junctions of two different wires. There is no standard symbol for this. A useful made-up symbol would be

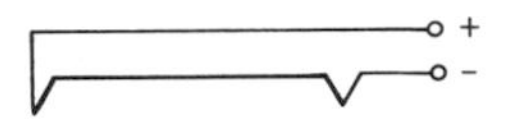

The symbol for a battery having a single cell is

The two vertical lines represent the two electrodes of the battery. The shorter one is always to the left of the longer one.

INCORRECT:

CORRECT:

The use of plus and minus symbols is optional.

CORRECT:

ALSO CORRECT:

Draw these with a fine felt-tipped pen with the aid of a straightedge.

The symbol for a battery having more than one cell is the single-cell symbol repeated two to four times. But the symbol with two repeats predominates:

There is no attempt represent the exact number of cells. Thus if there are six cells, six cells are not drawn, only multiple cells.

If plus and minus signs are used, they appear only one time, at the ends of the symbol:

The symbol for the vacuum tube consists of a circle within which are represented the anode and the cathode if the tube is a diode; and the anode, the cathode, and the grid if it is a triode:

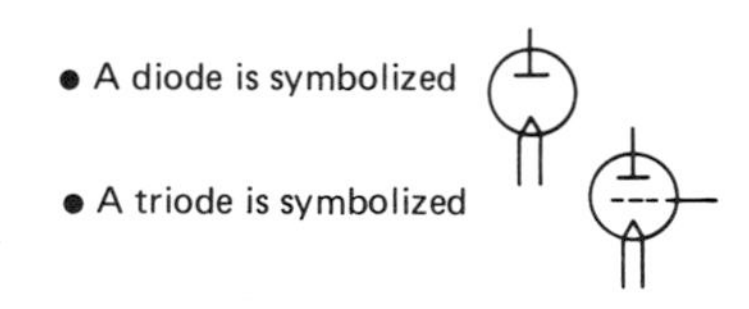

Tetrodes and pentodes are represented with added grids:

- A tetrode is symbolized
- A pentode is symbolized

These symbols are most commonly used in circuit diagrams, but they may appear within text.

The symbol for the semiconductor diode is the hand-drawn (with the aid of a straightedge)

It may appear in text or in circuit diagrams.

The arrowhead is always drawn solid.

The arrowhead stands for the emitter in the diode, and the straight line stands for the collector in the diode.

The symbol for the transistor is

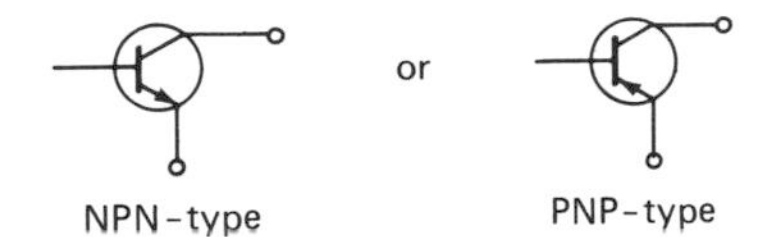

The ⊣ portion of the symbol is the base. Note that the vertical portion is often drawn with a heavier line than the horizontal portion.

The ⁄‾ portion of the symbol is the collector.

The remaining portion of the symbol, that for the emitter, is drawn with its arrow either touching the base or touching the outer circle. In the latter case this means that the transistor has an n-type emitter and collector and a p-type base; in the former case this means that the transistor has a p-type emitter and collector and an n-type base (Figure 5-20b).

5.8.3 Motors and Generator Symbols

The same symbol is used for the generator and the motor. It is

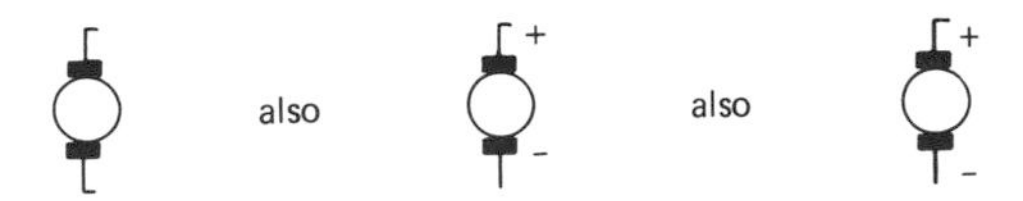

The circle represents the frame of the machine, and the heavy short lines represent brushes. Values of resistance and voltage may be included. The + and − may or may not be included.

5.9 MAGNETISM

Symbols and units of magnetism are tabulated in Table 5-3.

CHAPTER SIX

Chemistry

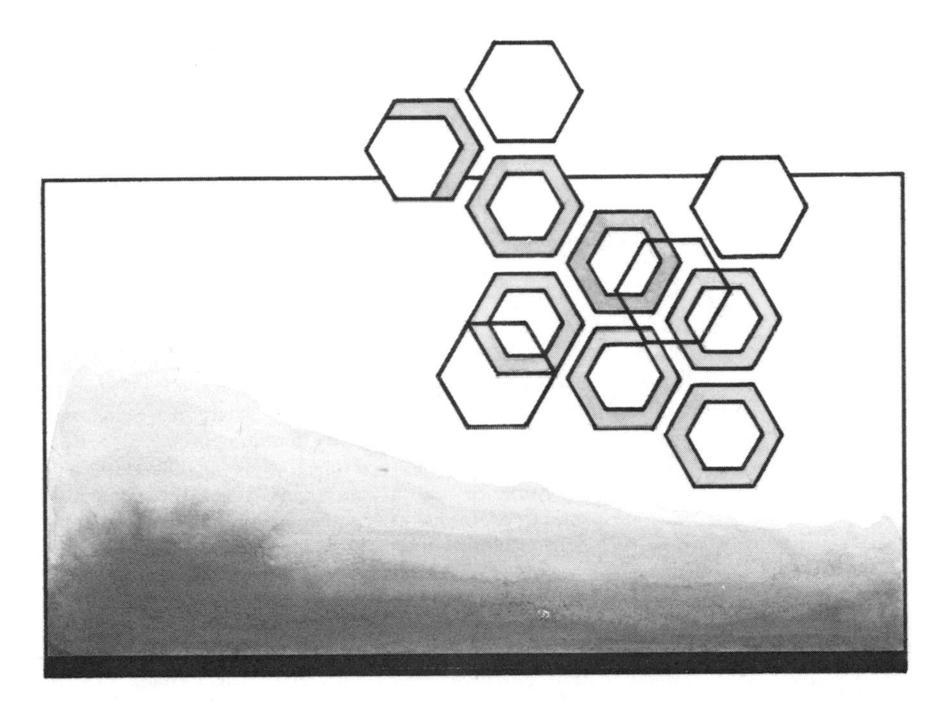

Basic Principles

This chapter treats the special grammar, vocabulary, and notation that must be used in the field of materials. Since every material in the world is a chemical substance, it is correct to call this chapter "Chemistry," although the text will often describe such diverse materials fields as metallurgy, ceramics, chemical engineering, electrochemistry, geology, petroleum engineering, and plastics. There is even a brief section on molecular biology, a discipline in which living matter is considered a chemical material.

6.1 ELEMENTS, COMPOUNDS, REACTIONS

6.1.1 The Chemical Elements

Every material in the world is made up of different combinations of only about 100 different basic (elemental) building blocks. These building blocks are known as the *chemical elements.*

NOTE: As of this printing, there are 107 chemical elements that have been identified (or produced artificially), but the ones most recently reported, 104, 105, 106, and 107, have not yet been given universally accepted names.

1. List of the Chemical Elements

These elements are listed, with their symbols, in Glossary H.

2. Names of the Chemical Elements

The chemical elements get their names from several different and interesting sources.

Some elements are named for people (curium, einsteinium, fermium, lawrencium, mendelevium, nobelium), some for places (americium, berkelium, californium, europium, francium, germanium, polonium), some for characters from mythology (mercury, niobium, promethium, thorium), and two are named for planets—although the planets' names came from mythology (neptunium and uranium). Many of the names are descriptive; for example, the name of the element radium was chosen by its discoverers, Marie and Pierre Curie, because it gives off radiation, and tungsten comes from the Swedish *tung,* which means "heavy," and *sten,* which means "stone." Some elements have two names. For example, tungsten is also called wolfram after one of the major minerals from which it comes, wolframite. There can also be regional variations in spelling such as *aluminum* in the United States and *aluminium* in Canada and the United Kingdom. Most of them end in "-um" or "-ium."

3. Relative Abundance of the Chemical Elements

Some elements are plentiful and some are rare.

The plentiful ones are cheap and the rare ones are expensive. Almost half of the planet Earth consists of oxygen, and approximately another quarter consists of silicon. The final quarter of our planet consists of all the other elements (Figure 6-1).

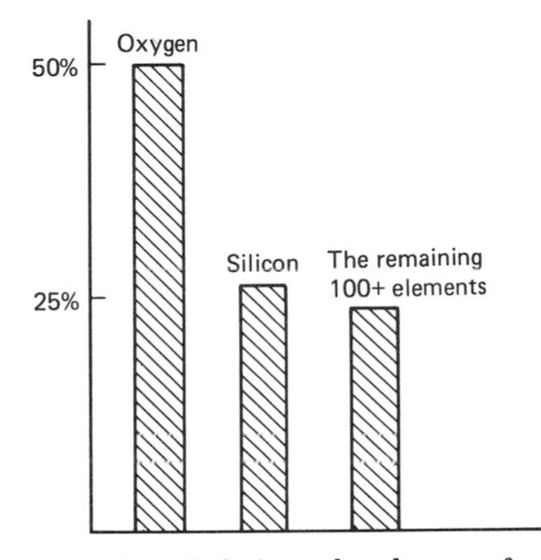

FIG. 6-1 Relative abundance of the chemical elements.

NOTE: Some elements are so rare that they are precious. The precious metals include gold, silver, and platinum. Cost and availability determine that when we wish to have a yellow metal for a doorknob, we use brass (an alloy of copper and zinc) rather than gold.

But you must know that all 107 elements exist and how to write them (at least the 103 that have names), because most of them will come up sometime during your career.

4. Only a Few Chemical Elements Common in Each Industry

How many elements are you likely to encounter over and over again on your job? Not many.

Although there are more than a hundred different chemical elements in the universe (see Glossary H), in any job it is unlikely that you will routinely encounter the names of more than about 20 of these. But each job will be different and will probably use a different 20. If you work for a steel company, you will deal with iron and its alloying elements like chromium, manganese, nickel, and a few others. If you work in an electronics plant, you will probably deal with silicon, copper, gold, and a few others. If you work in an oil refinery, you will probably deal with carbon, oxygen, hydrogen, sulfur, platinum, and a few others. And so on.

5. Symbols for Chemical Elements

Symbols for chemical elements are very rigidly prescribed and are used internationally. The symbols always consist of one or two letters (never more). The first letter is always capitalized; the second is always written lowercase. The symbols are always written on one line; the lowercase letter is not to be shown as a subscript. Periods are not used (except, of course, when the symbol ends a sentence).

- Mg is the symbol for magnesium.
- Pt is the symbol for platinum.
- B is the symbol for boron.
- H is the symbol for hydrogen.

NOTE: When in doubt, always check Glossary H for the correct symbol. Many of the symbols for the chemical elements are abbreviations of their English names, but not all are. Be careful because there is a snare here, and you could make a (logical) mistake. For example, Si is not silver; it is silicon. S is not silver either; it is sulfur. The symbol for silver is Ag, which comes from *argentum,* the Latin word for silver. In the same way the symbol for antimony is Sb, from the Latin *stibium.* The symbol for tungsten is W, which stands for wolfram, another name for tungsten that is commonly used in Europe and Asia but not in the United States.

6.1.2 Atoms and Molecules

Any tiny speck of a pure material may be made up of a pure element or of a combination of elements, in which case it is a chemical compound (Section 6.1.4). But it is interesting that each speck of element or compound in turn is made up of trillions and trillions of identical particles, each of which retains all the qualities of that particular element or compound.

1. The Atom

The smallest indivisible unit of an element is the *atom* (from the Greek word *atomos,* which means "indivisible"). (See Section 4.8.1 for a discussion of the structure of the atom.)

2. The Molecule

The smallest indivisible unit of a compound (Section 6.1.4) is the *molecule.*

NOTE: A molecule can in fact be broken up into its component atoms, but in that case it ceases to be a molecule or a compound.

6.1.3 Inorganic and Organic Materials

Most materials that contain the chemical element carbon are considered to be *organic* materials. This includes all matter that is or once was alive. So we ourselves (considered as materials—our flesh), the food we eat, and fuel—wood, coal, oil, gas—are included. So are plastics, which are manufactured from petroleum products, which in turn were produced by geological processes over millions of years from decayed animal and vegetable matter. And this means that even synthetic fibers like nylon, which is made from petrochemicals, have a "natural" source—as wool and cotton do. The chemistry of nylon and of all plastics is part of organic chemistry too.

Every other material in the world is *inorganic*.

Although most of the rules of chemical notation and usage pertain to both organic and inorganic chemistry, there are some differences that will be noted as they come up, particularly in Section 6.4.

6.1.4 Chemical Compounds

Some elements are familiar to us in their pure state, not combined with any other element. A diamond is pure carbon; nothing else is present. Mercury, such as you may see in a thermometer, is pure elemental mercury; nothing is added to it or combined with it. A balloon at a carnival usually contains the gas helium; just that, nothing else.

But most of the things we encounter and handle in the world are made up of combinations of elements. These combinations are *chemical compounds*. The smallest possible bit of a compound is a molecule.

The world contains millions of different chemical compounds that result from different combinations of the hundred or so elements. Can there be millions of different ways to combine a hundred things? Indeed there can.

1. It Takes Two (or More) Atoms to Make a Compound

A chemical compound is the result of the combination in some proportion of two or more atoms to make the basic molecule of that compound.

NOTE: These are usually different elements, but not necessarily—O_2 is a compound made of two oxygen atoms.

2. The Chemical Bond

The combining of atoms to form the molecules of a compound is characterized by a bonding process.

In other words, there is a distinct linking between the elements of a compound. These elements are not merely lying together side-by-side; they are in effect "glued together." That "glue" is known as the *chemical bond*. It is actually electric in nature.

3. Whole-Number Proportions of Elements

The elements in each compound are present in distinct whole-number proportions.

In the chemical compound chromium oxide, the ratio of chromium to oxygen is 2:3 (there are two chromium atoms for every three oxygen atoms).

It is written

$$Cr_2O_3$$

with the numerical-ratio designations shown as subscripts (see "Numerical Subscripts" in Section 6.1.4).

4. The Chemical Formula

We all have heard the expression *chemical formula.* A chemical formula is the notation for a chemical compound showing its ingredients and the numerical proportions of these. Think of the formula as just another language which describes what is in a chemical compound.

In the preceding section, the chemical compound chromium oxide was represented by the chemical formula Cr_2O_3.

NOTE: The use of symbols for elements and formulas for compounds is analogous to ordinary language. The element symbol is the "letter" and the formula is the "word." When we deal with chemical reactions (Section 6.1.7), in which compounds combine with each other, those are shown as equations, which are analogous to sentences.

This chemical "language" is international; it is understood by engineers and scientists from all nations. In this sense it is comparable to music. The words to a song may be different in different countries, but the page of musical notation is the same in every country and can be read with equal ease by musicians from every country.

In any compound, the elements are in certain exact proportions to each other. A numerical *subscript* after each element in a formula tells what proportion that element is of the whole.

NOTE: If the subscript is 1, it is left out. Thus

INCORRECT: K_1Cl_1

CORRECT: KCl

and thus

- In water, H_2O, a compound of hydrogen and oxygen, the ratio of hydrogen to water is 2:1 (two hydrogen atoms for each oxygen atom).
- In hydrogen peroxide, H_2O_2, the same elements are present as in water, but the proportions are different. The ratio is 2:2 (two atoms of hydrogen for two of oxygen).
- In carbon dioxide, CO_2, there is one carbon atom for every two oxygen atoms.
- In sulfuric acid, H_2SO_4, there are two atoms of hydrogen, one of sulfur, and four of oxygen.

5. The Radical

Some elements form little groups which occur repeatedly in many different compounds. Each such group then behaves almost as though it were a single element. These groups are called *radicals,* and multiples of them are recognized easily in chemical notation because they are usually set apart by parentheses.

In	$Ca(OH)_2$	calcium hydroxide
and	$Zn_3(PO_4)_2$	zinc phosphate

(OH) and (PO_4) are radicals.

a. Subscripts for radicals Radicals are treated in notation as though they were elements, and thus they can have their own subscripts, which are applied just as they would be to a single element.

Thus in the preceding examples, the subscripts tell us that in $Ca(OH)_2$ the ratio of (OH) to calcium, Ca, is 2:1, and that in $Zn_3(PO_4)_2$ the ratio of (PO_4) to zinc, Zn, is 2:3.

NOTE: To get the number of atoms in any compound containing radicals, multiply the subscripts for each radical by the subscripts for each atom within the radical.

This means that the number of atoms in the compounds $Ca(OH)_2$ and $Zn_3(PO_4)_2$ is as follows:

$$CaO_2H_2 \quad \text{and} \quad Zn_3P_2O_8$$

Both styles of formula show the same amounts of each element in each compound, but the CaO_2H_2 style is imprecise, because it fails to indicate that oxygen and hydrogen atoms are present in the form of OH radicals; therefore, the $Ca(OH)_2$ style is more common and useful.

b. Names for radicals Radicals sometimes have convenient names which are used in text.

Thus (OH) has two different names. It is called the *hydroxyl* radical or the *hydroxide* radical, and compounds containing it may be called *hydroxides.* The (SO_4) radical is known as the *sulfate* radical. There are hundreds of others such as *nitrate, chlorite, borate, thiosulfate,* etc.

NOTE: Suffixes and prefixes ("-ate," "-ite," "thio-" and others like them) have special meanings and are described in Section 6.1.6.

c. Leaving out parentheses For compounds in which radicals occur singly, for example, $H_2(SO_4)$, it is customary to omit the parentheses. Thus H_2SO_4 is the preferred form. But note that $Ca(OH)_2$ is not the same as $CaOH_2$. The former has two oxygen atoms and the latter only one. Parentheses are *essential* in multiples of radicals.

6. Waters of Hydration

When chemical compounds crystallize out of water solution (see Section 6.2.2), they often carry some molecules of the water with them. This water is bonded to the compounds and then makes up an integral portion of the material (the crystal). Note that the incorporated water is no longer wet (liquid), since the crystal is a solid.

Such "associated" water molecules are known as *water of hydration,* or *bound water.* The resulting compound is then called a *hydrate.*

An example of a hydrate is copper sulfate. When it is in water solution, it can be represented by the formula $CuSO_4$. When it is crystallized out of the water (most of the water is evaporated away), forming an array of blue crystals, it is in a form that can be represented by $CuSO_4 \cdot 5H_2O$. This indicates that the crystalline copper sulfate is a hydrate, with five molecules of bound water for every molecule of copper sulfate.

Hydration is always indicated by adding the number of water (H_2O) molecules to the end of the formula by means of a dot placed one-half index up ($CuSO_4 \cdot 5H_2O$).

NOTE: When the hydrated copper sulfate crystals are heated to very high temperatures—hundreds of degrees—the bound water becomes unbound and vaporizes away. What results is crystalline copper sulfate without water of hydration. It is then known as anhydrous copper sulfate and is represented by the formula $CuSO_4$. The term *anhydrous* is used to designate a material from which the water of hydration has been removed (or in which it was never present).

6.1.5 Kinds of Compounds—Acids, Bases, Salts

Compounds can be grouped into various classes. We will discuss three of the most important classes (there are others). The compounds in these classes differ from others in that they are more or less ionizable. For more about ions, see Section 6.3.2.

1. Acids

An *acid* is a compound that when added to a solution causes the concentration of hydrogen ions (H^+; see Section 6.3.2) in that solution to increase. In most cases the acid contains one or more hydrogen atoms that become hydrogen ions in water solution.

The ionizable hydrogen atoms are usually at the beginning of the written formula for inorganic acids (for example, HCl and H_2SO_4).

Besides the ionizable hydrogens, an acid contains one or more atoms of a different element or elements, which may be grouped as a radical in the formula.

HCl	hydrochloric acid
HNO_3	nitric acid
H_2SO_3	sulfurous acid

NOTE: There is a notation that shows how much hydrogen ion is present in a water solution. It is *pH* (Section 6.3.5).

2. Bases

A *base* (or an alkali) is a compound that when added to a solution causes the concentration of hydroxyl ions (OH^-) in that solution to increase. In most cases the base contains one or more hydroxyl radicals (OH) which become hydroxyl ions (OH^-) in water solution.

The ionizable OH is usually at the end of the written formula for inorganic bases—for example, NaOH and $Ca(OH)_2$.

Besides the ionizable hydroxyl radical, a base contains one or more atoms of a different element or elements, which may be grouped as a radical in the formula.

NH_4OH	ammonium hydroxide
$Ca(OH)_2$	calcium hydroxide

NOTE: Parentheses are used or not used for radicals depending on whether there are more than one (see "Leaving Out Parentheses" in Section 6.1.4).

NOTE: The *pH* is a measure of how much base is present in a water solution, just as it tells how much acid is present. This is because they balance each other; if the concentration of H^+ is high, the concentration of OH^- is low (see Section 6.3.5).

3. Salts

A *salt* is a compound that results from a reaction between an acid and a base. (The hydrogen atoms of the acid and the hydroxyl radicals of the base combine to form water.) A salt consists of two or more different atoms or radicals in any combination.

NaCl	sodium chloride (common table salt)
$CaSO_4$	calcium sulfate
$(NH_4)_2SO_4$	ammonium sulfate
Li_2MoO_4	lithium molybdate

6.1.6 Suffixes and Prefixes for Chemical Compounds

The chemist wants to be able to easily write or speak the name of a chemical compound in words as well as in symbols.

If the compound $Ba(MnO_4)_2$ is being discussed, it would be ridiculous for the speaker to have to say, "The salt that has a barium atom in it combined with two radicals made up of a manganese atom and four oxygen atoms." It is much easier if the chemist can just say, "Barium permanganate," knowing that the audience, also with some chemical training, understands that those two words indeed mean a salt of the formula $Ba(MnO_4)_2$.

Names are formed by using certain suffixes and prefixes that follow certain rules; this gives important information which describes each compound. A number of the more important suffixes and prefixes are described here.

NOTE: For more detailed information on suffixes and prefixes, you might find it useful to refer to the *Handbook of Chemistry and Physics* (Glossary Q), usually available in your department or in the company library.

1. The Suffix "-ide"

This suffix is used as the last syllable of the name of a compound of at least two parts (either elements or radicals). It describes the last part. You only need to know that it is a way to state that these elements or radicals are in combination with each other.

GaAs is gallium arsenide.

InSb is indium antimonide.

PbS is lead sulfide.

KI is potassium iodide.

LiBr is lithium bromide.

KCN is potassium cyanide.

NH_4Cl is ammonium chloride.

The names of the compounds in the foregoing examples are formed as follows. The first word is the name of the element or radical that comes first in the formula. The second word is a standard derivative based on the name of the second element or radical and ending with the suffix "-ide."

2. The *More* and *Less* Suffix Pairs

There are two sets of suffixes that describe the relative oxygen content (or *oxidation state*) of the radicals that are ingredients of certain acids and salts. These suffixes are "-ous" and "-ic" for acids and "-ite" and "-ate" for salts.

a. The suffixes "-ous" and "-ic" These generally refer to the *relative* oxygen content of the radical on the right (the anion—see Section 6.3.2) in the formula for the acid: "-ous" is for *less* oxygen and "-ic" is for *more* oxygen.

H_2SO_3 is sulfurous acid. It contains three oxygen atoms, one less than sulfuric acid.

H_2SO_4 is sulfuric acid. It contains four oxygen atoms, one more than sulfurous acid.

HNO_2 is nitrous acid.

HNO_3 is nitric acid.

NOTE: This practice of naming compounds using "-ous" and "-ic" has declined, but the usage remains sufficiently common that the processor of technical text should understand it.

NOTE: Many compounds have several names. For example, the simple compound NO can be called nitric oxide, mononitrogen monoxide, nitrogen monoxide, or nitrogen oxide. Some names are preferred, and usage changes with time. As with other cases of terminology, the important thing is always to be as consistent as possible within any one piece of work.

b. The suffixes "-ite" and "-ate" These refer to the relative oxygen content of the radical on the right (the anion—see Section 6.3.2) in the formula for the salt: "-ite" is for *less* oxygen and "-ate" is for *more* oxygen.

NOTE: For simplicity, these compounds are shown in their anhydrous form (see "Waters of Hydration" in Section 6.1.4), even though they more commonly occur as hydrates, for example, $Li_2SO_3 \cdot H_2O$.

Li_2SO_3, lithium sulfite, contains three oxygen atoms compared with four oxygen atoms for lithium sulfate.

Li_2SO_4 is lithium sulfate.

KNO_2 is potassium nitrite.

KNO_3 is potassium nitrate.

NOTE: As with "-ous" and "-ic," recent practice has been not to name compounds using this correlation of "-ite" and "-ate" with *less* and *more,* but the usage described is still common.

NOTE: The salts in the preceding example may be considered as corresponding to the acids which have the same radical. Thus lithium sulf*ate* is the lithium salt of sulfur*ic* acid, and lithium sulf*ite* is the lithium salt of sulfur*ous* acid. There is a correspondence between "-ate" and "-ic" and between "-ite" and "-ous."

3. Prefixes That Show Exact Quantities

The chemist often wishes to be more exact than is possible using added syllables that mean merely less or more. Then actual numbers can be used as prefixes, or conventional numerical prefixes can be incorporated into chemical expressions.

a. Number prefixes These are most frequently encountered in organic chemistry (see Section 6.4.7). The numeral precedes the term it refers to, separated by a hyphen. When there are two numerals together, they are separated by a comma.

2-chloro-1,3-butadiene
2-(3-chlorophenoxy)-propionic acid
6-chloropurine

These numbers concern the structural representation of organic compounds (see Section 6.4.2).

b. Word prefixes for numbers These are used in ordinary non-scientific text, but they may be incorporated into chemical expressions to show relative quantities of elements in chemical compounds (the term *stoichiometric* is used to refer to such relative quantities in any chemical compounds). They include such prefixes as "mono-," "bi-," and "hexa-." A list of such prefixes may be found in Glossary D.

Examples:

sodium bicarbonate
monosodium glutamate
potassium hexafluoroarsenate

Such prefixes are normally written without hyphenation—although there may be exceptions.

4. Some Miscellaneous Prefixes

Many more prefixes are used in forming chemical compound names, many with rather complex meanings. You need only understand that for the most part they relate to the relative proportions within a compound family of certain atoms or radicals in the compound—as in the phosphoric acid family (see the second and third examples below).

Some of the most common prefixes are

meta-	hypo-
pyro-	thio-
ortho-	per-

Examples:

$Na_2CO_3 \cdot H_2O$	sodium carbonate metahydrate
$H_4P_2O_7$	pyrophosphoric acid
H_3PO_4	orthophosphoric acid (also, phosphoric acid)
$Na_2S_2O_3$	sodium hyposulfite (also, sodium thiosulfate)
$KMnO_4$	potassium permanganate
H_2O_2	hydrogen peroxide

Glossaries D, E, and F contain many additional prefixes that are used in chemistry.

6.1.7 Chemical Reactions

Every chemical compound has a history. It became a combination of elements bonded together in definite numerical proportions because that many atoms, in those proportions, interacted (reacted) with each other in a process known as a *chemical reaction* to form the molecules of that compound. We can express a chemical reaction by writing it as an equation.

As has been noted in the section on mathematical equations (Section 2.3.3), an equation is a sentence which says, "This is the same as that." Chemical equations conform in some ways to this concept; they are also subtly different.

1. Equality of Both Sides of the Equation

In both chemistry and algebra, the expression on one side of the equation has the same value as the expression on the other side.

This may not be obvious in algebra (Section 2.3). Take the expression

$$x = 2y$$

We are confident that x is the same as $2y$ only because the equals sign is there.

But in chemistry the same symbols for the same elements normally appear on both sides of the equation, and in the same quantity (see "Balancing Chemical Equations" on p. 6.15). This means that in chemistry (with the exception of nuclear chemistry) basically no element transforms; rather, it just recombines. And what is on the left side is just what is on the right side. Count up the subscripts (see "Multiplying Coefficients by Subscripts Gives the Number of Atoms" on p. 6.15)—the atoms *balance*.

In the chemical reaction

$$4Cr + 3O_2 = 2Cr_2O_3$$

there are four chromium atoms and six oxygen atoms—equal quantities of each element—on both sides of the equals sign, but recombined (reacted).

NOTE: Of course, an author may have a reason to write an equation in which balance does not occur. Such a circumstance is not usual, but it does happen.

2. Signs of Operation

In algebra there are many signs of operation ($+$, $-$, $\div$, $\times$, $\cdot$, etc.). But in chemical equations, there is only one sign of operation. It is the plus sign ($+$).

NOTE: The equals sign ($=$) is not a sign of operation; it is a sign of identity.

In chemistry, things are (with a few rare exceptions) only added together. This means a chemical equation says only that "this and this and this, reacted together, will give that and that and that." There is no division and no multiplication (and subtraction occurs so rarely that it will not be treated here).

3. Signs That Show Equality

In algebra, the equals sign has some variations and nuances (see "The Family of Equals Signs" in Section 2.3.1). In chemistry, there are some special variations too. In addition to the equals sign (=), the arrow is used in different ways to show variations and nuances (of reactions).

It makes sense to use arrows in chemistry because they "point" the directions in which reactions take place.

a. The equals sign Reacting two atoms of hydrogen with one atom of oxygen produces water:

$$H_2 + \tfrac{1}{2}O_2 = H_2O$$

NOTE: This is the common way of writing $2H + O = H_2O$. The oxygen molecule has a coefficient of ½ since hydrogen (H_2) and oxygen (O_2) exist as two-atom molecules (see "Coefficients That Tell How Many Molecules There Are" on p. 6.15).

b. The arrow

Using the arrow, the reaction looks like this:

$$H_2 + \tfrac{1}{2}O_2 \rightarrow H_2O$$

The arrow, pointing from left to right, tells us that the reaction proceeds as written from left to right (that is, the water does not become $H_2 + \tfrac{1}{2}O_2$—rather, the opposite happens). Thus it tells us more than the simple equals sign would have. But it is not impossible for the reaction to be reversed. Water can decompose to form hydrogen and oxygen. Then the reaction looks like this:

$$H_2O \rightarrow H_2 + \tfrac{1}{2}O_2$$

c. The reverse arrow

Or the same information could be given by reversing the direction of the arrow:

$$H_2 + \tfrac{1}{2}O_2 \leftarrow H_2O$$

d. The arrow pair

The reaction we have been using as an example is called a *reversible reaction,* which means it can go either way, either combining oxygen and hydrogen into water or decomposing water into hydrogen and oxygen. The usual way to show a reversible reaction is with a pair of half arrows, the top one of which points to the right and the bottom one to the left:

$$H_2 + \tfrac{1}{2}O_2 \rightleftharpoons H_2O$$

e. The unequal arrow pair

It is possible to show that in certain reversible reactions, the reaction is more likely to occur in one direction than in the other; yet the reverse reaction still occurs to a slight degree. This is expressed by using a longer arrow to show the direction of the preponderant reaction. In that case the arrow in the direction of the preponderant reaction is usually drawn on top (both still as half arrows):

$$2H_2 + \frac{1}{2}O_2 \rightleftharpoons 2H_2O$$

4. Coefficients That Tell How Many Molecules There Are

In a chemical reaction it is necessary to know how many molecules of each kind are involved. This is done with a numerical coefficient preceding the group of symbols that stands for the molecule. This coefficient differs from the formula subscripts; those indicate only how many atoms of each element each molecule contains.

a. Multiplying coefficients by subscripts gives the number of atoms

In the chemical reaction

$$4Cr + 3O_2 = 2Cr_2O_3$$

there are a total of four chromium atoms on the left side (4 × 1) and on the right side too (2 × 2). There are six oxygen atoms on the left side (3 × 2) and on the right side too (2 × 3).

b. When radicals are present As described in "The Radical" on p. 6.6, the number of atoms of each element in a radical that has a subscript is found by multiplying that subscript by the subscript that follows the element symbol within the radical (remember, where there is no subscript, 1 is implied). Now if the whole compound has a coefficient, that number is in turn multiplied by all the previously multiplied numbers. Such manipulations tell the total number of each atom type in each compound.

Thus for $5Al_2(SO_4)_3$, there are 10 aluminum atoms (2 × 5), 15 sulfur atoms (3 × 1 × 5), and 60 oxygen atoms (3 × 4 × 5).

c. Balancing chemical equations It is important to recognize that when all the numbers are taken into account (all the molecular coefficients are multiplied by all the atomic or radical subscripts to which they apply), there must be exactly the same number of atoms of each sort on each side of the equation. That is one of the rules of chemical equations: they must *balance*.

NOTE: This is because atoms are not created or destroyed merely because they participate in (rearrange themselves in) a chemical reaction. This is the *law of conservation of matter.*

6.2 PHYSICAL CHEMISTRY

The fact that chemical compounds react with each other and that these reactions can be described with chemical notation in chemical equations was discussed in Section 6.1.7. Let us now go one step further and inquire about the effect of various physical (nonchemical) influences on those reactions. The fact is that chemical reactions (and the chemicals themselves) are very much affected by physical influences.

- The velocity and degree of completeness of chemical reactions is strongly affected by such conditions as temperature, pressure, and relative quantities of the reactants.
- Energy in all forms, but most importantly heat, is required for certain chemical reactions to take place. Conversely, certain reactions give off energy in the form of heat.

  The most common example is the heat given off by the chemical reaction known as *burning fuel.*

- The condition or state of matter of materials depends on the temperature and pressure.

  Thus the same material may be solid, liquid, or gaseous depending on the particular temperature and pressure at that time, as water is ice when "below freezing," water when "warm," and steam when "above boiling."

  The important thing to remember is that the material is unchanged *chemically.* The water remains water, with the formula H_2O, in spite of its drastic physical transformations. All this is diagrammed in Figure 6-2*a*.

The portion of chemistry dealing with these effects of changes in energy and changes in the environment on chemical reactions and materials is known as *physical chemistry.*

6.2.1 Changes of State

The change in physical condition of a material (where *physical condition* means whether it is a solid, a liquid, or a gas) without a change in its chemical composition (the chemical formula describing it is unchanged) is known as a *change in state;* to the scientist the physical condition of a material is its *state.*

The physical form of a material is also known as its *phase,* and changes of state are also known as *phase changes.*

1. Melting

This is the process whereby a solid, if sufficiently heated, becomes a liquid. The process is also called *fusion.*

NOTE: This should not be confused with *nuclear fusion,* which designates a radiochemical reaction.

2. Freezing

This is the process whereby a substance goes from its liquid phase to its solid phase by the removal of heat (decreasing the temperature). It is also called *solidification.*

3. Boiling

This is the process whereby a substance goes from its liquid phase to its gaseous phase as a consequence of the adding of heat. The same phase change occurs during *evaporation,* which may be considered to be non-violent, low-temperature boiling.

4. Condensation

This is the process whereby a substance in its gaseous state becomes a liquid or a solid because of the removal of heat.

This is what happens when moisture accumulates on the outside of a glass of iced tea on a humid summer day. The water in the air condenses on the cold glass. Or the vapor may skip the water phase and go directly to ice—as when frost accumulates on a window on a cold winter day.

NOTE: Boiling followed by condensation is the basis of the purification of materials by the process known as *distillation.*

5. Sublimation

This is the process whereby a solid material turns into a vapor without first becoming a liquid.

Sublimation occurs when wet clothes are hung on a line on a subzero day. The water in the clothes instantly freezes, but after a few hours, they are dry. The moisture has evaporated out of them without ever having gone through a liquid phase.

Figure 6-2*b* shows the different states of matter and the various processes (paths) whereby a single material can change from one state to another.

6. Representing Changes of State in Chemical Equations

A number of symbols can be added to ordinary chemical equations to show that physical (as well as chemical) changes have taken place.

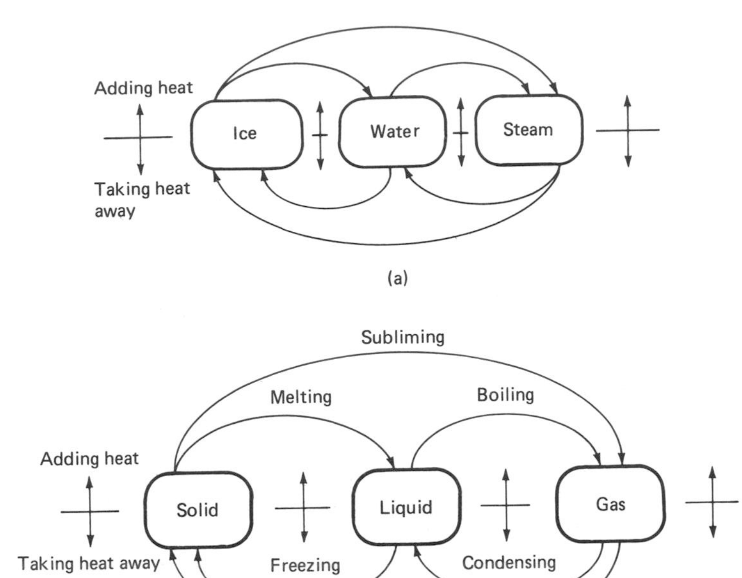

FIG. 6-2 The physical states of matter and the transitions from one to another. (*a*) The different physical states of water. (*b*) The general case for all materials.

These are letters in parentheses, vertical arrows, overbars, and underbars. They all refer to the states of matter.

- Liquid: $_{(l)}$ or (l)
- Solid: $_{(s)}$ or (s) or ↓ or
- Gas: $_{(g)}$ or (g) or ↑ or $\overline{}$

The (l), (s), and (g) designations are preferred.

In the reaction between silver nitrate ($AgNO_3$) and hydrobromic acid (HBr) in a water solution, particles of solid silver bromide will appear, as a cloud of suspended motes. These will sink to the bottom of the vessel to make a kind of silt. It can then be said that the silver bromide has *precipitated out of solution* and that the solid mass is a *precipitate* (see Section 6.2.2) of silver bromide. The reaction can be shown in any of the following four ways, although the first one, with the (s) at normal index level, is the most frequently used:

$$AgNO_3 + HBr \rightarrow AgBr(s) + HNO_3$$

$$AgNO_3 + HBr \rightarrow AgBr_{(s)} + HNO_3$$

$$AgNO_3 + HBr \rightarrow AgBr\downarrow + HNO_3$$

$$AgNO_3 + HBr \rightarrow \underline{AgBr} + HNO_3$$

Another reaction, which produces a gaseous product, involves the dissociation of sulfuric acid (H_2SO_4) when it is heated. In this case, instead of a solid precipitate that comes out of solution and sinks to the bottom of the vessel, bubbles of gas (sulfur trioxide; SO_3) form, rise to the surface of the liquid, and, breaking out of that surface, escape into the air above. This reaction may be shown in any of the following four ways, although the first one, with the (g) at normal index level, is the most frequently used:

$$H_2SO_4 \rightarrow H_2O + SO_3(g)$$

$$H_2SO_4 \rightarrow H_2O + SO_{3(g)}$$

$$H_2SO_4 \rightarrow H_2O + SO_3\uparrow$$

$$H_2SO_4 \rightarrow H_2O + \overline{SO_3}$$

6.2.2 Solutions

The process of *solution* is the physical combining of one substance with another to form a homogeneous mixture which has its own identity as a phase. This means that the mixing extends to the molecular or atomic level. A mixture formed this way is known as a *solution.*

NOTE: It is possible to have such mixtures (solutions) in any state of matter. Thus there can be solid solutions (such as alloys; see Section 4.7.2), liquid solutions (such as sugar in water), or gaseous solutions (such as smoggy air).

Here are some terms you will frequently encounter when dealing with solutions.

- The material into which another material is dissolved (disappears) is known as the *solvent.*
- The material which is dissolved (disappears) into the solvent is known as the *solute.*
- If only a small amount of solute is dissolved in the solvent, the solution is said to be *dilute.*
- If a great deal of solute is dissolved in the solvent, the solution is said to be *concentrated.*
- There is a limit to how much of a specific solute any solvent can contain (dissolve or retain in the dissolved state). Such a limit, for example, the maximum amount of solute A that can be dissolved in solvent B, is the *solubility*—in this example, the solubility of A in B.
- When there is so much solute that its limits of solubility in that solvent are exceeded, no more solute will dissolve, and the solvent is said to be *saturated* with that solute. If the amount of solvent in

this saturated solution is then somehow decreased (for example, by evaporation), then the limits of solubility of the solute will again be exceeded, and the excess solute will "come out of" the solution. This material, which usually drifts down through the solute and forms a silt on the bottom of the vessel, is known as the *precipitate*.

6.2.3 Thermodynamics

Changes in materials, whether chemical or physical, involve energy. Either energy must be invested from outside the system, or energy that is locked up in materials is released as a result of chemical reactions or physical changes. This study of energy-material interactions is known as *thermodynamics* (see Section 4.6); when applied to chemistry, it is called *chemical thermodynamics*.

1. Heat and Chemical Reactions—Endothermic and Exothermic

As can be seen from the prefix "thermo-," heat in some form is involved in thermodynamics. In any chemical reaction, either heat must be supplied from the outside or the reaction generates heat.

Reactions requiring outside heat are known as *endothermic*. Let us consider two examples.

- The process of boiling water (changing liquid water to steam)

$$H_2O(l) \rightarrow H_2O(g)$$

  is endothermic. Heat is applied to the bottom of a kettle. When the heat is turned off, the boiling stops.
- In the making of a stew, it is common to maintain the pot and its contents at a medium or low temperature for an hour or more. Chemical reactions involving the breakdown of protein molecules are going on during that period; we call this process tenderizing. The tenderizing process is endothermic. Turn the heat off too soon and the stew meat will not become tender.

Reactions that give off energy (heat)—sources of energy—are known as *exothermic*.

- When we burn wood or any other fuel, flames result. These give off heat. Burning is exothermic. We make use of the exothermic nature of the chemical reactions involved in burning to heat our houses.

2. Four Basic Thermodynamic Quantities

The forces and energies associated with chemical reactions can be described mathematically.

This means there are exact values associated with every chemical reaction. The thermodynamics expert knows at once, by looking at the pertinent thermodynamic values, just how easily any chemical reaction will proceed. It is then possible to predict how much heat must be supplied in an endothermic reaction and how much will be given off in an exothermic one.

Four thermodynamic values suffice to describe almost every chemical reaction. If you work with chemists or chemical engineers, you will constantly encounter them. They are

- Temperature T
- Heat content or enthalpy H
- Free energy G (or F or $F°$)
- Entropy S

NOTE: G is the preferred form for free energy. F and $F°$ are the older forms, now falling into disuse. The G stands for Gibbs free energy, named after J. W. Gibbs, an American physicist and a pioneer in the field of thermodynamics.

At any particular temperature T, every material has characteristic values for the other three factors, H, G, and S, and the chemist can look them up in tables. As the materials react with each other to form new materials, the thermodynamic values combine too. Not only has there been an overall chemical or physical change, but also there has been an overall thermodynamic change associated with it. It is these overall changes that tell the chemist how likely the reaction is to proceed. The changes are represented by the uppercase Greek letter *delta*.

- ΔH
- ΔG (or ΔF or $\Delta F°$)
- ΔS

NOTE: These factors apply in exactly the same way to all chemical and physical changes. Thus they apply to inorganic chemistry, organic chemistry, physical chemistry, electrochemistry, etc.

Thermodynamic notation in chemical reactions The energy and force relations may be stated along with the chemical reactions themselves, as a second statement after the equation.

In that case, thermodynamic equations follow the chemical reaction to which they pertain, after a semicolon.

NOTE: These quantities are usually expressed in calories per mole and in degrees Kelvin.

The equation is always understood to be in the form:

$$\Delta G = \Delta H - \Delta S T$$

Some examples:

$$H_2(g) + \tfrac{1}{2}O_2(g) \rightarrow H_2O(g); \Delta G = -60{,}180 + 13.9T$$

$$Ni(l) + \tfrac{1}{2}O_2 \rightarrow NiO(s); \Delta G = -60{,}750 + 25.1T$$

6.2.4 Catalysis

A *catalyst* speeds up the rate at which a reaction occurs without itself undergoing any chemical modification. The process is known as *catalysis.*

> What the catalyst usually does is to lower the free energy required to make the chemical reaction proceed, and this makes the reaction proceed more easily (faster) than it otherwise would have. The catalyst does not change the reaction in any other way.
>
> An example is the catalytic coating which is available in some ovens in kitchen ranges. That coating covers the inside walls of the oven giving them a velvety-black appearance. When food spatter hits these walls, it gradually disappears during the cooking cycle. That is because it is "burning." The burning is flameless because it takes place at cooking temperatures such as 350°F, which is so low that no flame exists. (Visible flames occur at temperatures above about 1000°F.)
>
> Remember, catalysts only speed up reactions; they don't change them. That means that some burning must normally occur at relatively low temperatures in the absence of a catalyst. But the rate of that burning is so infinitesimal that for all practical purposes the burning does not take place. What the catalyst does is to speed up the reaction so much that it becomes significant; we can see its effects. Thus significant low-temperature burning occurs only because the oven walls are catalytic. However, the process is still a relatively slow one, which is why such ovens require "continuous cleaning" (hence their name). Compare this with the "self-cleaning oven," which requires a special heating cycle for a few hours at 950°F to achieve what the catalytic oven achieves at normal cooking temperatures.

The interesting thing about catalysts is that they are not changed by the chemical reactions which they stimulate. After the reaction has taken place, the unchanged catalyst is ready to be used again.

6.3 ELECTROCHEMISTRY

All atoms are made of electrically charged particles, which is a way of saying that all molecules and all chemicals and all materials are also made of electrically charged ingredients. The portion of chemistry concerned with the electrical behavior of chemicals is called *electrochemistry.*

The professionals who work in this field are the *electrochemist* and the *electrochemical engineer.* The product most commonly associated with electrochemistry is the *battery* (see Section 6.3.4).

The process most commonly associated with electrochemistry is *electroplating.* But all chemists and physicists have occasion to discuss and write about the electrical properties of chemicals.

6.3.1 The Electrical Structure of Matter

This is a subject that pertains to both physics and chemistry. See "The Structure of the Atom" in Section 4.8.1.

1. Charged Particles

Figure 4-22 in Section 4.8.1 shows a typical atom, sodium, consisting of a centrally placed *nucleus* around which *electrons* circle. The nucleus consists of *neutrons* and *protons*. The numbers of neutrons, protons, and electrons are different for each chemical element (Table 4-7).

a. Electrons Electrons are small and mobile, and each one has one negative electric charge.

b. Protons Protons are relatively large and immobile (when compared with small and mobile electrons). Each one carries a positive electric charge.

c. Balance Electrons and protons balance in atoms as they normally occur in nature. Except under certain conditions (see Section 6.3.2), there are just as many electrons as protons in the atom. Thus most atoms are electrically neutral.

d. Neutrons Neutrons carry no electric charge and essentially do not participate in electrochemical phenomena.

2. Opposite Charges Attract—Static Electricity

We are all familiar with *static electricity* in our daily lives.

> When layers of clothing cling together it is because when the layers rub together, some of the (mobile) electrons get transferred from one layer to the other. Thus one layer has more electrons than it formerly had and is more negatively charged than it used to be, and the other layer has fewer electrons than it used to have and is less negatively charged than it used to be (which is the same as being more positively charged). It is these *opposite* electric charges that cause the clinging.

Opposite charges establish a force which drives them toward each other: *opposite charges attract.* Those charges are, in fact, *static electricity.*

3. The Role of Electric Charges in Forming Chemical Compounds

Static electricity may be said to cause the same attractions between particles on an atomic level as occur between the layers of clothing in the preceding example.

The attraction between positively charged and negatively charged atoms (*ions;* see Section 6.3.2) is responsible for the *ionic bond,* which holds many inorganic molecules together.

6.3.2 Ions

For many materials, the electric bonds that hold the compounds together are easily ruptured. This happens most commonly when these compounds are added to water.

When grains of salt are added to water, the solid grains seem to vanish. But the salt is still there. We know this because we now have salty water, which we can taste. As can be seen from Figure 6-3, the ionic bonds that made up the solid salt are ruptured when the salt is dissolved. Each atom of sodium is free to float around in the water, not tied to any particular chlorine atom. The same holds for all the chlorine atoms, which are no longer bonded to the sodium atoms.

This type of bond is called an *ionic bond.* When these ionic bonds are broken, the molecule does not separate into two neutral atoms (for simplicity we are assuming a two-atom molecule). Instead, one of the atoms splits off along with an extra electron (one more than the neutral atom

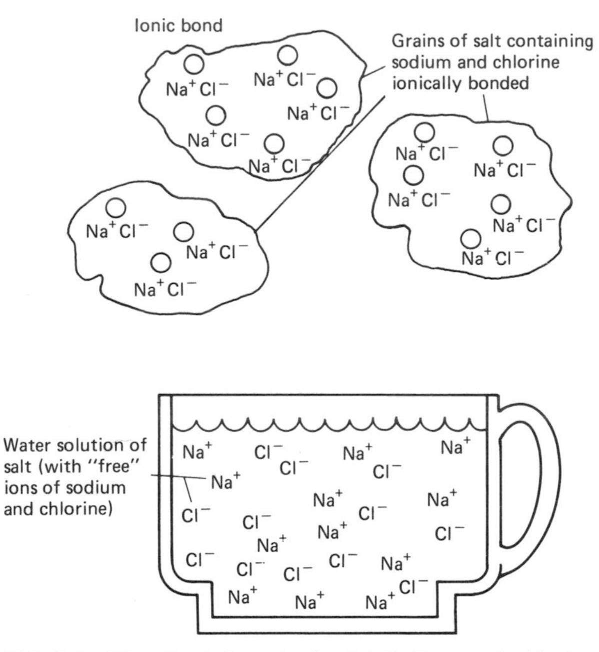

FIG. 6-3 The dissolving of salt. (*a*) Sodium and chlorine are held together by the electric force of the ionic bond to form molecules of salt (represented by the circles).
(*b*) The salt has dissolved in water causing the ionic bonds to rupture, with the result that each atom now carries an electrical charge—positive for the sodium and negative for the chlorine—and all atoms are now ions.

would have) and thus has a negative charge (see Section 4.8.1). The other atom is left with a "missing" electron and thus has a positive charge. These charged atoms are called *ions.*

1. Ions Have Electric Charge

All ions have electric charge. The charge may be either positive or negative.

2. Radicals May Be Ions

Groups of atoms that characteristically act together may also have charges, once dissolved in water, just as though they were single atoms. These are *radicals* (see "The Radical" in Section 6.1.4). They are ions too, since they bear electric charges.

3. Set Amount of Charge for Each Ion—They Balance

Every ion, whether it be a charged single atom or a radical, has a characteristic number of charges, either positive or negative.

Thus in table salt, NaCl, dissolved in water,

Na always has just one positive charge: Na^+

Cl always has just one negative charge: Cl^-

In sulfuric acid, H_2SO_4, dissolved in water,

H always has one positive charge: H^+

SO_4 always has two negative charges: SO_4^{2-}

NOTE: It is possible for a single atom to form more than one kind of ion. For example, iron may form ions with two positive charges, Fe^{2+}, or three positive charges, Fe^{3+}. For a description of ionic notation, see Section 6.3.3.

In any water solution the electric charges balance. There is always just as much negative charge as positive charge.

In the two examples, the single positive charge of the sodium ion, Na^+, is balanced by the single negative charge of the chlorine ion, Cl^-; the double negative charge of the sulfate ion, SO_4^{2-}, is balanced by two hydrogen ions, H^+, each of which has a single positive charge.

NOTE: Although ions occur normally in solutions, they may exist in the gaseous state. They conduct the electricity of lightning during a storm, and also conduct other electric sparks (see Fig. 5-10).

4. Valence

For single ions, the number of charges normally present is determined by the nature of that atom and gives rise to a property known as the *valence* or combining power of that atom. To a first approximation, the valence is equal to the number of charges, positive or negative, associated with that atom when it becomes an ion.

Thus

The valence of sodium is 1.

The valence of chlorine is 1.

The valence of calcium, which forms ions with two positive charges, is 2.

The valence of iron, which can form ions with two or three positive charges, is 2 or 3. (There is even an occasional ion of iron with six positive charges.)

NOTE: Ions with positive charges are called *cations;* those with negative charges are called *anions.*

6.3.3 Ionic Notation—Compounds and Equations

There is a special way to write the formulas for ionic compounds (compounds whose molecules include charged atoms or ions). This extends to chemical equations involving such compounds (electrochemical reactions).

1. Ionic Compounds

When a compound can dissolve in water and break up into its constituent ions, it can be written two ways. It can be shown as any other compound (see "The Chemical Formula" in Section 6.1.4), or notation can be added to indicate the identity and amount of charge of each ion.

Thus

Lithium molybdate as an ordinary compound: Li_2MoO_4

Lithium molybdate as an ionic compound:

$$2Li^+ + MoO_4^{2-}$$

$$\text{or } 2Li^+MoO_4^{2-}$$

NOTE: In this example, the molybdate radical may be shown in parentheses, if the author prefers:

$$(MoO_4)$$

$$(MoO_4)^2$$

NOTE: The use of the connecting + in the compound is optional with the author, depending on what is being conveyed.

The rules for expressing an ionic compound are as follows:

- Subscripts move to the front of the ion (or ionized radical) and act as though they were molecular coefficients, that is, they "multiply" each atom they precede.
- Every ion (electrically charged atom or radical) must be followed by its number of charges—positive or negative—written as a superscript.
- In an ionic molecule, the charges balance. There are exactly as many positive charges as negative charges.

For lithium molybdate, the two positive charges (associated with the two lithium ions) are balanced by the two negative charges (associated with the single molybdate radical).

- Those compounds that do not ionize are written as usual without appended charges. There is no reason why ionized and un-ionized compounds cannot appear together in the same chemical expression. They often do.

2. Ionic Equations

Equations that involve electrochemical reactions (ionic compounds) can be written either as normal equations or with ionic notation.

Some examples:

$$Zn + 2AgCl \rightarrow ZnCl_2 + 2Ag(s)$$

$$MnO_2 + 4HCl \rightarrow 2H_2O + MnCl_4$$

The same reactions written ionically would be

$$Zn + 2AgCl \rightarrow Zn^{2+} + 2Cl^- + 2Ag(s)$$

$$MnO_2 + 4H^+ + 4Cl^- \rightarrow 2H_2O + Mn^{4+} + 4Cl^-$$

NOTE: In preceding examples, the AgCl and the MnO_2 are given in nonionic form on the left side of the equation, indicating that they are solid (not dissolved) when the reaction begins. In fact, both compounds ionize (dissolve) to some degree in water, and that could have been shown with ionic notation had the author chosen to.

6.3.4 Free Electrons and the Battery

There are devices which can harness ionic chemical reactions in such a way that the energy inherent in the opposite charges of ions can be put to work. One example is the battery, which converts chemical energy to electric energy. Chemical reactions between the electrodes and the electrolyte that make up the battery produce *free electrons,* which are channeled off through a wire in the form of electric current.

1. Free Electrons

Electrons are called *free* when they are not attached to any specific atoms and can move freely toward whatever attracts them. The symbol for a free electron is e^-. A less common usage is the italic e^-. An older practice employs a ⊖, which is a hand-drawn circle with a horizontal diameter line.

NOTE: This symbol differs from θ, the Greek letter *theta*.

Equations must balance electrically, so a numerical coefficient may be employed in front of the free electron as though it were an ion.

Here is the ionic equation for producing electricity from a lead battery

$$Pb + SO_4^{2-} \rightleftharpoons PbSO_4(s) + 2e^-$$

2. Batteries

Section 5.3.3. and Figure 5-5 described the electrical performance of the battery. Here we consider it as a chemical system which produces free electrons.

A battery is made up pf certain chemicals (such as the lead sulfate in the example above) held in a tightly sealed package. Chemical reactions slowly take place at the electrodes inside the battery. As the electric charges slowly rearrange themselves, some electrons end up *free*. If a wire is then attached to the battery so that one end leads into the anode (where there is a chemical reaction that *gives off* free electrons) and the other end leads into the cathode (where there is a chemical reaction that takes *takes in* free electrons), there will be a continuous movement of free electrons through the battery and through the wire known as an *electric current*.

6.3.5 The Hydrogen Ion Concentration, pH

When an acid dissolves (ionizes), the hydrogen ion, H^+, with a single positive charge, is always present (Section 6.1.5). The hydrogen ion may be present in greater or lesser amounts (concentrations), and the chemist has devised a notation which indicates how much is present in any particular case. Once that is known, the chemist knows the chemical activity of that acid; the more hydrogen ions there are, the more acidic the solution.

The notation used to indicate the strength of an acid is the symbol pH.

Mathematically, pH is the logarithm of the reciprocal of the hydrogen ion concentration:

$$pH = \log\left(\frac{1}{H^+}\right)$$

which means that small numbers (low pH) tell us that relatively many hydrogen ions are present (the solution is very acidic), and large numbers (high pH) tell us that relatively few hydrogen ions are present.

The pH scale goes from 1 to 14, with 1 indicating the highest concentration of hydrogen ions (acidity), 7 indicating neutrality, and 14 indicating the lowest concentration of hydrogen ions (H^+). When the pH is greater than 7, the solution is not acidic at all, but the opposite of acidic, basic, and more hydroxyl ions (OH^-) are present than hydrogen ions (H^+).

NOTE: The product of the concentration of hydrogen ions times the concentration of hydroxyl ions is always constant (10^{-14}). Thus when the concentration of hydrogen ions is high, the concentration of hydroxyl ions must be low, and vice versa.

6.4 ORGANIC CHEMISTRY

This is the part of chemistry and chemical engineering that focuses on the chemical behavior of the compounds of the element carbon. There are two important qualities of carbon that set it off from other elements. One is that carbon is an essential chemical ingredient of everything that lives or has lived on earth. That includes our own bodies and almost everything we eat, wear, use as medicine, and use as fuel. The other is the distinctive way it participates in chemical bonds with itself and with other elements.

6.4.1 The Carbon Bond

Carbon combines with other elements in a special way. It always forms four *covalent bonds.*

Like ionic bonds (see Section 6.3.2), *covalent bonds* involve electric charges (electrons). However, unlike compounds "glued together" with ionic bonds, compounds with covalent bonds *share* electrons between different atoms, and the compounds do not split apart into ions when dissolved in water.

1. Bonds Show Structure

What is important for the person working with organic chemistry text to know is that the arrangements of the carbon bonds are usually *shown*, because they are different for each carbon compound. For this reason you may find yourself constantly diagraming the *structure* of each compound.

This is very different from writing down a simple formula of an inorganic compound with element symbols and pertinent subscripts, as we have done for, let us say, lead sulfate, $PbSO_4$. Rather, for the organic compound, we "draw a picture" of it.

Let us look at *propylene,* for instance. Propylene, written as though it were an ordinary inorganic compound, is C_3H_6. Such a representation of propylene is not wrong, but several different chemicals may have the same formula. For example, cyclopropane is also C_3H_6 (see Section 6.4.4). Thus organic chemists usually show something about structure in the formula. So they would more likely represent propylene as CH_3CHCH_2. This indicates to the reader that there are groups that have their own identity within the compound (different from the radicals discussed in Section 6.1.4); the groups are CH_3, CH, and CH_2. Written structurally, the formula of propylene becomes

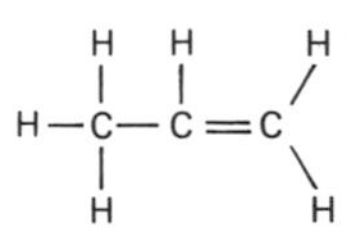

NOTE: Each carbon atom has four bonds: count them.

NOTE: An author may sometimes use a formula that gives a little more information than CH_3CHCH_2 but not quite as much as the structural formula (see Section 6.4.5). Thus an author might write $CH_3CH{=}CH_2$ or even $CH_3{-}CH{=}CH_2$. All these forms are correct. The choice usually depends on the emphasis of the discussion. As with other cases of terminology, consistency is preferred (one normally should not use $CH_3CH_2CH_2CH_3$ and $CH_3{-}CH_2{-}CH_3$ in the same discussion). If you think an author is being inconsistent, it wouldn't hurt to ask which form should be used.

NOTE: In fact other chemical compounds can be depicted structurally, as the compounds of carbon are. This is most frequently done with compounds of the element silicon, which is a "cousin" of carbon in that it too forms four covalent bonds. But its chemistry is not as commonly encountered as carbon chemistry is. Some silicon compounds that are structurally similar to carbon compounds are called *silicones.* Do not confuse *silicon* with *silicone.*

Bonds are shown structurally primarily by straight lines that interconnect the atoms that make up the compound. But that's not all. There are variations on the straight-line bond indication that provide additional information about the structural situation.

Other bond symbols include, but are not limited to,

Dots

Dashed lines

Lines with arrows

Wedge-shaped lines

The ways in which bonds are diagrammed provide information about such things as tendencies of electrons or atoms to change location, and in which direction, and positions of atoms or groups above or below a certain structural plane. There are also notations for "weak" bonds and for multiple bonds. There are compounds in which the atoms are arranged in patterns such as hexagons, known as *rings.* The technical secretary and writer must learn how to present all these.

NOTE: Chemical bonds are longer than the hyphen or the equals sign on a typewriter or word processor. If true bonds are not available in the font you are using, draw them in by hand.

2. Depicting Bonds

a. Other elements In addition to carbon, the most commonly encountered elements in organic chemistry are nitrogen, oxygen, hydrogen, sulfur, and the halogens.

NOTE: The *halogens* are a family of chemical elements with similar chemical behavior. They include fluorine, chlorine, bromine, and iodine.

Carbon forms four bonds, nitrogen usually three, sulfur two, oxygen two, and hydrogen one. (Sulfur and nitrogen can have other numbers of bonds. You should follow the copy you get or ask your author, if in doubt.) Other elements occur in organic compounds as well and are also depicted structurally with the valences (numbers of bonds; see "Valence" in Section 6.3.2) that are associated with each.

b. Dots Some atoms have locations that could participate in bonding but are not bonded at the moment. Such sites are commonly shown as two dots since they represent a pair of electrons that could constitute part of a bond.

In ammonia, which is often shown structurally like organic compounds, the nitrogen atom has four bond sites but utilizes only three of these. Thus NH_3 can be shown structurally as

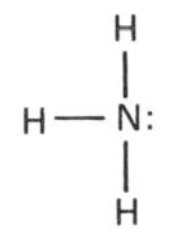

c. Curved arrow Sometimes the chemist wishes to show that when a reaction takes place, electrons in effect move from one atom to another. Such relocations are shown with a curved arrow:

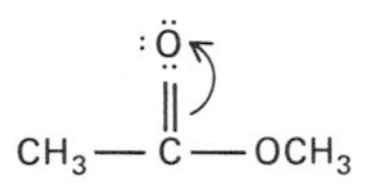

d. Weak bonds Some bonds are "weak," which means easily ruptured. A weak bond is shown as a dashed line rather than as a solid line. A dotted line is also acceptable.

Weak bonds often occur between hydrogen and oxygen.

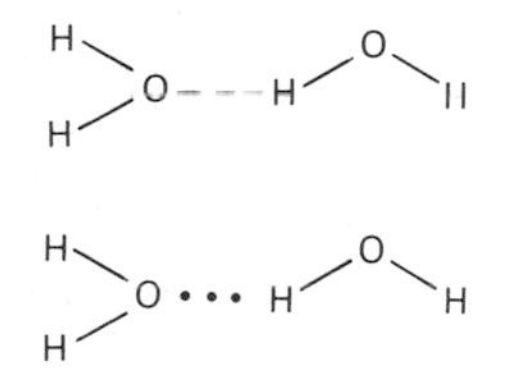

NOTE: Other ways to show bonds are described in the following sections, which deal with more complex structural arrangements, such as *rings*.

6.4.2 Ionic Notation in Structural Formulas

Organic compounds may ionize in water solution as inorganic compounds do (Section 6.3.2), and there are organic acids and bases. Charged atoms or groups of atoms may then be shown with ordinary superscript plus or minus signs like inorganic ions, or these plus or minus signs may be incorporated into structural diagrams.

Acetic acid, CH_3COOH, ionizes with an acetate radical and a hydrogen ion and may be shown in ionic notation (Section 6.3.3)

$$CH_3COO^- \ H^+$$

or structurally

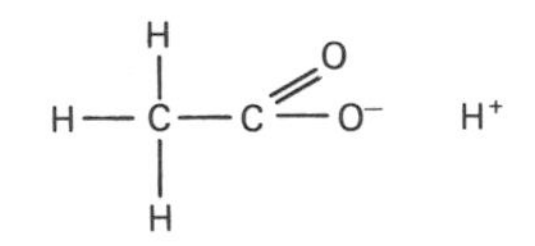

6.4.3 Hydrocarbons

There are thousands of known compounds of the two elements carbon and hydrogen. They are known as *hydrocarbons*. Hydrocarbons are best understood when they are written structurally. Many hydrocarbons can then be seen to be long chains of repeated units. One common unit is

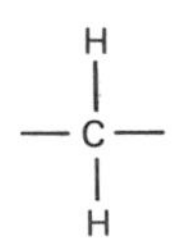

NOTE: Two of the carbon bonds in this formula don't seem to be attached to any atoms. Of course they are. But sometimes the chemist writes only the atoms that are important for a particular discussion. In this case, various atoms or groups of atoms could be attached at the unspecified bonds. Sometimes symbols such as R or X are used to represent the same thing: RCH_2R'.

Hydrocarbons range from very small compounds, with only one of these units, to very large molecules, with hundreds of units.

Thus

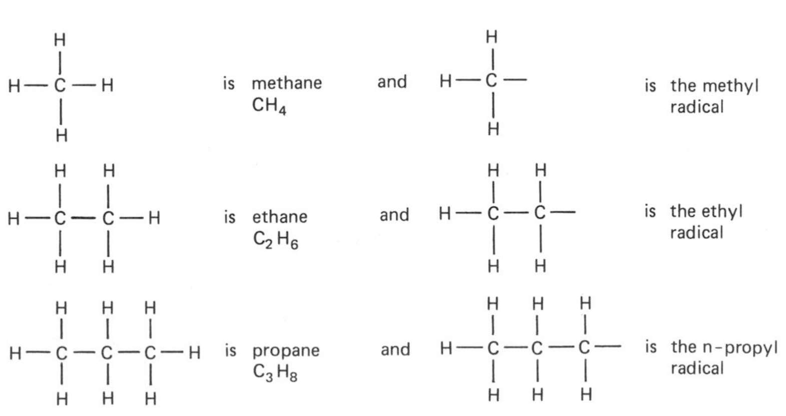

The compounds go on:

C_4H_{10}	butane
C_5H_{12}	pentane
C_6H_{14}	hexane

and so on to the point where hundreds of atoms are involved in a single molecule. This family of compounds is known as *alkanes* or *paraffins.*

NOTE: The general formula here is C_nH_{2n+2}, where n can be a very large number. In any family of hydrocarbons, each compound has its own name and corresponding group with its corresponding name.

There are other hydrocarbon families with different structural arrangements; they also build into chains by adding similar structural units. These families are *alkenes* (or *olefins*), *alkynes,* and *cycloalkanes* (which are also *paraffins*), and each family has its own general formula. You will often see the term *aliphatic* applied to noncyclic families.

6.4.4 Isomers

This word comes from the Greek *isos,* which means "same," and *meros,* "part." It designates compounds that have the same chemical composition but different structure.

It is interesting that isomers are quite different from each other in physical properties in spite of having the same kinds of atoms in the same numbers. The differences derive from how these atoms are arranged—structurally. Thus

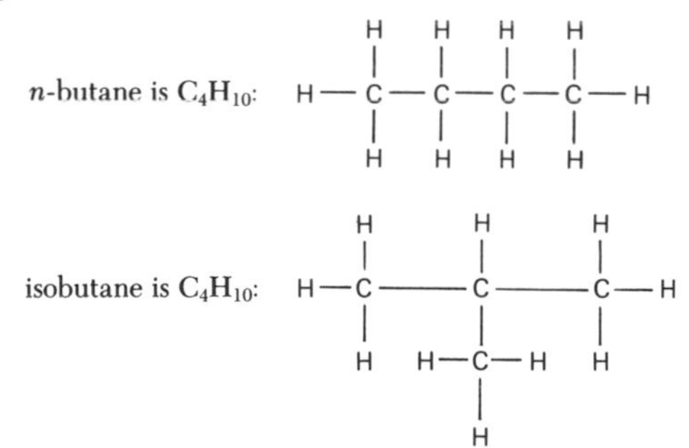

One physical difference between these two compounds, for example, is their boiling points: n-butane boils at $-0.5°C$ and isobutane boils at $-11.7°C$.

6.4.5 Condensed Structural Formulas

In order to save space and minimize complexity, the chemist will often write in the nonstructural manner a group of atoms which is well-understood structurally (by other chemists), using structural representation only for certain important groups.

Thus the formula for methyl butyl ketone may be written to show its most important structural ingredient, a group known as a *carbonyl,*

$$-\overset{\displaystyle O}{\overset{\|}{C}}-$$

The structure of the other organic groups may not need to be shown. Thus an acceptable way to write this formula is

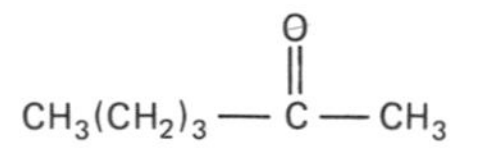

NOTE: Hydrocarbon units can be grouped in parentheses, with subscripts used, just as parentheses and subscripts are used with inorganic radicals (see "The Radical" in Section 6.1.4). Thus $CH_3CO(CH_2)_3CH_3$ is a much more economical way of writing $CH_3COCH_2CH_2CH_2CH_3$. Also note that it is not necessary to use structural representation for the carbonyl group. Either form, as well as the form in which all groups are shown structurally, is acceptable.

NOTE: Chemists also use abbreviations such as Me (for methyl), Et (for ethyl), etc. When in doubt about the meanings of such abbreviations, the best procedure would be to look them up in a reference such as *The Handbook of Chemistry and Physics* (Glossary Q).

6.4.6 Aromatics and the Benzene Ring

There is a group of organic compounds known as the *aromatics* because certain members of the group have strong, often pleasant aromas. Oil of wintergreen and vanilla are examples. Aromatics are characterized by the fact that they contain six-membered rings with three double bonds. The atoms of the simplest one are arranged in a particular structure known as the *benzene* ring.

Benzene, with the formula C_6H_6, may be written structurally

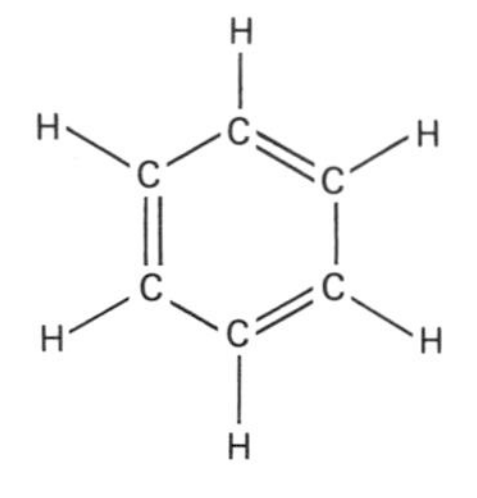

1. The Benzene Ring Symbol

Chemists have determined that the carbon atoms of benzene are arranged in a hexagonal pattern. Therefore, benzene rings may be shown as simple hexagons with alternating double sides.

It is understood that in each ring there is a carbon atom at each corner, and each carbon atom has four bonds, one of which is to a hydrogen atom, as in the preceding example. The double sides represent the double bonds.

a. Hexagon with a circle It is also common for benzene rings to be represented as follows:

When various chemical groups are substituted for one or more of the hydrogens on the benzene ring, other specific aromatic chemicals are formed. These can also be represented using the benzene hexagon notation.

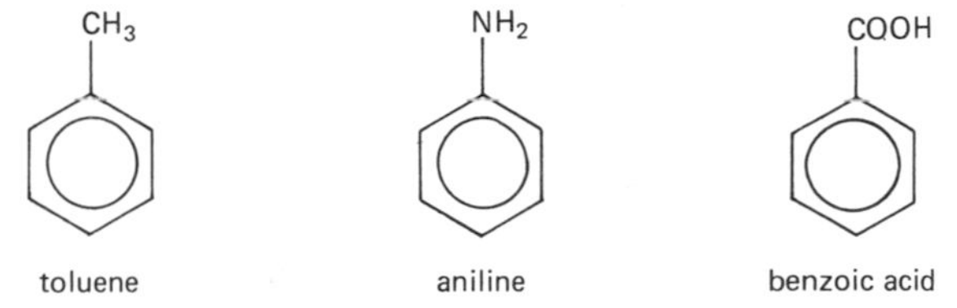

NOTE: Like benzene, benzene derivatives can be written without all the hydrogen atoms. Unless otherwise indicated, the hydrogen atoms are understood to be there. Thus in the example above, each of the benzene rings is assumed to have five hydrogen atoms at its five vacant points.

b. Positioning of added atoms or atom groups The position of the added atoms or atom groups tells the chemist something important about the compound. Such positioning is described in text with prefixes (*ortho-*, *meta-*, and *para-*). They describe the following types of structures:

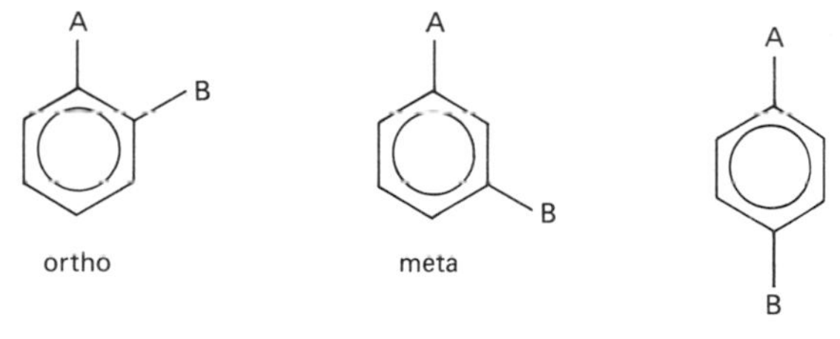

NOTE: These prefixes may be abbreviated *o-*, *m-*, and *p-*. Thus *para*-dichlorobenzene may also be written *p*-dichlorobenzene. It is shown structurally as

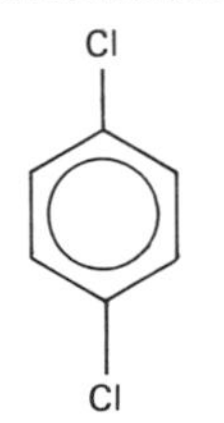

2. Other Ring Structures

These are used to represent other compounds. There is a large variety of possible structures.

Some examples:

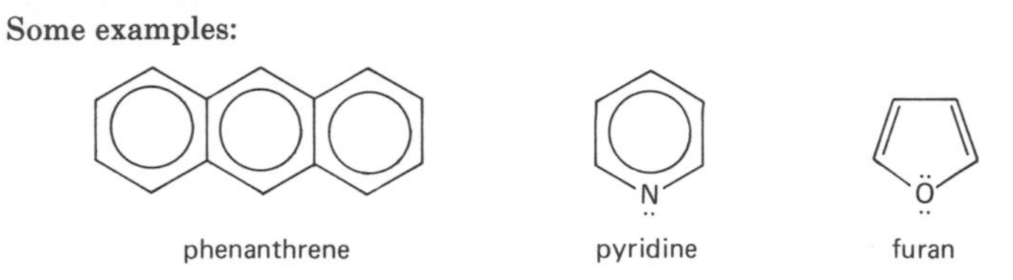

There is no end to the possible complexity of some of the compounds you may be asked to diagram.

For example, reserpine:

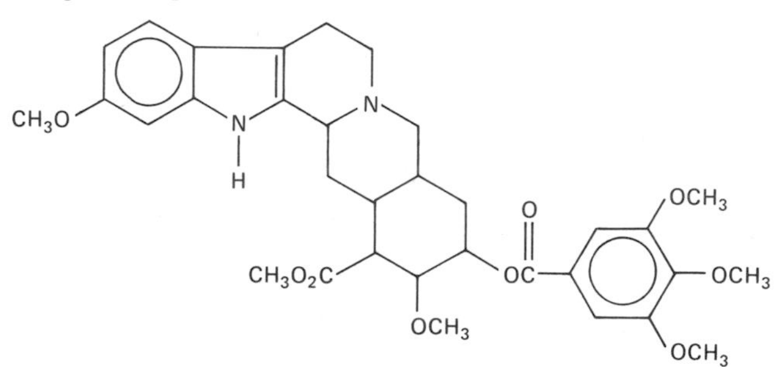

and chlorophyll *a*:

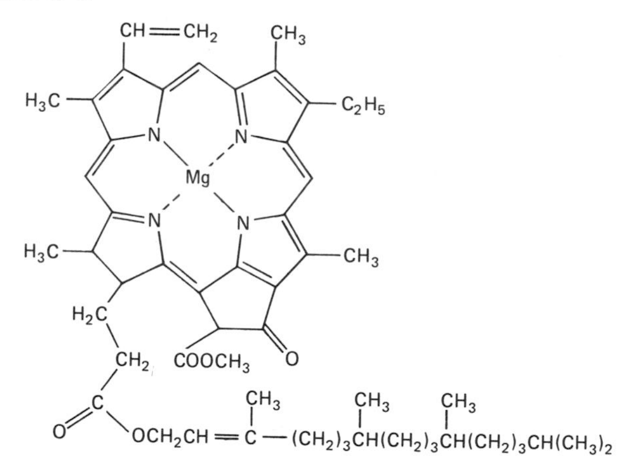

6.4.7 Nomenclature

As in other scientific disciplines, the organic chemist has invented a nomenclature that eases the task of talking and writing about the subject.

1. Suffixes In organic chemistry, suffixes are frequently used to designate functional groups (groups that tend to have similar properties even when located in quite different compounds; for example, an acid group is a functional group).

Table 6-1 lists the most commonly encountered suffixes.

TABLE 6-1
SUFFIXES FOR SOME COMMON STRUCTURAL GROUPS

Family Name (example)	*Suffix*	*Structure*	*Example*
Alkane	-ane	RCH_2CH_2R'	butane
Alkene	-ene	$RCH{=}CHR'$	butene
Alkyne	-yne	$RC{\equiv}CR'$	pentyne
Alcohol	-ol	ROH	ethanol
Aldehyde	-al	$\overset{H}{RC}{=}O$	formaldehyde
Ketone	-one	$R\overset{\overset{O}{\parallel}}{C}R'$	acetone
Carboxylic acid	-ic acid	$R\overset{\overset{O}{/\!/}}{C}{-}OH$	acetic acid

2. Locants Locants are numbers or letters which designate the *location* of an atom or of a well-defined group in a molecule.

By *location* is meant the structural position in a molecule. For example, *para*-dichlorobenzene (see Section 6.4.6) is also called 1,4-dichlorobenzene. This indicates that the chlorine atoms are located on carbon atoms 1 and 4 in the benzene ring. The location of a group can strongly affect the properties of a material.

Examples:

1,i-cyclohexadiene

α-oxoglutaric acid

3. Prefixes Prefixes may designate quantities or qualities.

Examples:

dichloroacetic acid

tert-butyl alcohol

cis-cinnamic acid

Table 6-2 lists locants and prefixes. Notice that some should be italicized.

TABLE 6-2
LOCANTS AND CHEMICAL PREFIXES

Combining Form		*Adjective Form*	*Capitalization in Labels and at Beginning of Sentence*
n-		normal	*n*-Butane
sec-		secondary	*sec*-Butyl alcohol
tert- or *t-*		tertiary	*tert*-Butyl alcohol
cis-		cis	
trans-		trans	
sym- or s-		symmetrical	
asym-		asymmetrical	
unsym- or uns-		unsymmetrical	
syn-		syn	Cis form
anti-		anti	*cis*-Cinnamic acid
endo-		endo	
exo-		exo	
erythro-		erythro	
threo-		threo	
meso-		meso	
scyllo-		scyllo	
gem-		geminal	Geminal form
vic-		vicinal	*gem*-Dimethyl
m- or *meta-*	Not invariably abbreviated, but should be consistent	meta	
o- or *ortho-*		ortho	*p*-Aminobenzoic acid
p- or *para-*		para	*para*-Chlorophenol
β-		*β or* beta (*β* oxidation)	*β*-Alanine
D-		D	D-Glucopyranose
D(+)-		D(+)	D(+)-*α*-Phenethyltrimethylammonium
L-		L	
DL-		DL	
d-		*d or* dextro	
l-		*l or* levo	*d*-Tartaric acid
dl-		*dl*	
(+)-		dextrorotatory	
(−)-		levorotatory	(−)-Ephedrine
(*R*)- (rectus)		*R*	1(*S*),3(*S*)-Dimethylcyclohexane
(*S*)- (sinister)		*S*	
(*E*)- (entgegen)		*E*	(*Z*)-1-Bromo-1,2,-dichloroethene
(*Z*)- (zusammen)		*Z*	
bis			
tris			
cyclo		cyclo	Tris(*p*-nitrophenyl)methyl bromide
di			
bi			Isovaleric acid
mono		mono	
ixo		iso	

TABLE 6-2
LOCANTS AND CHEMICAL PREFIXES (continued)

Combining Form	*Adjective Form*	*Capitalization in Labels and at Beginning of Sentence*
2,2-dimethyl-3-chlorobutane		2,2-Dimethyl-3-chlorobutane
bicyclo[2.2.1]heptane	bicyclo compound [2.2.1] system	Bicyclo[2.2.1]heptane
N-, *O*-, *S*-, etc. (substituent on nitrogen, oxygen, sulfur, etc.)	the *N* position	*N*,*N′*-Bis(*p*-hydroxyphenyl)-2-pentene-1,5-diamine

Reprinted with permission from Marie Longyear, *The McGraw-Hill Style Manual*, McGraw-Hill, New York, 1982.

6.4.8 The Creation of New Organic Compounds

The organic chemist and chemical engineer have created many of the materials we use in our daily lives by putting together—*synthesizing*—new organic compounds. Much of chemical engineering involves the development of new synthesizing processes.

1. Mers and Polymers

A compound can be formed by *linking* many similar units together. These units are called *mers*.

> *Mer* comes from the Greek word *meros*, meaning "part." In this case it means a basic structural unit. It is also used as a combining form, "-mer," in constructing chemical terminology (see Section 6.4.4 and below).

The compound so formed is known as a *polymer* (from the Greek "many parts"), and the process of linking (with covalent bonds) similar units (mers) is known as *polymerization*. If you work with organic chemists, much of your text will deal with this subject.

2. Giant Molecules

Polymers may become so complex (be formed of so many mers) that the resulting molecules are called *giant molecules*. Nature engages in this polymerization process too. Fats, carbohydrates, proteins, and nucleic acids are such giant molecules; thus much living matter consists of such molecules.

> Some common synthesized materials consist of giant molecules, for example, rubber and most plastics.

The chemist who creates giant molecules designs them to have different mechanical properties.

One example is elasticity. When elastic compounds are formed into fibers and woven into cloth, the result is stretch fabrics. (See "Elasticity," Section 4.2.3, and Figure 4-2.)

6.5 MOLECULAR BIOLOGY AND GENETIC ENGINEERING

It is not within the scope of this book to treat the life sciences such as medicine and the various areas of biology. To do this would require an additional volume of the same size as this one. But these days, biology and chemistry often overlap, especially in the areas of *biochemistry* and *molecular biology*. Both biochemists and molecular biologists study the molecules of which living cells are composed.

Molecular biology today is where solid-state physics was in the 1950s, when solid-state physics gave birth to the semiconductor explosion which has given us transistors, computers, and many other devices and systems which are the subject of much of this book. Molecular biology has given birth to *genetic engineering,* which is just beginning to "explode." Genetic engineering will be the source of much research and business in the United States from here on, and many of the readers of this book will find themselves working with text in this field.

6.5.1 Keyboarding Similarities with Chemistry and Physics

Most of the material you will work with in the fields of molecular biology and genetic engineering will resemble what has been described in the section on organic chemistry. You will also use most of the other techniques of chemical notation described in this chapter.

Substantial text will resemble physics text (especially in the field called *biophysics*). Reference to Chapter 4 should be useful, especially Section 4.7 and Section 4.8.

6.5.2 Nomenclature, Notation, Biological Concepts

What is different in the fields of molecular biology and genetic engineering is that they have developed (and are continuing to develop) their own systems of nomenclature and their own notation practices (including methods of diagramming molecules and reactions). Fortunately, these are similar to the practices of chemistry and physics dealt with here, so that although you must learn them on the job, you will undoubtedly find that they are only "more of the same."

Basic Practice—Techniques and References

6.6 CHEMICAL SIGNS AND SYMBOLS

There are very well defined ways to represent chemicals and their interactions with each other. Descriptions of these are given in the following sections, along with information about some of the special symbols, prefixes, and suffixes used in constructing the names of chemicals.

Organic chemistry requires the greatest care and good judgment on the part of the technical secretary and technical editor because organic chemistry makes heavy use of *structural formulas*. Rules for constructing such formulas on a typewritten page are outlined.

When the editor has any doubt about the meaning of what is being reproduced and the reasons for a particular practice, he or she is advised to refer to the relevant section of the first part of this chapter.

6.6.1 The Chemical Elements

At this writing 107 chemical elements have been identified. Names have been given to 103 of them; numbers 104, 105, 106, and 107 remain unnamed. They are listed in Glossary H with their abbreviations.

Two elements have two names. Both names are included in Glossary H, with the preferred one indicated.

6.6.2 Writing the Chemical Elements

Chemical elements appear in narrative as words or symbols, and in notation as the same symbols.

1. As Words When used in written text, the names of the elements are merely nouns and are written without capitalization.

The ring is made of gold.

The bumper is plated with chromium.

2. As Symbols In chemical notation, such as in chemical equations, symbols are always used. These are very rigidly prescribed in form and are used internationally. Symbols always consist of one or two letters (never more).

- K is the symbol for potassium.
- Na is the symbol for sodium.

Always check Glossary H for the correct symbol.

Note that the first letter is always capitalized; the second is always written lowercase and always on the same line, not as a subscript. Periods are not used.

The author may also choose to use symbols in written text instead of writing out the name of the element. However, consistency is preferred. It is not considered good usage to mix symbols and words in the same passage. Thus "the potassium atom and the sodium atom will be discussed" or "the K atom and the Na atom will be discussed" are both acceptable. But "the potassium atom and the Na atom will be discussed" is not good usage.

6.6.3 Graphic Symbols

Glossary I tabulates the symbols and schemes of notation used in chemical text. These relate particularly to chemical reactions and chemical bonds.

6.6.4 Chemical Compounds

Chemical compounds may be expressed as words and as formulas.

1. As Words In written text, the name of a compound is usually spelled out.

Thus

potassium chloride	(a compound of potassium and chlorine)
iron oxide	(a compound of iron and oxygen)

NOTE: In each case there are simply two nouns; neither is capitalized.

2. As Formulas A chemical formula represents a compound by showing the symbols of its components

Thus

potassium chloride	KCl
iron oxide	FeO

NOTE: Spaces are never left between chemical symbols in a formula.

INCORRECT:	K Cl	and	Fe O
CORRECT:	KCl	and	FeO

These examples show compounds in which the ratios of the different elements are 1:1, or one of each. When the proportions are more complex, numerical subscripts show this.

carbon dioxide	CO_2
water	H_2O
aluminum oxide	Al_2O_3

3. Parentheses Groups of atoms called *radicals* are denoted by the use of parentheses.

$$Zn_3(PO_4)_2$$

4. Leaving out the Parentheses With compounds in which radicals occur singly, as $H_2(SO_4)$, it is customary to omit the parentheses—H_2SO_4 (preferred).

NOTE: Rules for writing formulas have been made by the International Union of Pure and Applied Chemistry (IUPAC), and these rules are being followed here.

Thus

NaOH is correct and prescribed by IUPAC.

Na(OH) is correct, but poor form.

$MgSO_4$ is correct and prescribed by IUPAC.

$Mg(SO_4)$ is technically correct, but not usual and not recommended.

CAUTION: It is important that you never make the error of omitting the parentheses in multiples of radicals, where they are required. For example, $Ca(OH)_2$ (where the subscript 2 is a multiple) is a very different compound from $CaOH_2$. The first compound has two oxygen atoms, but the second has only one oxygen atom.

5. Use of the Dot to Show Water of Hydration Compounds which have attached water of hydration make use of a dot, placed one-half index up, before the water:

$$CuSO_4 \cdot 5H_2O$$

6. Number and Letter Prefixes (Locants) When these are used the numeral or letter precedes the term it refers to, separated by a hyphen. When two numbers or letters occur together, a comma is used between them without space,

2-chloro-1,3-butadiene

2-(3-chlorophenoxy)-propionic acid

β-chloropropionitrile

N-dimethylformadine

7. Numerical Prefixes Prefixes such as "mono-," "bi-," "hexa-," and so forth are normally written without hyphenation, although there may be exceptions.

sodium bicarbonate

monosodium glutamate

potassium hexafluoroarsenate

Some prefixes may occur within an expression, in which case they usually precede a group enclosed in parentheses or brackets.

tetrakis(triphenylphosphine) palladium

Lists of such prefixes may be found in Table 6-2 and Glossary D.

6.6.5 Signs That Show Equality

1. The Equals Sign If your keyboard has no equals sign, it may be typed using two hyphens. Leave one space around all signs of operation.

$$H_2 + \frac{1}{2}O_2 = H_2O$$

2. The Arrow The same reaction may be written with an arrow, with a space separating the ends of the arrow from both sides of the equation.

$$H_2 + \frac{1}{2}O_2 \rightarrow H_2O$$

NOTE: If your keyboard does not have arrows, draw them neatly with a fine fiber-tipped pen and with the aid of a ruler. Some precautions for drawing similar arrows used in vector algebra are given in "Writing Vectors" in Section 2.6.5.

Refer to "Signs That Show Equality" in Section 6.1.7 for instructions on other types of arrows such as double arrows, reverse arrows, and unequal arrows.

Figure 6-4 illustrates the writing of chemical equations.

6.6.6 Representing Changes of State in Chemical Equations

A number of symbols can be added to ordinary chemical equations to show that physical (as well as chemical) changes have taken place. These are letters in parentheses, vertical arrows, overbars, and underbars. They all refer to the states of matter.

- Liquid: $_{(l)}$ or (l)
- Solid: $_{(s)}$ or (s) or ↓ or $\underline{\quad}$
- Gas: $_{(g)}$ or (g) or ↑ or $\overline{\quad}$

The (l), (s), (g) designations—at index level—are preferred.

PRODUCTION OF SULFURIC ACID (CATALYTIC PROCESS):

a) $\frac{1}{2}O_2 + 2VO_2 \rightarrow V_2O_5$

b) $V_2O_5 + SO_2 \rightarrow 2VO_2 + SO_3$

c) $SO_3 + H_2SO_4 \rightarrow H_2S_2O_7$

d) $\underline{H_2S_2O_7 + H_2O \rightarrow H_2SO_4 + H_2SO_4}$

e) $SO_3 + H_2O \rightarrow H_2SO_4$

REDUCTION OF MnO_4^- IN ALKALINE SOLUTION:

a) Na_2SO_3:

$$2MnO_4^- + 3SO_3^{2-} + H_2O \rightarrow 2MnO_2\downarrow + 3SO_4^{2-} + 2OH^-.$$

b) $MnCl_2$:

$$2MnO_4^- + 3Mn^{2+} + 4OH^- \rightarrow 5MnO_2\downarrow + 2H_2O.$$

FIG. 6-4 Inorganic and ionic equations. This is an actual page of text that could be typed by a technical secretary. *(Courtesy of Stohler Isotope Chemicals, Waltham, Mass.)*

NOTE: Some publications prefer italic for the letter symbols, for example, AgBr(*s*) or $AgBr_{(s)}$ or $SO_3(g)$. As in other instances of terminology, the important thing is to be consistent within any one manuscript. When in doubt, leave it roman. It's easier and neater for the editor to add the marks for italic than to delete them.

6.6.7 Thermodynamic Notation in Chemical Reactions

Thermodynamic values may be stated with chemical reactions. They follow the chemical reaction to which they pertain, after a semicolon.

Four thermodynamic quantities are used for this purpose.

- Temperature T
- Heat content or enthalpy H
- Free energy G (or F or $F°$)
- Entropy S

NOTE: *G* is the preferred form for free energy. *F* and *F°* are the older forms, now falling into disuse.

Changes in these quantities are shown using the Greek letter *delta,* ΔH, ΔG, etc.

The equation is always understood to be in the form

$$\Delta G = \Delta H - \Delta ST$$

For example,

$$H_2(g) + \tfrac{1}{2}O_2(g) \rightarrow H_2O(g); \Delta G = -60{,}180 + 13.9T$$

6.6.8 Expressing Ions

Charges associated with ions are always designated by a plus sign, +, for positive charges and a minus sign, −, for negative charges. If there is more than one charge, there are two choices. Either a numeral is used with the charge sign ($^{3+}$), or the sign is repeated ($^{+++}$). The former is preferred.

These are always written as superscripts following the symbol for the chemical element or radical to which they pertain.

Thus

The sodium ion is written Na^+.

The chlorine ion is written Cl^-.

The nitrate ion is written NO_3^-.

The sulfate ion is written SO_4^{2-}(preferred) or SO_4^{--}.

The iron ion is written Fe^{2+} (for valence of 2).

The iron ion is written Fe^{3+} (for valence of 3).

NOTE: When the ion radical appears in parentheses with its own numerical subscript, then the ionic superscript is applied as follows:

$$(SO_4)_3{}^{2-}$$

Ions may be used in ionic equations. Figure 6-4 illustrates such usage.

6.6.9 Symbols for Free Electrons

Free electrons are represented by

e$^-$ (preferred)

e^-

or ⊖

The symbol ⊖ differs from the Greek letter *theta*, *θ*, in that the circle should be completely round rather than oval like the capital O or the theta. Practice drawing this before commiting it to the typed page.

6.6.10 The Carbon Bond

With organic chemistry text, compounds are shown structurally.

Except for the most complex compounds, the secretary or the reproduction department is expected to know how to do this.

1. The Bonds Bonds are almost always short, straight lines hand-drawn with a fine fiber-tipped pen and a straightedge.

Bonds are normally between three-sixteenths and three-eighths of an inch long. The length depends on your judgment of what looks neat and readable. Bonds are usually positioned either horizontally or vertically, but some formulas require that they be drawn at 45-degree angles, and some at any convenient angle.

2. Multiple Bonds Carbon bonds may be single, double, or triple. In the latter two cases, the bonds are drawn as parallel duos or trios.

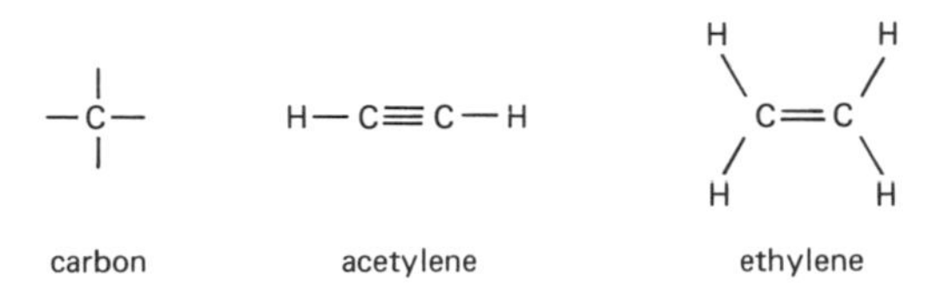

3. Dots Dots, usually in pairs, may appear at bond sites. (See the example in the following subsection.)

4. Curved Arrow Symbols These show the movement of electrons or atoms when a reaction takes place. Draw such arrows freehand, smoothly and neatly.

:Ö:

‖

CH_3—C—OCH_3

5. Ring Symbols These are generally in the form of hexagons.

You may be required to draw the hexagon many times. Learn how to do it neatly and in a way such that each one you draw looks like the others.

Templates exist for the hexagons used to make benzene rings, and they should be used whenever possible. If templates are not available, the rings can be carefully hand-drawn using a fine felt-tipped pen.

Other ring variations must also be produced exactly. Use templates whenever possible.

Refer to Section 6.4.6 for additional information on drawing ring symbols.

NOTE: There is no rule that says how much space a structural formula should occupy on the typewritten page. If it seems logical to stretch it to fill the whole width of the paper, that should be done.

6. Boldface Symbols Structural diagrams make use of the wedge and of particular bonding lines, both drawn with heavy lines.

Ureidosilanes. Of the several reactive silane comonomers used in the preparation of poly(carboranesiloxanes), the bis(ureido) silanes were the preferred monomer, Peters et al. (1977). These reagents are powerful silylating reagents which do not form acid or basic byproducts during reaction. Although a large structural variety of bis(ureido) silanes were prepared, the particular ureido group chosen to provide the most easily synthesized and purified crystalline solids was the N-phenyl-N', N'-tetramethylene structure, VII, in which the group R could be varied at will. These are prepared by reaction of the corresponding bis(pyrrolidino) silanes with phenyl isocyanate, eq. 10.

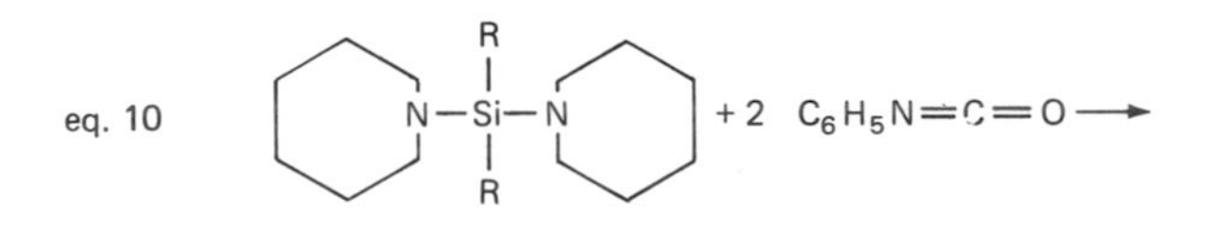

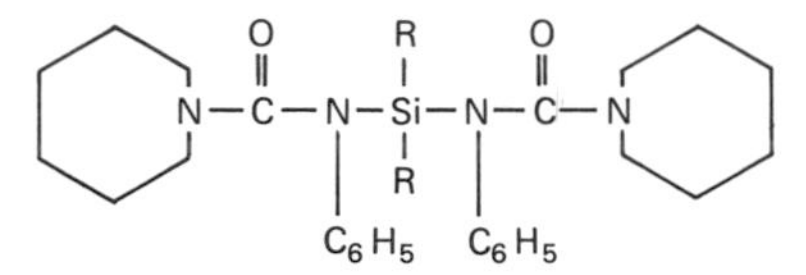

Carborane-siloxanes were first prepared by the ferritic chloride catalyzed polymerization of dichlorodimethylsilane, with bis(methoxydimethylsilyl) m-carborane, eq. 11, Schroeder et al. (1966).

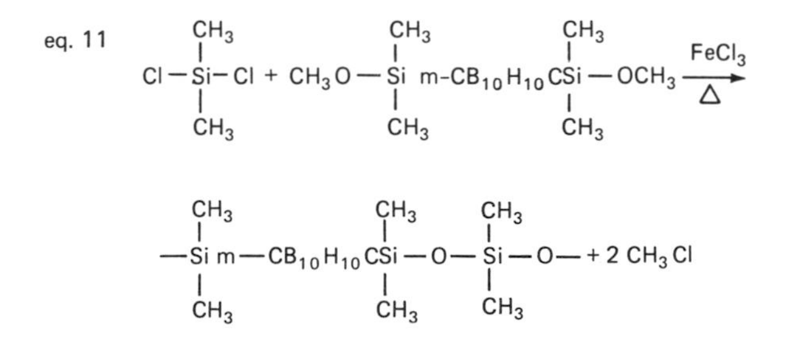

FIG. 6-5 Organic chemistry. This is an actual page of text typed by a technical secretary. *(Courtesy of General Electric Company, Pittsfield, Mass.)*

Example of a formula using wedges:

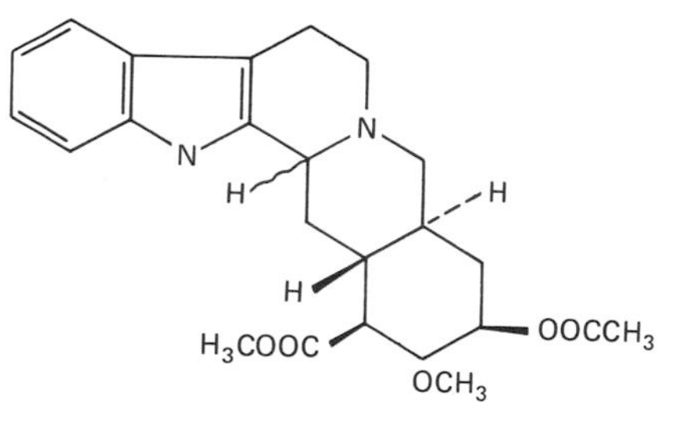

Example of a formula using boldface bond symbol.

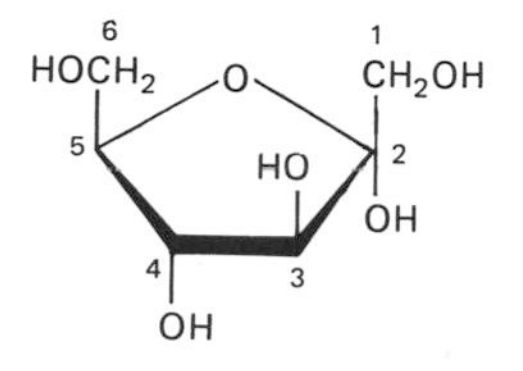

As always, draw these symbols neatly with a straightedge and a fine felt-tipped pen.

6.6.11 Depicting Organic Chemical Reactions

In general, the rules for organic chemical equations are the same as those for inorganic chemical equations (see Section 6.1.7). The difference lies in the fact that the person reproducing them has a real challenge in the task of layout.

The following equation is an example:

$$CH_3-\overset{\overset{\ddot{O}:}{\|}}{C}-CH_3 + OH^- \rightleftharpoons CH_3\ \underset{OH}{\overset{\ddot{O}:^-}{C}}-CH_3 \xrightleftharpoons{+H_2O} CH_3-\underset{OH}{\overset{OH}{C}}-CH_3 + OH^-$$

NOTE: The bent-over double arrow shown in the example is an expedient which is allowed when space is a problem. The best practice is to minimize its use.

Figure 6-5 shows an example of a page of organic equations such as you might encounter.

PART TWO

Reference

Part 2 presents reference material in the form of organized lists, arranged for convenient and rapid reference.

GLOSSARIES

The glossaries are organized into three groups. The first group is a reference for the *language matters* of Chapter 1. Here one can find information on alphabets and on the construction of scientific words.

The second group is a reference for the five chapters that describe the five major disciplines of mathematics, computers, physics, electricity and electronics, and chemistry. These put into one location the most often-needed information on how to depict symbols both in text and in graphics. Although some of the content of these glossaries is repeated in the individual chapters, the convenience of having it all assembled in one place is felt to be worth an occasional repetition. This set of glossaries also includes lists of the most used physical units and abbreviations.

The third group of glossaries lists miscellaneous reference material, primarily books and manuals that will be useful in order to put to rest any final doubts about how to spell or how to present any questionable item of scientific or technical text.

The glossaries are as follows:

General reference

Glossary O	Abbreviations of Scientific and Technical Organizations
Glossary P	Names of Scientists and Engineers
Glossary Q	Selected Science and Technology References
Glossary R	General References

DICTIONARY OF TECHNICAL TERMS

Following the glossaries is a dictionary containing several thousand of the most commonly encountered scientific and engineering terms. For a brief explanation of how the dictionary is arranged, see p. R.47.

The Greek Alphabet

Name of Letter	*Greek Alphabet*		*Latin Equivalent*
Alpha	Α	α α*	a
Beta	Β	β	b
Gamma	Γ	γ	g
Delta	Δ	δ ∂*	d
Epsilon	Ε	ϵ	e
Zeta	Ζ	ζ	z
Eta	Η	η	ē
Theta	Θ	θ ϑ*	th
Iota	Ι	ι	i
Kappa	Κ	κ	k
Lambda	Λ	λ	l
Mu	Μ	μ	m
Nu	Ν	ν	n
Xi	Ξ	ξ	x
Omicron	Ο	ο	o
Pi	Π	π	p
Rho	Ρ	ρ	r
Sigma	Σ	σ	s
Tau	Τ	τ	t
Upsilon	ϒ	υ	u
Phi	Φ	ϕ φ*	ph
Chi	Χ	χ	kh
Psi	Ψ	ψ	ps
Omega	Ω	ω	ō

*Old-style character.

Variations in the Latin Alphabet

Danish has three additional letters:

Å å, Ææ, Ø ø

Finnish has two umlauted vowels:

Ä ä, Ö ö

French uses two wedge-shaped accent marks, the *acute,* which points to two o'clock, and the *grave,* which points to ten o'clock.

église (acute)
crème (grave)

French also has the cedilla, which makes *c* soft and the circumflex, which indicates that in Latin an *s* followed the accented vowel.

garçon (cedilla)
fenêtre (circumflex)

German has one additional letter, the double s.

ß

German may also apply umlauts to certain vowels:

Ä ä, Ö ö, Ü ü

Hungarian uses several varieties of accented vowels:

Á á, É é, Í í, Ó ó, Ö ö, Ő ő, Ú ú, Ü ü, Űű

Italian may use the grave accent, as in French.

Norwegian has the same three additional letters as Danish.

Polish requires the following special characters:

Ą ą, Ć ć, Ę ę, Ł ł, Ń ń, Ó ó, Ś ś, Ź ź, Ż ż

Portuguese has three special characters:

Ã ã, Ç ç, Õ õ

Spanish has one special character:

Ñ ñ

Spanish also repeats the question mark and the exclamation point at the beginning of a question or an exclamation—upside down:

> ¿Qué pasa, amigo?
> Por favor, señor ¿donde está la biblioteca municipal?
> Alguien viene. ¡Vamonos!

Swedish has the following special characters:

Å å, Ä ä, Ö ö

Turkish requires the following special characters:

Â â, Ç ç, Ğ ğ (*or* Ğ ğ), İ ı, Ö ö, Ş ş, Û û, Ü ü

Preferred Plural Forms

Singular	*Plural*
	-a
formula	formulas
	-is or -ix
analysis	analyses
appendix	appendixes
axis	axes
basis	bases
hypothesis	hypotheses
matrix	matrices
synthesis	syntheses
thesis	theses
	-um, -us, or -on
criterion	criteria
datum	data
erratum	errata
focus	foci
fulcrum	fulcrums
medium	media
nucleus	nuclei
phenomenon	phenomena
radius	radii
stylus	styli

Numerical Prefixes

Numerical prefixes	
One	mon-, mono-, primo-, uni-
Two	bi-, di-, do-, duo-, sec-
Three	ter-, tri-
Four	quad-, quadro-, tert-, tetra-
Five	pent-, penta-
Six	hex-, hexa-, sex-, sexa-
Seven	sept-, septi-, hept-, hepta-
Eight	oct-, octa-, octo-
Nine	nona-
Ten	dec-, deca-, deka-
Twelve	dodeca-

*See Table 2-3 for numerical equivalents of SI prefixes.

GLOSSARY E

Descriptive Prefixes

Prefix	Meaning
aceto-	The acetate radical
aero-	Relating to flight
amino	Containing the radical NH_2
anti-	Against, opposite
asym-	Asymmetrical
calc-, calci-, calco-	Pertaining to calcium
chemi-	Chemical
chloro-	Chlorine
cis-	This side of
co-, con-	Together
cryo-	Low-temperature
cyclo-	Circle
demi-	Partly
e-	Not; out of
electro-	Electricity
endo-	Taking in
epi-	On, over, above
equi-	Equal
ex-	Not; out of
exo-	Giving off
extra-	Outside
ferri, ferro-	Iron
fluo-, fluor-	Giving off light
gem-	Geminal
gon-	Angle
heli-, helio-	The sun
hemi-	Half
hemo-	Blood
hetero-	Different
homo-, homeo-	Same
hydro-	Water
hygro-	Humidity
hyper-	Greater than
hypo-	Above
in-	In, included
infra-	Less than; below
intra-	Within
intro-	Between
is-, iso-	The same
leuco-, leuko-	White
lum-	Light
macro-	Large
magneto-	Magnet

Prefix	Meaning
mal-	Bad
meso-	Middle
meta-	After, later
mid-	Middle
mini-	Small
mis-	Wrong
morph-, morpho-	Shape
mult-, multi-	Many
ne-, neo-	New
neuro-	Nerve
non-	Not
ob-	Opposing
organo-	Organic
orth-, ortho-	Straight
ox-, oxy-	Oxygen
paleo-	Ancient
para-	Similar
per-	Through; containing most
peri-	Encircling
petr-, petri-	Stone
phil-	Having affinity
phon-, phono-	Sound
phot-, photo-	Light
pneumo-	Air
poly-	Many
post-	After
pre-	Before
pro-	In favor of; in front of
proto-	First
pyr-, pyro-	Fire
quasi-	Similar
radi-	Radial
re-	Again
retro-	Back
rheo-	Flow
semi-	Partial
spectro-	Spectrum
stereo-	Solid; three-dimensional
strati-	Layered
sub-	Under
sulfa-	Sulfur
super-	More than
supra-	Above
sur-	Over
sym-	Symmetrical
syn-	With
synchro-	Synchronous
taut-	Same
tax-, taxo-	Classification
tel-, tele-	Across space
thermo-	Heat
thio-	Containing divalent sulfur atom
trans-	Across
ultra-	Beyond
un-	Contrary to
unsym-, uns-	Unsymmetrical
vari-	Varied

GLOSSARY F

Suffixes

Suffix	Meaning
-age	Quantity of
-ance	Quantity of
-ane	An organic functional group
-ase	Enzyme
-ate	A radical of a salt having more oxygen than an "-ite" salt
-ene	An organic functional group
-fuge	Flowing
-gon	Angle
ic	An acid having more oxygen than an "-ous" acid
-ics	Practice of
-ide	A two-element compound
-ile	Relating to
-ion	Ion
-ite	A radical of a salt having less oxygen than an "-ate" salt
-ium	Designates a chemical element
-ive	Tending toward
-ivity	Amount of units of a given quality
-mer	Organic bonding
-modality	Class
-ode	An electronic path
-oid	Similar to
-ol	An organic functional group
-osis	A process
-otic	Pertaining to
-ous	An acid having less oxygen than an "-ic" acid
-phil, -philic	Having affinity for
-phobic	Having an aversion for
-phoresis	Relating to transmission
-phyte	Plant
-scence	Quality of
-scope	Viewing instrument

-tron	Electronic device
-tuple	Multiplying
-ular	Relating to
-um	*See* -ium
-ure	Process
-urgy	Art of
-yl	An organic functional group
-yne	An organic functional group

GLOSSARY G

Mathematical Signs and Symbols*

Symbol	*Definition*
$+$	Plus (sign of addition)
$+$	Positive
$-$	Minus (sign of subtraction)
$-$	Negative
$\pm\ (\mp)$	Plus or minus (minus or plus)
$\times$	Times, by (multiplication sign)
$\cdot$	Multiplied by
$\div$	Sign of division
$/$	Divided by
$:$	Ratio sign, divided by, is to
$::$	Equals, as (proportion)
$<$	Less than
$>$	Greater than
$\ll$	Much less than
$\gg$	Much greater than
$=$	Equals
$\equiv$	Identical with

Symbol	*Definition*
$\sim$	Similar to
$\approx$	Approximately equals
$\cong$	Approximately equals, congruent
$\leq$	Equal to or less than
$\geq$	Equal to or greater than
$\neq$	Not equal to
$\rightarrow\ \doteq$	Approaches
$\propto$	Varies as
∞	Infinity
$\sqrt{\ }$	Square root of
$\sqrt[3]{\ }$	Cube root of
$\therefore$	Therefore
$\parallel$	Parallel to
$(\ )[\ ]\{\ \}$	Parentheses, brackets, and braces, quantities enclosed by them to be taken together in multiplying, dividing, etc.

$\overline{AB}$	Length of line from A to B
π	(Pi) = 3.14159+
$°$	Degrees
$'$	Minutes
$''$	Seconds
$\angle$	Angle
dx	Differential of x
Δ	(Delta) difference
Δx	Increment of x
$\partial u/\partial x$	Partial derivative of u with respect to x
$\int$	Integral of
$\int_b^a$	Integral of between limits b and a
$\oint$	Line integral around a closed path
Σ	(Sigma) summation of
$f(x)$, $F(x)$	Functions of x
$\exp x = e^x$	(e = naperian log base) (abbreviation for e^x)
∇	(Del or nabla) vector differential operator
∇^2	Laplacian operator
£	Laplace operational symbol
4!	(Factorial 4) = $1 \times 2 \times 3 \times 4$
$\|x\|$	Absolute value of x
$\dot{x}$	First derivative of x with respect to time
$\ddot{x}$	Second derivative of x with respect to time
$\mathbf{A} \times \mathbf{B}$	Vector product; magnitude of **A** times magnitude of **B** times sine of the angle from **A** to **B**; $\overline{AB} \sin \overline{AB}$
$\mathbf{A} \cdot \mathbf{B}$	Scalar product of **A** and **B**; magnitude of **A** times magnitude of **B** times cosine of the angle from **A** to **B**; $AB \cos \overline{AB}$

*Reprinted with permission from the *McGraw-Hill Dictionary of Scientific and Technical Terms*, 3d ed., McGraw-Hill, New York (1984).

The Chemical Elements

Element	*Symbol*	*Element*	*Symbol*
Actinium	Ac	Copper	Cu
Aluminum	Al	Curium	Cm
Americium	Am	Dysprosium	Dy
Antimony	Sb	Einsteinium	Es
Argon	A	Erbium	Er
Arsenic	As	Europium	Eu
Astatine	At	Fermium	Fm
Barium	Ba	Fluorine	F
Berkelium	Bk	Francium	Fr
Beryllium	Be	Gadolinium	Gd
Bismuth	Bi	Gallium	Ga
Boron	B	Germanium	Ge
Bromine	Br	Gold	Au
Cadmium	Cd	Hafnium	Hf
Calcium	Ca	Helium	He
Californium	Cf	Holmium	Ho
Carbon	C	Hydrogen	H
Cerium	Ce	Indium	In
Cesium	Cs	Iodine	I
Chlorine	Cl	Iridium	Ir
Chromium	Cr	Iron	Fe
Cobalt	Co	Krypton	Kr

Element	*Symbol*	*Element*	*Symbol*
Lanthanum	La	Rhenium	Re
Lawrencium	Lw	Rhodium	Rh
Lead	Pb	Rubidium	Rb
Lithium	Li	Ruthenium	Ru
Lutetium	Lu	Samarium	Sm
Magnesium	Mg	Scandium	Sc
Manganese	Mn	Selenium	Se
Mendelevium	Md	Silicon	Si
Mercury	Hg	Silver	Ag
Molybdenum	Mo	Sodium	Na
Neodymium	Nd	Strontium	Sr
Neon	Ne	Sulfur	S
Neptunium	Np	Tantalum	Ta
Nickel	Ni	Technetium	Tc
Niobium*	Nb	Tellurium	Te
Nitrogen	N	Terbium	Tb
Nobelium	No	Thallium	Tl
Osmium	Os	Thorium	Th
Oxygen	O	Thulium	Tm
Palladium	Pd	Tin	Sn
Phosphorus	P	Titanium	Ti
Platinum	Pt	Tungsten†	W
Plutonium	Pu	Uranium	U
Polonium	Po	Vanadium	V
Potassium	K	Xenon	Xe
Praseodymium	Pr	Ytterbium	Yb
Promethium	Pm	Yttrium	Y
Protactinium	Pa	Zinc	Zn
Radium	Ra	Zirconium	Zr
Radon	Rn		

*A less preferred name for niobium is columbium.
†A less preferred name for tungsten is wolfram.

Graphic Symbols of Chemistry*

*International Graphic Symbols, reprinted with permission from H. Dreyfuss, *Symbol Source Handbook,* McGraw-Hill, New York (1972).

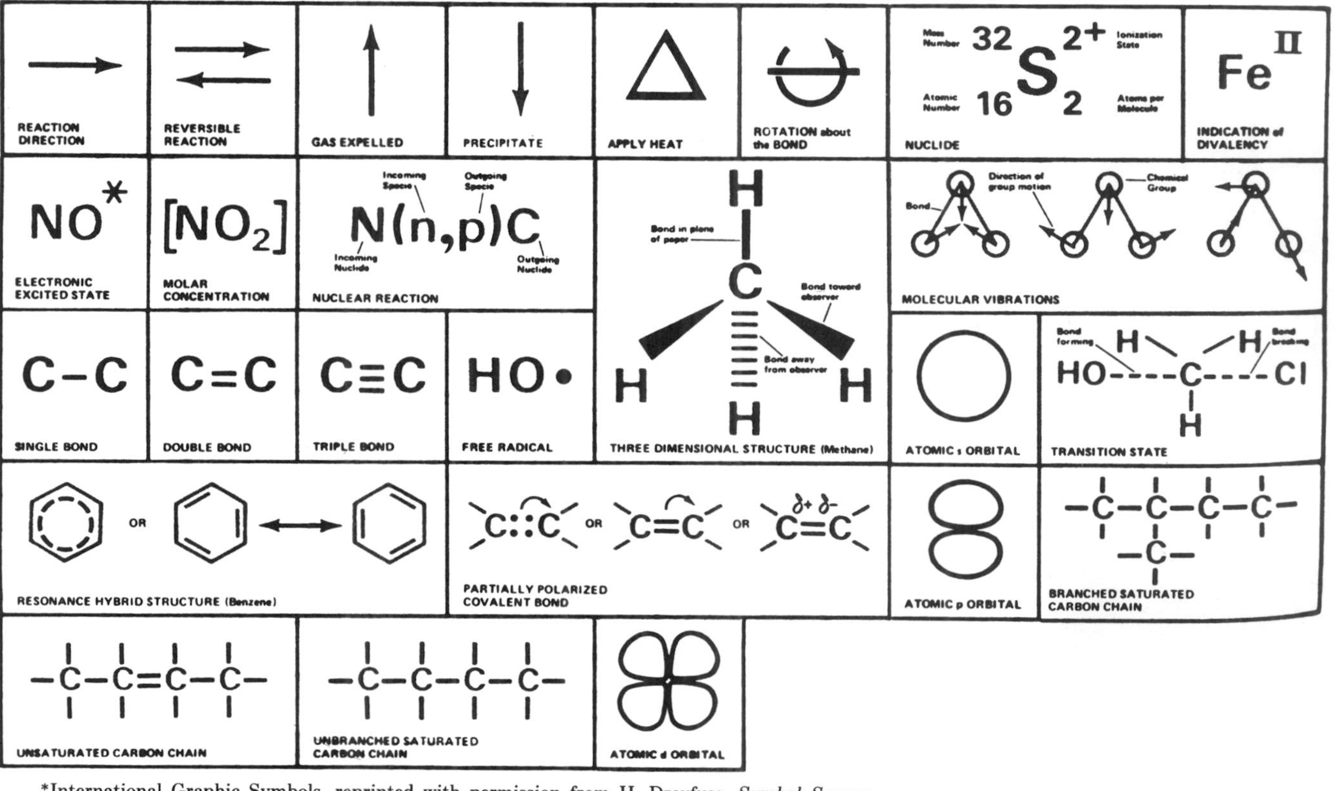

*International Graphic Symbols, reprinted with permission from H. Dreyfuss, *Symbol Source Handbook*, McGraw-Hill, New York (1972).

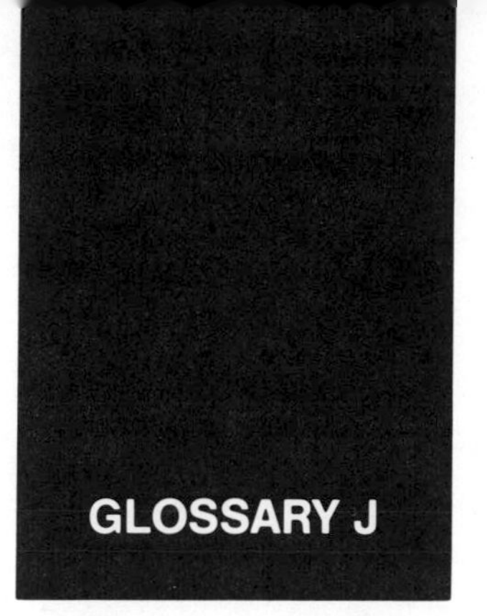

Schematic Electronic Symbols*

*From R. F. Graf, *Modern Dictionary of Electronics,* 4th ed., Indianapolis, Howard Sams (1972).

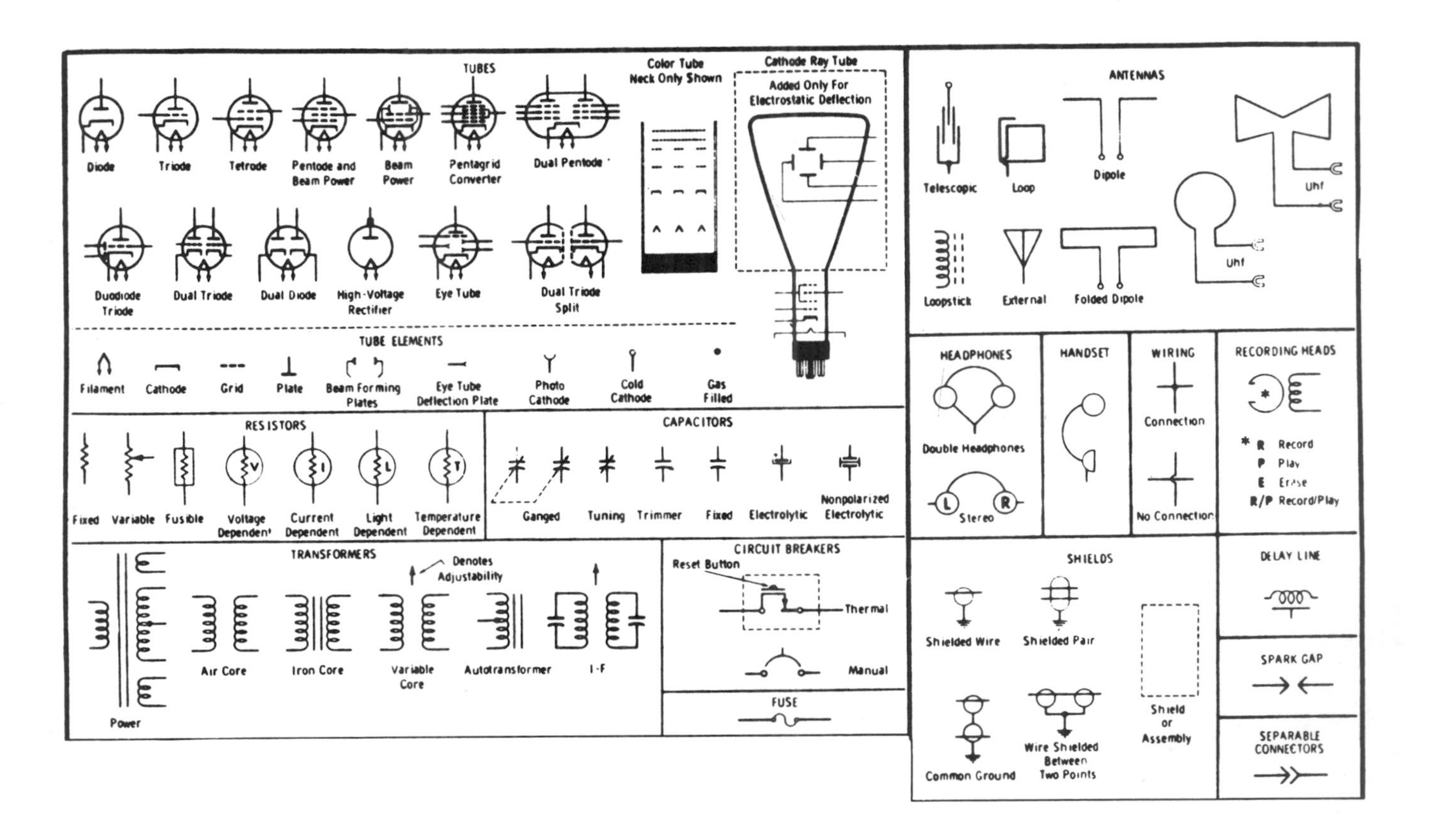
TUBES
Diode
Triode
Tetrode
Pentode and Beam Power
Beam Power
Pentagrid Converter
Dual Pentode
Duodiode Triode
Dual Triode
Dual Diode
High-Voltage Rectifier
Eye Tube
Dual Triode Split
Color Tube Neck Only Shown
Cathode Ray Tube
Added Only For Electrostatic Deflection
TUBE ELEMENTS
Filament
Cathode
Grid
Plate
Beam Forming Plates
Eye Tube Deflection Plate
Photo Cathode
Cold Cathode
Gas Filled
RESISTORS
Fixed
Variable
Fusible
Voltage Dependent
Current Dependent
Light Dependent
Temperature Dependent
CAPACITORS
Ganged
Tuning
Trimmer
Fixed
Electrolytic
Nonpolarized Electrolytic
TRANSFORMERS
Denotes Adjustability
Power
Air Core
Iron Core
Variable Core
Autotransformer
I-F
CIRCUIT BREAKERS
Reset Button
Thermal
Manual
FUSE
ANTENNAS
Telescopic
Loop
Dipole
Uhf
Uhf
Loopstick
External
Folded Dipole
HEADPHONES
Double Headphones
L
R
Stereo
HANDSET
WIRING
Connection
No Connection
RECORDING HEADS
* R Record
P Play
E Erase
R/P Record/Play
SHIELDS
Shielded Wire
Shielded Pair
Shield or Assembly
Common Ground
Wire Shielded Between Two Points
DELAY LINE
SPARK GAP
SEPARABLE CONNECTORS

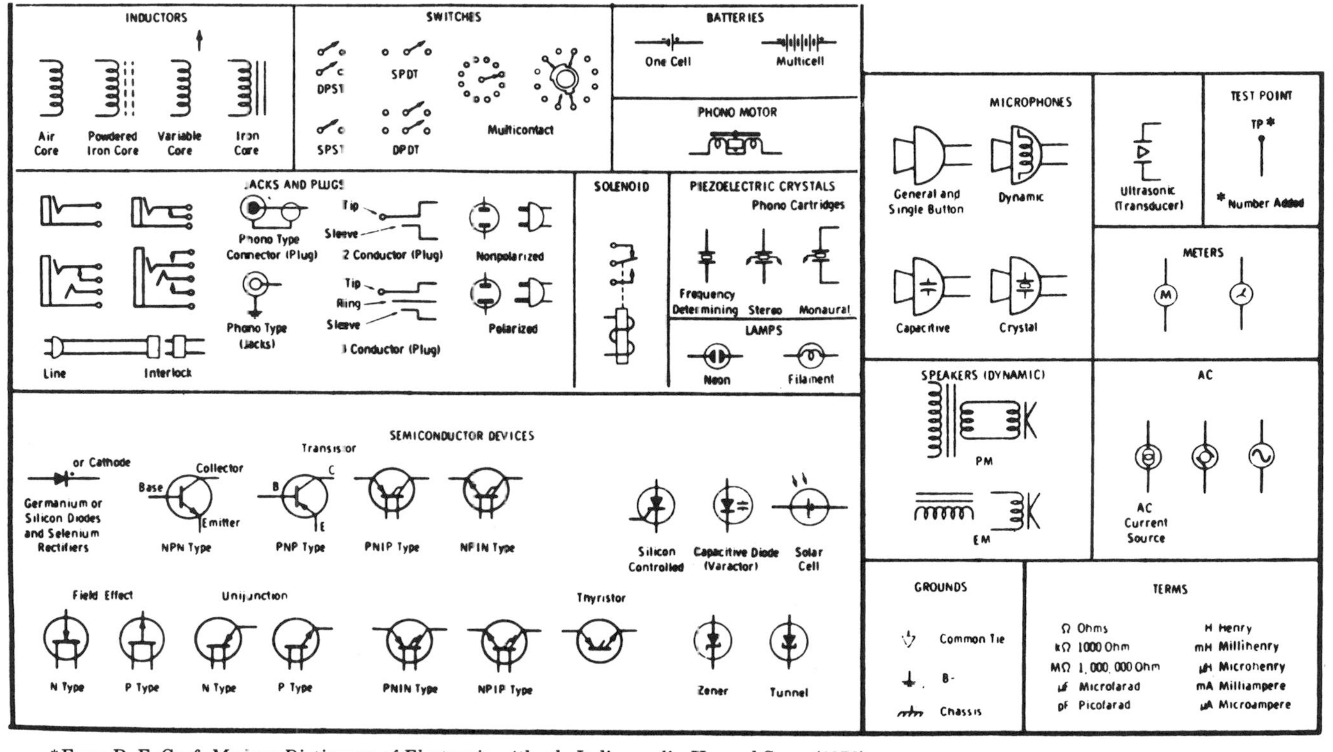

* From R. F. Graf, *Modern Dictionary of Electronics,* 4th ed., Indianapolis, Howard Sams (1972).

Biology Symbols*

*Reprinted with permission from H. Dreyfuss, *Symbol Source Handbook,* McGraw-Hill, New York (1972).

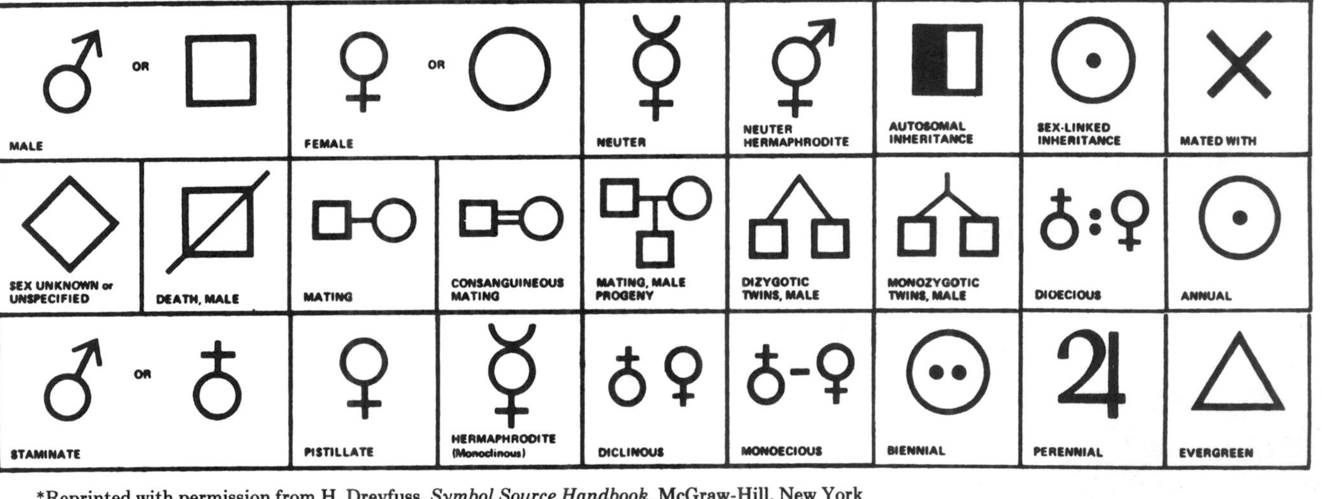

*Reprinted with permission from H. Dreyfuss, *Symbol Source Handbook*, McGraw-Hill, New York (1972).

Computer Flowchart Symbols*

*Reprinted with permission from H. Dreyfuss, *Symbol Source Handbook,* McGraw-Hill, New York (1972).

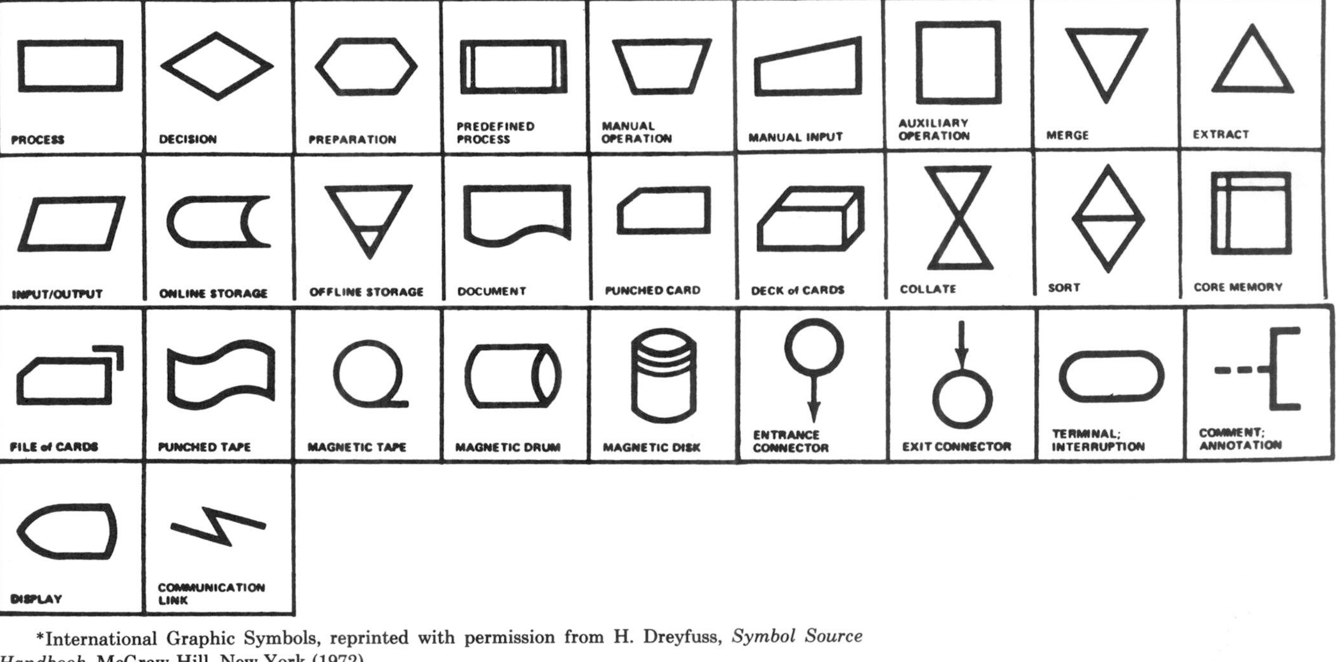

*International Graphic Symbols, reprinted with permission from H. Dreyfuss, *Symbol Source Handbook*, McGraw-Hill, New York (1972).

Commonly Encountered Physical Units*

Unit	*Symbol*
ampere	A
ampere-hour	Ah or A·h
angstrom	Å
astronomical unit	AU
atmosphere	atm
atomic mass unit (unified)	u
bar	bar
barn	b
barrel	bbl
baud	Bd
billion electron volts	GeV
British thermal unit	Btu
calorie	cal
candela (preferred to candle)	cd
centimeter	cm
centipoise	cP
centistokes	cSt
circular mil	cmil
coulomb	C
cubic foot per minute	ft^3/min
cubic foot per second	ft^3/s
curie	Ci
cycle (mechanical)	cycle (also c)
cycle per second (see hertz)	
decibel	dB
degree (plane angle)	. . .°
degree (temperature):	
degree Celsius	°C
degree Fahrenheit	°F
kelvin	K
degree Rankine	°R
dyne	dyn
electron volt	eV
erg	erg
farad	F
foot	ft
foot per second	ft/s

Unit	*Symbol*
foot-pound	ft·lb
footcandle	ftc
footlambert	ft-L
gallon	gal
gauss, pl. gauss	G
gilbert	Gb
grain	gr
gram	g
hectare	ha
henry, pl. henrys	H
hertz	Hz
horsepower	hp
hour	h
inch	in
international unit	IU
joule	J
kelvin	K
kilogram	kg
kilowatt-hour	kWh or kW·h
knot	kn
lambert	L
liter or litre	liter or L or l
lumen	lm
lux	lx
meter or metre	m
mho (see siemens)	
micrometer (formerly micron, μ)	μm
mile	mi
mile per hour	mi/h
millimeter of mercury	mmHg
millimicron (see nanometer)	
minute (time)	min
mole	mol
nanometer (preferred to millimicron, mμ)	nm
nautical mile	nmi
newton	N
newton-meter	N·m
oersted	Oe
ohm	Ω
ounce (avoirdupois)	oz
parsec	pc
pascal	Pa
phot	ph
pint	pt
poise	P
pound	lb
pound per cubic foot	lb/ft^3
pound-force	lbf
pound-force per square foot	lb/ft^2
pound-force per square inch	lb/in^2 or psi (not preferred)
quart	qt
radian	rad
radiation absorbed dose	rad or rd (to avoid confusion with radian)
revolution per minute	r/min
revolution per second	r/s
roentgen	R
second (time)	s
siemens (preferred to mho)	S
square meter	m^2
steradian	sr
stokes	St
tesla	T
ton (2000 lb)	ton
ton, metric (1000 kg or 1 Mg)	t
torr, pl. torr	torr
var	var
volt	V
volt-ampere	VA or V·A
watt	W
watt-hour	Wh or W·h
weber	Wb
yard	yd

*Adapted from *ECS (Editing and Composition Standards)*, McGraw-Hill, New York (1976).

GLOSSARY N

Abbreviations

alternating current	ac	chemically pure	cp
altitude	alt	coefficient (in subscript)	coef
amplitude modulation	AM	concentration	concn
analog-to-digital	A/D	constant	const
anhydrous	anhyd	continuous wave	cw
antilogarithm	antilog	cosecant	csc
approximate (in subscript)	approx	cosine	cos
aqueous	aq	cotangent	cot
atomic weight	AW or at wt	cubic	cu
		cubic centimeter	cm^3
audio frequency	af	curl	∇x
automatic gain control	AGC	data out	DO
average (in subscript)	av	decibel	db
barometer	bar	degree Celsius	°C
barrel	bbl	deoxyribonucleic acid	DNA
Baumé	Be	deviation	dev
body-centered-cubic	bcc	diameter	diam
boiling point	bp	digital-to-analog	D/A
calculated (in subscript)	calc	dilute	dil
cathode-ray tube	CRT	direct current	dc
center of mass	c.m.	divergence	div
centimeter	cm	edition	ed.
centimeter-gram-second (system)	cgs	electromotive force (plural is emf's)	emf

electron paramagnetic resonance	EPR or epr
electron spin resonance	ESR or esr
equation	Eq.
equations	Eqs.
equivalent weight	EW or equiv wt
error function	erf
estimated standard deviation	e.s.d.
experiment(al) (in subscript)	expt(l)
exponential	*e*, exp
face-centered-cubic	fcc
field-effect transistor	FET
field-emission spectroscopy	FES
figure	Fig.
figures	Figs.
footcandle	ftc
footlambert	ft-L
formula weight	FW
frequency modulation	FM
full-scale	F.S.
gradient	grad
hexagonal-close-packed	hcp
high-frequency	HF
hour	h
hyperbolic cosecant	csch
hyperbolic cosine	cosh
hyperbolic cotangent	coth
hyperbolic secant	sech
hyperbolic sine	sinh
hyperbolic tangent	tanh
inch	in or in.
infrared	IR or ir
input/output or input-output	I/O
inside diameter	i.d.
insoluble	insol
integrated circuit	IC
intermediate frequency	IF
light-emitting diode	LED
limit	lim
logarithm	log
logarithm (natural, base *e*)	ln
low frequency	LF
magnetomotive force	mmf
maximum	max
melting point	mp
metal-oxide semiconductor field-effect transistor	MOSFET
meter-kilogram-second (system)	mks
minimum	min
minute	min
mole percent	mol%
molecular weight	MW or mol wt
multiple feedback	MF
noise figure	N.F.
nuclear magnetic resonance	NMR or nmr
number	No.
numerical aperture	N.A.
observed (in subscript)	obs
outside diameter	o.d. or OD
parts per million	ppm
percent volume per volume	%v/v or %vol/vol
percent weight per volume	%w/v or %wt/vol
percent weight per weight	%w/w or %wt/wt
phase modulation	PM
potential difference	PD
pound-foot	lb-ft
pound-inch	lb-in
pounds per square inch	psi or lb/in^2
probable error	pe
precipitate	ppt
radio frequency	rf
ribonucleic acid	RNA
room temperature	RT
root-mean-square	rms
scanning electron microscopy	SEM
secant	sec
second	s
section	Sec.
sections	Secs.
semiconductor	SC
silicon controlled rectifier	SCR

sine	sin	unijunction transistor	UJT
single-pole single-throw (switch)	SPST	valence band	VB
solution	soln	versus	vs
specific gravity	sp gr	voltage-controlled current source	VCCS
square	sq	voltage-controlled oscillator	VCO
standard temperature and pressure	STP	voltage-controlled voltage source	VCVS
Système International	SI	voltage-to-frequency converter	VFC
tangent	tan	volume	vol. or vol
temperature	temp	volume percent	vol%
theory, theoretical (in subscript)	theor	watt-hour	Wh or W·h
total (in subscript)	tot	weight	wt
trace	tr, Tr	weight percent	wt%
ultrahigh-frequency	UHF		
ultraviolet	UV or uv		

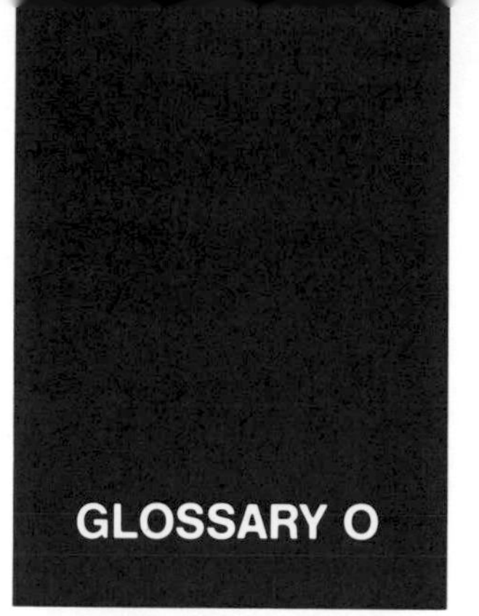

Abbreviations of Scientific and Technical Organizations

AAAS — American Association for the Advancement of Science
AAPM — American Association of Physicists in Medicine
ACA — American Crystallographic Association
ACS — American Chemical Society
AFS — American Foundrymen's Society
AGA — American Gas Association
AIAA — American Institute of Aeronautics and Astronautics
AICE — American Institute of Consulting Engineers
AIChE — American Institute of Chemical Engineers
AIME — American Institute of Mining, Metallurgical, and Petroleum Engineers
AIP — American Institute of Physics
AISI — American Iron and Steel Institute
API — American Petroleum Institute
ASHRAE — American Society of Heating, Refrigerating, and Air Conditioning Engineers (formerly ASHAE and ASH&VE)
ASM — American Society for Metals; American Society for Microbiology
ASTM — American Society for Testing and Materials

ECS	Electrochemical Society, Inc.
EIA	Electronic Industries Association
EJC	Engineers Joint Council
EPRI	Electric Power Research Institute
GSA	Geological Society of America
ICE	Institution of Civil Engineers
IEEE	Institute of Electrical & Electronics Engineers (successor to AIEE)
IGT	Institute of Gas Technology
IMPI	International Microwave Power Institute
INA	Institution of Naval Architects
ISA	Instrument Society of America
MAA	Mathematical Association of America
NIH	National Institutes of Health
NIOSH	National Institute of Occupational Safety and Health
OSA	Optical Society of America
SAMPE	Society of Aerospace Material & Process Engineers
SERI	Solar Energy Research Institute
SPHE	Society of Packaging and Handling Engineers
SWE	Society of Women Engineers

Names of Scientists and Engineers (with common usages)

Ampère (ampere)	Bessemer process
Ångström (angstrom)	Bohr atom
Archimedes principle	Boltzmann (Stefan-Boltzmann law)
Arrhenius	Boole (Boolean algebra)
d'Arsonval meter	Bose
Auger analysis	Boyle's law
Avogadro's number	Bragg angle
Babbage	Bridgman technique
Bain (bainite)	Brillouin zone
Bardeen	Brinnel hardness
Barkhausen effect	de Broglie
Baumé effect	Bunsen burner or bunsen burner
Bernoulli theorem	Brown (brownian motion)
Berzelius	Carnot cycle
Bessel function	Cavendish

Celsius scale
Cerenkov effect
Clausius-Clapeyron equation
Clerk-Maxwell
Compton
Colpitts oscillator
Cotrell process
Coulomb (coulomb)
Curie (curie)
Czochralski technique
Dalton
De Forest
Diesel engine or diesel engine
Dewar flask or dewar flask
Dirac
Doppler effect or doppler effect
Edison
Euclidean geometry
Euler
Faraday (farad)
Fermi level
Fresnel lens
Fahrenheit scale
Fourier series
Frauenhofer lines
Galvani (galvanic)
Gauss (gaussian curve)
Gay-Lussac
Gibbs
Gilbert (gilbert)
Haber process
Haldane
Hall effect
Hamilton (hamiltonian)
Hartley oscillator
Heaviside layer
Heisenberg uncertainty principle
Helmholtz
Henry's law
Hertz
Hooke's law
Huxley
Huygens
Jacobi
Jeans
Joliot-Curie
Josephson junction
Joule (joule)
Kapitza
Kelvin scale
Kirchhoff's laws
Lambert (lambert)
Langmuir
Laplace transform
Lavoisier
Laue pattern
Legendre
Lenz's law
Leibniz
Lewis
Lissajous figures
Lorentz
Luria
Mach number
Marconi
Marten (martensite)
Maxwell equations
Mendeleev
Michelson
Miller indices
Mohr's salt
Mohs' scale of hardness
Moseley's law
Mössbauer effect
Nernst mantle
Nicol prism
Newton (Newtonian physics)
Nobel prize
Oersted (oersted)
Ohm (ohm)
Pascal
Pasteur
Pauli exclusion principle
Peltier effect

Pitot tube
Planck's constant
Poisson's ratio
Poynting vector
Pythagoreas (Pythagorean theorem)
Rabi
Raman effect
Raoult's law
Reynold's number
Riemann (Riemannian geometry)
Röntgen (also Roentgen)
Rutherford
Rydberg's constant
Salk vaccine
Schottky diode
Schrödinger
Seaborg
Seebeck effect
Shockley
Snell laws
Solvay process
Sommerfeld
Steinmetz
Stirling engine
Stokes' law
Szilard
Tesla coil
Thomson
Torricelli barometer
Townes
Urey
Van Allen belt
Van de Graaff generator
van der Waals force
van't Hoff
Venturi tube (also venturi tube)
Volta (volt)
Watson
Watson-Watt
Watt (watt)
Weber (weber)
Wheatstone bridge
Wigner theorem
Young's modulus
Zeeman effect
Zener diode or zener diode

Selected Science and Technology References

Abbreviations for Use on Drawings and in Text, ANSI Y1.1, American Society of Mechanical Engineers, New York (1972).

Authors' Guide, Science Magazine (a publication of the American Association for the Advancement of Science), Washington, D.C. (1979).

Code Name Handbook, 11th ed., DMS, Greenwich, Conn. (1982).

Day, Robert, *How to Publish a Scientific Paper,* ISI Press, Philadelphia (1979).

Dreyfuss, H., *International Graphic Symbols,* McGraw-Hill, New York (1972).

A Guide for Better Technical Presentations, IEEE Press, New York (1975).

Government Acronyms and Alphabetical Organizational Designations Used in DTIC, Defense Technical Information Center, Alexandria, Va. (1983).

Handbook for Authors, American Chemical Society, New York (1978).

Haskins, Craig, and Daniel L. Plung, eds., *A Guide for Writing Better Technical Papers,* IEEE, New York (1982).

IEEE Standard Dictionary of Electrical and Electronic Terms, ANSI/IEEE Std. 100-1977, IEEE, distributed by John Wiley & Sons, New York (1978).

International Code Name Handbook: Aerospace/Defense Technology 1983, DMS, Greenwich, Conn. (1982).

McGraw-Hill Dictionary of Scientific and Technical Terms, 3d ed., McGraw-Hill, New York (1984).

McGraw-Hill Encyclopedia of Science and Technology, 5th ed., McGraw-Hill, New York (1982).

Style Manual, American Institute of Physics, New York (1978).

Thesaurus of Engineering and Scientific Terms, Engineers Joint Council, New York (1969).

Weist, Robert C., ed., *Handbook of Chemistry and Physics,* 61st ed., CRC Press, Boca Raton, Florida (1981).

General References

The Chicago Manual of Style, 13th ed., University of Chicago Press, Chicago (1982).

Encyclopedia of Associations, Gale Research, Detroit (updated yearly).

Fernald, James, *English Grammar Simplified,* revised by Cedric Gale, Barnes & Noble, New York (1979).

Longyear, Marie, ed., *The McGraw-Hill Style Manual,* McGraw-Hill, New York (1983).

Sabin, William A., *The Gregg Reference Manual,* 5th ed., Gregg, New York (1977).

Strunk, William, and E. B. White, *Elements of Style,* 3d ed., Macmillan, New York (1978).

Today's Office, Hearst Business Communications, Inc., UTP Division, New York (magazine).

Zinsser, William, *On Writing Well: An Informal Guide to Writing Nonfiction,* 2d ed., Harper & Row, New York (1980).

Dictionary of Technical Terms

This dictionary is a selection of several thousand of the most commonly encountered scientific and engineering words and terms.

It is far from exhaustive. For example, the *McGraw-Hill Dictionary of Scientific and Technical Terms* contains well over 100,000 entries. This dictionary is not exhaustive in another sense: although a good technical dictionary often devotes more than 100 words to a definition, this one is very terse. Most definitions are one sentence or less. But it should be of value for those who use this book for exactly these reasons.

The point is, most people who reproduce scientific and technical text—unlike scientists and engineers—require "just enough" scientific information that when the meaning of a text is in danger of becoming obscure or lost, they will have a quick reference to set them back on the right course. No more than that is required. Indeed, more information might cause needless complication. So that is the goal of this dictionary—to provide "just enough" information to be useful to text processers, and no more.

THE SELECTION

Any selection is a process of leaving out and leaving in. What has been left out is some of the material that is treated in other parts of the book. The glossaries of Part 2 should be convenient references for such information as Greek letters, prefixes and suffixes, the chemical elements, physical units, abbreviations, names of noted scientists, and so on. Some of the most important are included in the dictionary as well. Also, most names such as those of chemical compounds—which number in the thousands—have been left out. Including them would have converted this convenient little selection into a section larger than the rest of the book.

What do you do when the word you want is not in this dictionary or in any of the other glossaries or tables? Look it up in an ordinary dictionary; or, if it is not to be found there, look it up in a technical dictionary such as one of those listed in Glossary Q.

What has been left in is what we feel is necessary to provide crucial concepts. Notable here are many words that have commonly accepted English meanings, but which have far different connotations in technical contexts. So do not be surprised to learn that a *die* is a small semiconductor fragment and that it is the singular form of *dice* (and that dice in this sense has no connection with those little cubes with black dots on them used for gambling).

WORD DIVISION

Rules of word division are often confusing, especially since they may be different for typists in offices and for typesetters in printing establishments. Correct usage of technical terms is also confusing because new ones appear every day, as new scientific facts are gathered, as new scientific theories are promulgated, and as new technologies are developed—and as each generates its own jargon. The writers have attempted to follow the practice of *The Gregg Reference Manual* (Glossary R) for "new" expressions.

Word division in the dictionary is indicated as follows:

- The places where words may be broken for typing puposes are shown by a centered dot.
- Hyphenated words are indicated by long dashes where the hyphens appear.
- Expressions containing more than one word are indicated by leaving conventional spaces between them.

DICTIONARY

A

ab– *prefix* used for electromagnetic units. *See* stat-.

ab·am·pere *n.* The electromagnetic unit of current, equal to 10 amperes. *Abbr.* aA

ab·cou·lomb *n.* The electromagnetic unit of charge, equal to 10 coulombs. *Abbr.* aC.

ab·er·ra·tion *n.* In an optical system, an irregularity that causes an imperfect image.

ab·la·tion *n.* The melting or vaporization of nose-cone material in aerospace vehicles and projectiles.

abort *v.* To stop an experiment or process short of completion.

ab·scis·sa *n.* The horizontal coordinate of a point on a graph, known as the x coordinate.

ab·sorb *v.* To assimilate a substance.

ab·sorp·tion *n.* The taking in of matter by other matter (such as the taking in of water by a sponge).

ac *See* alternating current.

ac·cep·tor *n.* An impurity in a semiconductor crystal that will pick up an electron.

ac·cess *v.* To locate desired data in computer memory storage.

ac·cu·mu·la·tor *n.* A circuit in a computer that stores figures for processing.

achro·mat·ic *adj.* Capable of transmitting light without decomposing it into its constituent colors.

achro·mat·ic lens *n.* A combination of two or more lenses designed to eliminate color fringes.

ac·id *n.* A class of chemical compounds usually containing the hydrogen ion (H^+) plus a negatively charged radical; thus, $H^+NO_3^-$ or nitric acid. Acids are corrosive, have a sour taste, and turn litmus paper red. In solution, acids release hydrogen ions. They are donors of protons. *See* base.

acid·u·late *v.* To make acidic.

ac·tin·ic *adj.* Pertaining to the property of radiant energy by which chemical changes are produced (such as the changes made by light in a photographic medium).

ac•tiv•ity *n.* A measure of the tendency of a substance to react chemically with other substances.

acute an•gle *n.* An angle of less than 90 degrees.

ADA *n.* A computer programming language. *Also* Ada. Named for Ada Byron, Countess Lovelace (1816–1852), mathematician and assistant to Charles Babbage (1791–1871), builder of an early calculating machine.

ad•dend *n.* A number to be added to another number.

add•er *n.* A computer circuit in which two or more signals are combined.

ad•dress *n.* A label that identifies a particular location in the memory storage of a computer.

ad•dress reg•is•ter *See* address.

adi•a•bat•ic *adj.* Pertaining to a physical change in which no heat is transferred.

ad•sorb *See* adsorption.

ad•sorp•tion *n.* The taking up of molecules of one substance by the surface of another substance. *v.* ad•sorb.

aer•o•sol *n.* A suspension of particles in a mist.

af•fect *v.* To cause to change.

af•fin•ity *n.* A measure of the tendency of chemical elements to combine.

age *v.* To expose to a given set of conditions for extended time, as to age wine at a certain temperature and humidity.

age–hard•en•ing *v.* To harden an alloy by a low-temperature heat treatment.

ail•er•on *n.* The movable part of an airplane wing that controls roll.

air–core coil *n.* An electrical coil wound around an air space rather than around a magnetic core.

air•foil *n.* A surface designed to interact with air during movement to create lift.

air•frame *n.* The essential structure of an aircraft, distinct from the covering, or skin.

al•che•my *n.* The medieval craft of chemistry, now discredited. *See also* transmutation.

AL•GOL *n.* Algorithmic Oriented Language, a computer programming language.

al•go•rithm *n.* A set of well-defined rules for solving a problem.

al•i•phat•ics *n.* Pertaining to a class of organic chemical compounds in which the carbon atoms are linked in open chains. *See* aromatics.

al·ka·li *n.* A highly basic class of chemical compounds usually containing the hydroxyl radical (OH^-).

al·lo·trope *n.* One of several crystal forms of the same material, such as graphite, charcoal, diamond (all of which are pure carbon).

al·lot·ro·py *See* allotrope.

al·lu·vi·al *adj.* Pertaining to silt, gravel, and so forth deposited by running water.

al·ni·co *n.* A class of magnetic alloys of iron containing aluminum, nickel, and cobalt.

al·pha iron *n.* A crystal form in steel. *Also known as* α iron.

al·pha·nu·mer·ic *adj.* Pertaining to a computer word containing both letters (alpha-) and digits (numeric).

al·pha par·ti·cle *n.* A positively charged particle consisting of two protons and two neutrons; emitted by certain radioactive substances. *Also known as* α particle.

al·pha ray *n.* A stream of alpha particles. *Also known as* α ray.

al·ter·nat·ing cur·rent *n.* A flow of electrons which regularly reverses direction. *Abbr.* ac

al·ter·na·tor *n.* The rotating member of a generator that produces alternating current.

ALU *n.* Arithmetic logic unit (in a computer).

alu·mi·na *n.* Aluminum oxide, Al_2O_3.

alu·min·i·um *n.* The British spelling of aluminum.

amal·gam *n.* **1.** A mercury alloy. **2.** A mixture.

amal·ga·mate *v.* To combine.

am·bi·ent *adj.* Pertaining to the surrounding environment.

ami·no *adj., prefix* Relating to or containing the group NH_2 or a derivative.

ami·no ac·ids *n.* The building blocks of all proteins. *See* protein.

am·me·ter *n.* An instrument that measures the flow of electric current, in amperes.

am·mo·nia *n.* The chemical compound NH_3.

am·mo·ni·a·cal *adj.* Containing or having the properties of ammonia. *See* ammonia.

amor·phous *adj.* Without shape; having no crystal structure.

am·per·age *n.* The strength of electric current.

am·pere *n.* The unit of electric current. *Abbr.* A or amp

am·pere-hour *n.* A unit of the quantity of electricity. *Abbr.* Ah or A·h

am•pere–turn *n.* A unit of magnetomotive force generated in a loop carrying a 1-ampere current. *Abbr.* amp-turn

am•pho•ter•ic *adj.* Capable of reacting either as an acid or as a base.

am•pli•fy *v.* To increase or enlarge.

am•pli•tron *n.* A kind of microwave tube.

am•pli•tude *n.* The measure of a wave or vibratory movement taken from the midpoint to one extreme.

am•poule *See* ampule.

am•pule *n.* A small vial. *Also* am•poule.

an•a•log *adj.* Able to represent information as a continuous proportion. *Also* an•a•logue.

an•a•log com•put•er *n.* A computer that measures. It operates on data in the form of continuously varying quantities of physical information.. *Compare* digital computer.

an•a•logue *See* analog.

anal•y•sis *n.* The process of studying the nature of a thing by breaking it down into its component parts.

an•a•lyze *v.* To determine a thing's nature by means of analysis. *See* analysis.

an•as•tig•mat•ic lens *n.* A lens that corrects astigmatism. *See* astigmatism.

AND *n.* A computer function by which an output signal is produced only if all input channels carry a signal. A computer gate that performs the AND function.

an•e•cho•ic *adj.* Without echoes, absorbent of sound.

ang•strom *n.* A unit of length equal to one ten-billionth of a meter. One angstrom equals one-tenth of a nanometer (nm), a unit of measure preferred for most uses. *Abbr.* Å

an•hy•dride *n.* A compound from which chemically bound water has been removed.

an•hy•drous *adj.* Free from chemically bound water.

an•ion *n.* A negatively charged ion.

an•i•so•trop•ic *adj.* Having different qualities, such as conductivity of heat in different directions (said of materials).

an•neal *v.* To heat, then cool, a metal in order to render it less brittle.

an•no•ta•tion flag *n.* A computer flowchart symbol.

an•nu•lus *n.* A ring-shaped figure, or a ring of tissue. ***adj.*** an•nu•lar.

an•ode *n.* **1.** The positive terminal of a battery. **2.** The electrode which collects electrons in an electron tube. *See* cathode.

an·o·dize *v.* To form an oxide layer on a metal electrolytically.

anom·a·ly *n.* A case not explained by known rules or conditions. ***adj.*** anom·a·lous.

ANSI stan·dard *n.* A materials standard established by the American National Standards Institute.

an·ti·body *n.* A blood-plasma protein that protects the body against an invading substance by combining with the substance and thus inactivating it. *See* antigen.

an·ti·gen *n.* A substance which, when it invades a living organism, generates antibodies. Antigens may be viruses, bacteria, or foreign tissue. *See* antibody.

an·ti·log·a·rithm *n.* A second number whose logarithm is the first number. *Abbr.* an·ti·log.

APL *n.* A Programming Language, a high-level language used to program computers with large memories.

ap·pli·ca·tion soft·ware *n.* Any computer program designed for a specific purpose.

ar·chi·tec·ture (computer) *n.* The structure and arrangement of a computer's components.

ar·gon *n.* A chemical element which is an inert gas. *Symbol* Ar

a·rith·me·tic–log·ic unit (ALU) *n.* That part of a computer's central processing unit that manipulates data according to a set of instructions.

arm·a·ture *n.* The rotating element of a motor or generator.

aro·ma·tics *n.* A class of organic chemical compounds in which the carbons are linked in rings. *See* aliphatics.

ar·ti·fi·cial in·tel·li·gence *n.* A branch of computer science that employs computers to solve problems that appear to require human intelligence and ability to learn from previous errors. *Abbr.* AI

ASCII *adj.* Pertaining to American (National) Standard Code for Information Interchange, a data-processing standard for word size, 7 bits. *Pronounced* askey. *See* EBCDIC.

as·sem·bler *n.* A program that translates an assembler-language program into machine language. *See* machine language.

as·sem·bler lan·guage *n.* A low-level language that uses symbols and abbreviations to represent operations.

astig·ma·tism *n.* A defect in an optical system that causes a point to appear to be a line, and lines to blur.

as·tral *adj.* Pertaining to the stars.

asym·met·rical *adj.* Not balanced; askew.

asymp·tote *n.* A straight line on a graph that a curve approaches, but never quite reaches. ***adj.*** asymp·tot·ic.

-ate *suffix* In chemistry, indicates the highest oxidation state of a salt.

at•mo•sphere *n.* **1.** The gas envelope surrounding an object. **2.** A unit of pressure, equal to the weight of air at sea level. *Abbr.* atm **3.** The gases, principally nitrogen and oxygen, that surround Earth. Proceeding out toward space, they are divided into layers: the troposphere (to about 15 miles), the stratosphere (to about 25 miles), the ionosphere (to about 250 miles), and the exosphere (250 miles and beyond, into space).

at•om *n.* The smallest particle of any chemical element.

atom•ic num•ber *n.* The number of protons in an atom's nucleus, different for each element. The atomic number of hydrogen is 1, of oxygen 16, of uranium 92.

at•om•ize *v.* To break up into tiny particles.

at•ten•u•ate *v.* Of an electronic signal, to diminish. *See* gain.

atto- *prefix* One-billion-billionth. *Symbol* a

at•tri•bute *n.* A unique quality of a material.

at•tri•tion *n.* A wearing or grinding away.

aus•tem•per•ing *n.* A heat treatment for austenitic steel.

aus•ten•ite *n.* A crystalline form of iron.

aus•ten•it•ic steel *n.* An alloy whose structure is typical of austenite.

avi•on•ics *n.* The science of electronics as used in flight. Derived from aviation electronics.

Avo•ga•dro's num•ber *n.* The number of molecules in a mole of any substance. Equal numbers of moles of all substances contain the same number of molecules (1 mole = 6.02×10^{23} molecules, or 602 followed by 23 zeros).

ax•i•al *adj.* Along an axis.

ax•i•om *n.* An assumption upon which a theory is based.

ax•is *n.* The centerline of any system. ***pl.*** ax•es.

az•ide *n.* Compounds derived from hydrazoic acid HN_3. Inorganic azides include lead azide $Pb(N_3)_2$, important in the manufacture of explosives.

az•i•muth *n.* A horizontal direction on the Earth's surface.

B

Bab•bitt met•al *n.* An alloy primarily of tin and lead, used in bearings.

back-end proc•es•sor *n.* A computer which serves as an interface between a larger central processing unit and data bases stored in storage devices.

back·ward wave os·cil·la·tor *n.* An electronic device which amplifies microwave signals. *Abbr.* BWO

baf·fle *n.* A deflector.

bain·ite *n.* Steel formed by austempering. *See* austempering.

bake·out *n.* The degassing of surfaces of a vacuum system by heating.

bal·last *n.* **1.** A heavy weight used for stabilization. **2.** Coarse gravel used in concrete.

bal·lis·tics *n.* The study of the motion and behavior of objects in flight.

band *n.* A limited range of electromagnetic-wave frequencies assigned to a certain type of radio service.

ba·rom·e·ter *n.* An instrument that measures atmospheric pressure.

bas·al *adj.* **1.** Pertaining to the base. **2.** Pertaining to the minimal level of metabolism.

ba·salt *n.* A rock of volcanic origin.

base *n.* **1.** A chemical substance capable of receiving a hydrogen ion from another substance. A base will neutralize an acid, turn litmus paper blue, and accept protons. **2.** The number of which a logarithm is the exponent.

BA·SIC *n.* A computer programming language, Beginner's All-Purpose Symbolic Instruction Code.

ba·sic·i·ty *n.* The level of alkalinity.

batch *n.* A limited quantity. *adj.* Referring to a process performed in separate steps rather than continuously.

bath *n.* A medium for regulating temperature, or the vessel containing the medium, in which a process is carried on.

bat·tery *n.* **1.** A device which delivers direct current electricity as a result of electrochemical action. **2.** An array of structures used together.

baud *n.* A unit equal to the number of code elements per second. Named for Emile Baudot, French engineer (1845–1903), inventor of a 5-bit code widely used in printing-telegraphic devices.

Bau·mé *n.* A hydrometer scale for liquids used in describing their density. After Antoine Baumé (1728–1804), French pharmacist, its inventor. *Abbr.* Bé

baux·ite *n.* An ore of aluminum.

BCD sys·tem *n.* Binary-coded decimal system. In data processing, a system of number representation. Each digit of a number in the decimal (base 10) system is represented by a binary (base 2) number.

bea·con *n.* A radio transmitter or light beam used for identification of a position.

bead *n.* The melted region that makes up the path of a weld.

beak·er *n.* A cuplike piece of chemical ware.

beam *n.* A horizontal structural member.

bear·ing *n.* A machine part consisting of elements that slide, turn, or oscillate on each other.

beat *n.* The difference frequency between two frequencies.

bell jar *n.* A removable dome-shaped cover which creates a chamber in which vacuum or other gases may be maintained.

bel·lows *n.* An extendable part, usually made of foldable pleats of metal, plastic, or paper. Used to suck in air through a valve and blow it out through a tube.

bell–shaped curve *See* Gaussian distribution.

ben·zene *n.* A flammable hydrocarbon.

ben·zene ring *n.* The six-carbon hexagonal structural arrangement found in benzene and in other organic compounds formed from benzene.

be·ryl·li·um *n.* A metal used in beryllium-copper alloys, which have many industrial and scientific applications. *Symbol* Be

be·ta par·ti·cle *n.* A radioactively emitted electron or positron. *Also* β particle.

be·ta ray *n.* A stream of beta particles. *Also* β ray.

bev·el *n.* An edge plane at any angle but 90 degrees to primary surfaces.

bi·as *n.* A voltage difference. Dc voltage applied to a transistor control electrode to establish the desired operating point.

bi·ax·i·al *adj.* Having two axes.

bi·fi·lar *adj.* Marked with two fine lines.

bi·fi·lar lens *n.* A lens with fine wires or hairs mounted in the eyepiece, as an aid to focusing.

bi·fur·cate *v.* To separate into two sections.

bi·mod·al *adj.* Having two statistical modes. *See* mode.

bi·na·ry *adj.* Pertaining to a system that exists in two conditions, such as OFF/ON. It is usually represented in computer code by 0 and 1. ***n.*** A number system to the base 2.

bin·oc·u·lar *adj.* Pertaining to an optical system with an eyepiece for each eye.

bi·no·mi·al *adj.* A mathematical expression consisting of two terms, such as $5a + b$.

bi·po·lar *adj.* Referring to a physical entity or chemical compound having two ends (poles) bearing a positive or negative charge, respectively.

bi·re·frin·gence *n.* Splitting of a light beam to produce two rays.

bisque *n.* Ceramic ware that has been fired once but is still unglazed and therefore porous.

bit *n.* **1.** The basic unit of computer information. **2.** A binary digit.

bi·va·lent *adj.* Having a valence of 2.

blackbo·dy *n.* A surface which absorbs all radiant energy that strikes it and reflects none.

bleed·er cur·rent *n.* Current drawn off to lessen the effect of load changes.

bomb *n.* A chamber in which high-pressure experiments may be run.

Boo·le·an al·ge·bra *n.* An algebraic system that represents a two-valued (YES/NO) logic. Its operations are AND, OR, and NOT. It is widely applied in computer architecture. Invented by the English mathematician George Boole (1815–1864).

boom *n.* A movable strut used to hoist and swing cargo on a ship or on construction equipment.

bore *n.* The interior diameter of a gun or tube. *v.* To drill a hole.

bort *n.* Crude diamond, existing as imperfect crystals, used in industry as an abrasive. The "bread and butter" of the diamond trade.

boss *n.* A rounded, convex projection.

boule *n.* An unpolished manufactured crystal. *Also called* ingot.

bound wa·ter *n.* Molecules of water bonded to other molecules in a compound. *Also called* water of hydration. *See* hydrate.

branch *n.* A computer instruction which causes a transfer of control to a subprogram.

branch struc·ture *n.* A program that includes a provision for moving from one subprogram to another.

braze *v.* To join two pieces of metal with a brazing alloy or a high-temperature solder.

break·down *n.* A large, sudden rise in electric current that occurs after a small increase in voltage.

bridge *n.* An electrical shunt path.

Bri·nell test *n.* A test of the hardness of a metal. The results are given as a Brinell number.

Brit·ish ther·mal unit *n.* A heat unit (the amount of heat required to raise 1 pound of water 1°F). Equal to about 1054 joules. *Abbr.* Btu

broad·band *adj.* Relating to a spread of electromagnetic energy over a wide range of frequencies.

broad·band am·pli·fi·er *n.* An amplifier that responds with a minimum of distortion over a wide range of radio frequencies.

brush *n.* A contact that completes the circuit between a stationary and a moving element of a motor or generator.

buf•fer *v.* To add a chemical compound to a solution to stabilize a level of acidity. *n.* **1.** A chemical used to stabilize a level of acidity. **2.** In a computer, an area of main memory that temporarily stores data fetched from internal storage until the arithmetic-logic unit is ready to process it.

bug *n.* A computer error. *See* debug.

built–up frac•tion *n.* A fraction written with the numerator and the denominator placed one over the other with a horizontal fraction bar between them $\left(\frac{1}{2}\right)$. *Compare* shilling fraction.

Bun•sen burn•er *n.* A small gas burner used in chemical tests and experiments.

bu•rette *n.* A glass tube, marked off in fine divisions, used to measure very precise quantities of liquids in chemical analysis. *Also* buret.

burr *n.* A sharp projection left on a piece of metal by machining.

bus *n.* The circuit connection that carries electrons between electrical components.

butt *v.* To place two objects together, end-to-end.

butt weld *n.* The weld region of two pieces welded end-to-end.

BWO *n.* Backward wave oscillator (a microwave tube).

byte *n.* A fixed number of bits operated on by a computer as a unit. *See* bit.

byte–ad•dress•a•ble *adj.* Referring to a computer whose storage locations may be addressed by the byte.

C

cache *n.* A very-high-speed computer storage device.

CAD *n.* Computer-Aided Design (of mechanical and electrical systems).

CAD/CAM *n.* Computer-Aided Design/Computer-Aided Manufacture, the use of computer technology as an aid to automating design and manufacture and combining these two actions.

cal•cine *v.* To heat material to a high temperature, just below the point of melting or fusion in order to effect a change, such as oxidation or pulverization.

cal•i•ber *n.* A measure of the inside diameter of a cylinder.

cal•i•brate *v.* To set an instrument to conform with a standard.

cal•i•per *n.* An instrument for measuring inside and outside diameter. *Usually* cal•i•pers

call *v.* To transfer program control to a subroutine.

ca•lor•ic *adj.* Relating to a quantity of heat.

cal•o•rie *n.* **1.** A unit of heat. **2.** A unit of heat-producing energy of food when oxidized in the body; equivalent to one "large" (kilogram) calorie. *See* kilogram calorie. There are also "small" (gram) calories. (Because of confusion about the different values assigned to the calorie, the joule is the preferred unit of heat measurement. *See* joule;Joule's equivalent). *Abbr.* cal

cal•o•rim•e•ter *n.* A heat-measuring device.

CAM *n.* Computer-Aided Manufacture. *See* CAD/CAM.

cam *n.* A rotating mechanism which provides a programmed sequence of movement, such as opening and closing valves in an automotive engine.

cam•ber *n.* A deviation from a straight line.

can•de•la *n.* A unit of light intensity. *Abbr.* cd

canned pro•gram *n.* A computer program prepared by and available from an outside supplier.

can•ti•le•ver *n.* A horizontal beam, fixed at one end.

ca•pac•i•tance *n.* The ability of a capacitor to store electric charge. *Abbr. C*

ca•pac•i•tor *n.* An electrical component that stores an electric charge.

ca•pac•i•ty *n.* *See* capacitance.

cap•il•lary *n.* A narrow tube.

CAR *n.* Computer-*a*ided *r*etrieval.

car•at *n.* A unit of weight for precious stones, 200 milligrams. *See* karat.

car•bide *n.* **1.** A binary compound of carbon and a more electropositive element. **2.** A tungsten, titanium, or tantalum carbide, heat- and wear-resistant, used for cutting tools.

car•bon•yl *n.* The chemical radical CO (carbon plus oxygen).

car•bu•rize *v.* To cause carbon atoms to diffuse into a metal at high temperature, especially to surface harden steel by this process.

card read•er *n.* In a punched-card system, a machine that reads data from the punched cards.

car•te•si•an co•or•di•nates *n.* Mutually perpendicular coordinates in a graph.

cas•cade *n.* A series of circuit breakers in which each unit controls the operation of the one that follows.

case *n.* The surface layer of a ferrous alloy, harder than the core.

cas•sette ***n.*** Used in connection with computers, a plastic cartridge containing magnetic tape that passes between two reels.

ca•tal•y•sis ***n.*** The process of speeding a chemical reaction by the presence of a nonconsumed substance. ***v.*** cat•a•lyze.

cat•a•lyst ***n.*** A substance that speeds up chemical reactions. *See* catalysis.

cat•e•nary ***n.*** The shape of a curve made by a cord held up at both ends. Gateway Memorial Arch, St. Louis, Missouri, is an example of an inverted catenary curve in architecture.

cath•ode ***n.*** **1.** The negative terminal of a battery. **2.** The electron emitter in a vacuum tube or a semiconductor device. ***adj.*** cath•o•dic. *See* anode.

cath•ode–ray tube ***n.*** A vacuum tube in which cathode rays are projected onto a screen. *Abbr.* CRT

cat•ion ***n.*** A positively charged ion, attracted to the negative electrode, the cathode.

cat whis•ker ***n.*** A sharply pointed, flexible wire that makes contact with the surface of a semiconductor crystal.

C band ***n.*** The 3.7–5.1-centimeter wavelength, within microwave frequency range (0.3–30 centimeters).

cell ***n.*** The smallest functional unit of an organism or a structure.

cel•lu•lar ***adj.*** Relating to a cell or cells.

Cel•si•us ***adj.*** Relating to a temperature scale in which the range between the boiling and the freezing points of water is divided into 100 degrees. Named for its inventor, the Swedish astronomer Anders Celsius (1701–1744). *Abbr.* C

ce•ment ***n.*** In concrete, the bonding material that forms a matrix around pebbles and sand.

ce•ment•ite ***n.*** A brittle, crystalline compound, iron carbide, found in steel.

centi– ***prefix*** One-hundredth. *Symbol* c

cen•ti•grade ***adj.*** Relating to a temperature scale very close to Celsius, which is the preferred word. *See* Celsius.

cen•ti•me•ter ***n.*** A measure of length, one-hundredth of a meter. *Abbr.* cm

cen•tral proc•es•sing unit ***n.*** The portion of a computer which controls the execution and interpretation of instructions. Includes arithmetic-logic and control sections. *Abbr.* CPU

cen•trif•u•gal ***adj.*** Referring to a force directed out from the center.

cen•tri•fuge ***n.*** A machine which rotates materials at high speed for the purpose of separating their constituents by centrifugal force.

cen·trip·e·tal *adj.* Referring to a force directed toward the center.

ce·ram·ist *n.* One who practices ceramics technology. *Also* ce·ram·i·cist.

cer·met *n.* A material consisting of bonded particles of metal and ceramic.

cgs *adj.* Abbreviation for the centimeter-gram-second system of units. *See* metric system.

cham·fer *v.* To bevel an edge on a machined part.

chan·nel *n.* **1.** A groove. **2.** A radio-frequency band. **3.** A track on a tape or a drum. *v.* To direct down a designated path.

char·ac·ter *n.* In a computer, an alphabetic or a numeric symbol. Also a special character such as #, ?, or].

Char·py im·pact *n.* A measure of the fracture strength of a metal.

chas·sis *n.* The metal structure on which an electronic circuit system is constructed.

che·late *v.* To react chemically in such a way as to sequester metal atoms.

chip *n.* **1.** A semiconductor die on which an integrated circuit has been constructed. **2.** A packaged functional semiconductor circuit.

choke *n.* A coil used in an electric circuit to impede frequencies above a selected frequency. *v.* To constrict.

chop·ping *v.* The act of interrupting an electric current at regular intervals.

chord *n.* A line segment that intersects an arc at its two terminations.

chro·mat·ic *adj.* Relating to color.

chro·ma·tog·ra·phy *n.* The act of separating materials by percolation through a selectively adsorbing medium.

chro·nom·e·ter *n.* An extremely accurate instrument for measuring time.

cin·der *n.* The residue or slag after burning or melting.

ci·pher *n.* **1.** Zero. **2.** A code.

cir·cle *n.* A closed curve. *See* conic section.

cir·cuit *n.* **1.** A path that returns to its origin. **2.** An array of electric components that performs a given electrical function.

cir·cuit break·er *n.* A large electric switch.

cir·cuit card *n.* A printed circuit which has its terminations at one edge for easy insertion into a larger circuit system.

cir·cum·scribe *v.* To enclose.

cite *v.* To designate a reference, usually an authoritative one.

clad *adj.* Laminated, with the layers well bonded.

clad•ding *n.* The surface laminate in a clad structure.

clamp•ing cir•cuit *n.* A circuit used in a television receiver to maintain the average light value of the image.

cleave *v.* To split a crystal along one of its planes.

clock *n.* In a computer, an electronic system that generates the periodic signals used to control the timing of the central processing unit.

clone *n.* A group of cells produced asexually from a single ancestor cell. *v.* To produce such cells.

co•ax•i•al *adj.* Having a common axis.

co•ax•i•al cable *n.* A communications transmission line in which the core and the outer metal tube that sheaths it serve as separate conductors.

CO•BOL *n.* *C*ommon *b*usiness-*o*riented *l*anguage, a computer language.

CO•DA•SYL *n.* *C*onference of *d*ata *sy*stem *l*anguages.

code *n.* A set of rules for expressing information, such as Morse code, or computer machine language.

co•ef•fi•cient *n.* A factor in a product, thus, in $A \cdot B$, A is the coefficient of B and B is the coefficient of A.

co•er•cive force *n.* The magnetic field which must be applied cyclically to a magnetic material in order to demagnetize it.

co•ex•ten•sive *adj.* Sharing the same space on the same time.

co•here *v.* To stick together.

co•her•ent *adj.* Pertaining to waves that are in phase and therefore may produce interference effects.

coil *n.* **1.** A spiral winding. **2.** An electric inductor.

coin *v.* To stamp a pattern with a die.

co•in•ci•dent *adj.* Occupying the identical space or time span.

cold work•ing *n.* The deformation of metal below the recrystallization temperature.

col•lin•e•ar *adj.* Lying along the same line.

col•late *v.* To gather data, either mechanically or by hand, into a specific order.

col•lat•er•al *adj.* Accessory.

col•lec•tor *n.* The electrode that receives electrons.

col•li•mate *v.* To confine rays to a parallel path.

col•loid *n.* Ultrafine particles dispersed in a medium.

Col•pitts os•cil•la•tor *n.* An electron-tube oscillator.

co•lum•nar *adj.* Pertaining to arrays of vertical members.

com·mand *n.* A specific instruction in a computer program. *v.* To give that instruction.

com·mu·ta·tor *n.* A wiping contact in an electric motor or generator. *See* brush.

com·pact *v.* To condense.

com·pass var·i·a·tion *See* magnetic declination.

com·pil·er *n.* Computer software that translates high-level language to machine language.

com·plex num·ber *n.* An expression made up of real numbers and the imaginary number that is the square root of -1.

com·pres·sion *n.* Reduction in volume of a substance due to pressure. *adj.* Compressive.

com·put·er word *See* word.

con·cave *adj.* Having a curve which is rounded inward.

con·cav·i·ty *n.* A rounded indentation.

con·cen·trate *v.* To make less dilute. *n.* That which has been made less dilute.

con·cen·tra·tion *n.* The strength of a solution with reference to a specified ingredient.

con·cen·tric *adj.* Having the same center, a series of circles one surrounding the others.

con·crete *adj.* Real. *n.* A composite of particles bonded by cement.

con·den·sate *n.* A liquid which is deposited from a gas phase.

con·dens·er *n.* *See* capacitor.

con·di·tion·al op·er·a·tion *n.* In a computer program, an operation that is to be performed only if certain conditions exist.

con·duct *v.* To serve as a medium for the flow of a substance or of electrons.

con·duc·tance *n.* The measure of the ability of a circuit to conduct electricity. *Unit* mho (symbol ℧) or siemens (symbol S).

con·duc·tiv·i·ty *n.* The ability of a material to conduct electricity. *Symbol* σ.

con·gru·ent *adj.* Coinciding in size and form, as of two triangles that match when superimposed.

con·ic sec·tion *n.* The plane formed by slicing a cone. The four conics are the circle, ellipse, parabola, and hyperbola.

con·ju·gate *adj.* Joined together; related.

con·junc·tion *n.* **1.** A coming together. **2.** In mathematics, the connection of two statements by the word "and."

con·sole *n.* The part of a computer used by the operator to control the machine. It may consist of a keyboard only or an entire work station.

con•stant *n.* An unchanging value.

con•straint *n.* A limit imposed upon a system, either physical or mathematical.

con•tents *n.* The information stored at any address in computer memory.

con•tig•u•ous *adj.* Sharing a boundary.

con•tin•u•um *n.* A whole, which cannot be divided into parts except artificially.

con•tour *n.* An outline. *v.* To outline an object.

con•tract *v.* To shrink.

con•vec•tion *n.* The heat-activated motion of a gas or a fluid.

con•verge *v.* To come together.

con•ver•gence *n.* A coming together, as of the paths of the electron beams in a three-gun color television tube.

con•vex *adj.* Having a curve which is rounded outward.

con•vo•lu•tion *n.* A fold, coil, or twist.

cool•ant *n.* A flowing fluid which draws away heat (for example, from a cutting tool while it is being operated).

co•or•di•nates *n.* A set of numbers that designates the position of a point in space. *See* abscissa; ordinate.

co•or•di•nate plot•ter *See* XY plotter.

co•pol•y•mer *n.* A large molecule resulting from the linking of many smaller molecules of at least two different types.

core *n.* The center of an electric coil or transformer. It may be air or a magnetic material.

core stor•age *n.* In a computer configuration, the main memory area of the central processing unit.

cor•ol•lary *n.* A deduction that follows from previously established propositions. *adj.* Accompanying.

co•ro•na dis•charge *n.* A glowing discharge region that surrounds a conductor at high voltage.

cor•pus•cu•lar *adj.* Existing in separate particles.

co•se•cant *n.* *See* trigonometric ratios. *Abbr.* csc

co•sine *See* trigonometric ratios. *Abbr.* cos

co•tan•gent *See* trigonometric ratios. *Abbr.* cot

cot•ter pin *n.* A metal rod used as a fastener.

cou•lomb *n.* A unit of electric charge. *Abbr.* C

count•er *n.* **1.** A computer element that keeps track of the number of occurrences of an event in a program. **2.** An electronic circuit that produces an output pulse after a certain number of input pulses have been received.

coun·ter·sink *v.* To drill a larger hole over a smaller hole in a surface to accommodate a screw head flush to the surface.

co·va·lent bond *n.* An atomic bond based on electron sharing.

CPU *n.* Central Processing Unit in a computer.

crash *n.* A hardware or software failure in a computer.

craze *v.* To produce an array of fine cracks on a surface.

creep *n.* The slow distortion of a material under stress. *v.* Of soil, to move imperceptibly downward under stress.

crossed–field am·pli·fi·er *n.* An efficient, wide-bandwidth microwave electron tube. *Abbr.* CFA

cross hair *n.* A fine line scribed in the lens of an optical system, for accurate sighting. May also be made of hair or silk.

cross–link *v.* To establish chemical bonds between molecules of polymers.

cross talk *n.* Electric interference in a communication line.

CRT *n.* Cathode-ray tube. The receiver of a television set and the screen of a computer terminal are familiar examples.

crude *adj.* Not refined or processed, as for example crude oil.

cry·o·gen·ics *n.* The technology of very low temperatures.

cry·o·stat *n.* An apparatus used to create low-temperature environments.

cryp·to·gram *n.* Information written in code.

crys·tal *n.* A solid consisting of atoms in regular repeated geometrical arrangements.

crys·tal lat·tice *n.* The geometrical arrangement in a crystal.

crys·tal sys·tems *n.* Seven groupings that define crystals by their geometry, which in turn is influenced by their atomic structure. In order of decreasing symmetry, these groups are: cubic, hexagonal, tetragonal, trigonal, orthorhombic, monoclinic, and triclinic. Thus a crystal that belongs to the cubic system is called a cubic crystal. These geometries are consistent for a crystal of a particular mineral, whether it is microscopic in size or several feet long. The plane along which a crystal will cleave is determined by its symmetry, and this is an aid in identification. In nature, it is rare to find perfect specimens.

cu·bic *See* crystal systems.

cull *n.* A reject. *v.* To separate and retain the good portions of a mix while rejecting the bad.

cu·pric *adj.* Relating to copper.

cu•rie *n.* A unit of radioactivity. After Marie Curie (1867–1934), Polish-French chemist. *Abbr.* Ci

curl *n.* A mathematical operation in vector algebra. *Symbol* ∇

cur•rent *n.* A flow of electrons.

cur•sor *n.* An indicator on a computer terminal's video screen.

cusp *n.* The area where two curves intersect; shaped like one tip of a crescent

cut•off *n.* The electrical condition in which an active electronic device ceases to pass current.

cy•ber•net•ics *n.* The science that attempts to integrate matters of communication and control in machines and living organisms.

cy•cle *n.* A complete sequence of values or operations that repeats. *v.* To repeat. *adj.* cy•clics

cy•cloid *n.* The curve described by a point on the circumference of a wheel as it rolls in a straight line.

Czo•chral•ski tech•nique *n.* A way to grow crystals by seeding a melt with a single crystal of silicon or other semiconductor material.

D

D/A *adj.* Digital-to-analog.

dai•sy wheel *n.* A flat plastic wheel, containing numbers, letters, and other characters, used in a printer attached to a computer. It gives letter-quality hard copy.

damp *v.* To lessen the amplitude of a vibration.

d'Ar•son•val gal•va•nom•e•ter *n.* An instrument for measuring a small electric current.

DASD *n.* *D*irect-*a*cess *s*torage *d*evice.

da•ta base *n.* A collection of data; in computer technology, one that is organized for easy access.

da•tum *n.* A piece of information, usually quantitative. *pl.* data. Especially in computer texts, the plural form is often used with a singular verb.

de•bug *v.* To detect, locate, and repair errors in computer programs or hardware.

de•bug•ger *n.* A program employed to facilitate debugging.

deca– *prefix* Ten. *Also* deka *Symbol* da

de•cant *v.* To pour out a liquid-solids mixture, leaving a solid residue behind.

de•cel•er•ate *v.* To lessen velocity.

deci– *prefix* One-tenth. *Symbol* d

de•ci•bel *n.* In acoustics, a unit that measures power. *Abbr.* dB

dec•li•na•tion *n.* **1.** In navigation or astronomy, the angular distance to a star, measured along a great circle from the celestial equator. **2.** *See* magnetic declination.

dec•re•ment *n.* A negative increment; the decrease in a variable.

de•code *v.* To convert a coded message to uncoded ("clear") language.

de•duce *v.* To draw conclusions by applying a general principle.

deep draw•ing *n.* The process of making deep indentations in metal sheet.

de•flec•tion yoke *See* yoke.

de•gas *v.* To drive out gases from a liquid or a solid.

deka– *prefix* Ten. *Also* deca *Abbr.* da

de•lay line *n.* An electronic circuit that introduces lag in wave phases.

del *n.* The vector differential operator. *Symbol* Δ

del•i•ques•cence *n.* The taking up of water from the air.

del•ta *n.* A three-branched electric circuit. *Also,* del•ta star, del•ta wye.

de•mod•u•late *v.* To remove a modulating signal from a carrier wave.

den•drite *n.* A crystal with a treelike form.

de•nom•i•na•tor *n.* The term below the line in a fraction. *See* numerator.

den•si•tom•e•ter *n.* An instrument used to measure light intensity.

den•si•ty *n.* The mass, by a unit volume, of a material.

de•press *v.* To flatten or indent.

de•riv•a•tive *n.* **1.** The result of differentiation. Derivatives are used in calculating maximum anticipated profits from sales, the highest point of a missile's trajectory, and in many other applications. *See* differentiation. **2.** The result of a chemical process.

de•sign *n.* A plan based on a concept. *v.* To formulate such a plan.

de•sorb *v.* To release absorbed or absorbed gas from the surface of a material.

de•tail *v.* To add fine elements to a design drawing.

de•tec•tion *n.* The recovery of a modulating signal from a carrier signal.

de•ter•mi•nant *n.* An array of numbers (columns and rows) used to solve linear equations and to study linear transformations.

deu•te•ri•um *n.* An isotope of hydrogen. *Abbr.* D or ^{2}H *See* hydrogen, isotopes of.

de•vise *v.* To conceive, to invent.

Dew•ar flask *n.* A container contracted like a Thermos bottle, used to keep liquids very cold.

dew point *n.* The temperature at which atmospheric moisture condenses.

di•ag•no•sis *n.* **1.** Identification of a disease from its symptoms. **2.** Discovery of the cause of an error in a computer program, or in a process.

di•ag•nos•tics *n.* Error messages printed by a computer.

di•ag•o•nal *n.* A straight line joining opposite corners of a polygon with an even number of sides. *adj.* Pertaining to a diagonal.

di•al•y•sis *n.* The process of separation of substances in solution by selective diffusion through a membrane. *v.* di•a•lyze.

di•a•mag•net•ic *adj.* Referring to a magnetic property of a material. Materials having this property are repelled by a magnet and will align at right angles to a magnetic field.

di•a•met•ric *adj.* Along the diameter of a circle; completely opposite to.

di•a•mond *n.* The hardest substance known in nature, a mineral composed of crystalline carbon. The great majority of diamond is used as an industrial abrasive (*see* bort). Rare, nearly flawless crystals are cut as gems. When these are tinted by mineral impurities so that their appearance is enhanced, their value may increase greatly. The principal source for high-quality diamond is South Africa.

di•a•phragm *n.* A thin, flexible membrane.

di•a•ther•my *n.* The generation of high-frequency currents that are sensed as heat within the body, for therapeutic purposes.

di•a•tom•ic *adj.* Having two atoms.

di•chot•o•my *n.* A division into two parts.

di•chro•ism *n.* The property of causing color changes in a material that reflects certain colors of light while allowing others to pass through.

die *n.* **1.** A tiny piece of semiconductor material. *pl.* dice. **2.** A tool or mold used to form impressions or to cut patterns.

die at•tach *v.* To bond a semiconductor die to a substrate.

di•e•lec•tric *n.* A nonconductor of electricity, or a material in which voltage can be stored. *adj.* Pertaining to such a nonconductor.

die•sel en•gine *n.* A fuel-injection, internal-combustion engine.

dif•fer•en•tial *n.* **1.** A small increment. **2.** A system of gears and two shafts that allows one shaft to turn faster than the other.

dif•fer•en•ti•a•tion *n.* An operation in calculus that asks, "What is the rate of change of any quantity that changes or varies?" *v.* dif•fer•en•ti•ate.

dif•frac•tion *n.* The dispersal of light through fine slits, so that the beams interfere to produce spectra.

dif•frac•tion grat•ing *n.* A device that separates a beam of electromagnetic energy (such as light or x-rays) into its different wavelengths, producing a spectrum. Nature's diffraction grating is the raindrops, which separate sunlight into a spectrum called the rainbow.

dif•fuse *adj.* Widespread.

dif•fu•sion *n.* The spontaneous movement of one species within another, as when molecules of a gas disperse in air. *v.* dif•fuse.

di•ges•tion *n.* The breakdown and dissolution of a substance by a chemical solution.

dig•it *n.* A numeric character: 0, 1, 2, . . . , 9.

dig•i•tal *adj.* Referring to the use of numbers in defining data.

dig•i•tal com•put•er *n.* A computer that performs logic operations on numbers rather than physical quantities. *See* analog computer.

dig•i•tize *v.* To convert an analog measurement to a digital value.

di•he•dral *n.* A figure described by two intersecting planes.

di•late *v.* To spread or enlarge an opening.

di•la•tom•e•ter *n.* An instrument to measure a change of volume in a liquid.

di•lute *adj.* Referring to a low concentration of a substance in a solvent. *v.* To make less concentrated by increasing the amount of solvent.

dim•mer switch *n.* An electrical switch that allows the level of light to be gradually raised or lowered, as distinguished from the ordinary off-on switch. *See* rheostat.

di•mor•phic *adj.* Having two distinct forms.

di•ode *n.* An electronic device with two electrodes, often made of semiconductor material.

di•op•ter *n.* A measure of the power of a lens. *Abbr.* D

di•ox•ide *n.* A compound with two oxygen atoms, for example CO_2, carbon dioxide.

di•pole *n.* A system that is charged at two points.

di•rect ac•cess stor•age de•vice *n.* A computer storage device in which the time required to read or write data is not dependent on the location of the data.

di•rect–cou•pled *adj.* Joined without intermediate connections.

di•rect cur•rent *n.* A flow of electrons in one direction only. *Abbr.* dc

dis•as•so•ci•ate *v.* To separate into parts. *Also* dis•so•ci•ate.

disc *See* disk (preferred).

dis·charge ***v.*** To drain a storage device, such as a battery, of a voltage difference. ***n.*** The act of discharging.

dis·crete ***adj.*** Separate.

dis·crim·i·na·tor ***adj.*** An electronic circuit in which differing signals increase or decrease the level of the voltage.

disk ***n.*** A magnetized platter on which information is stored. *See* floppy disk; hard disk.

disk drive ***n.*** On a computer, a device that causes a magnetic disk to spin, thus allowing data to be entered onto or retrieved from the disk.

dis·kette ***n.*** A small, inexpensive disk used in computer data storage.

disk pack ***n.*** A group of hard disks, sealed in a pack and allowing direct access to data by the read/write head.

dis·lo·ca·tion ***n.*** A crystal defect.

dis·or·dered ***adj.*** Irregular.

dis·pense ***v.*** To provide measured quantities.

dis·perse ***v.*** To cause to scatter. ***adj.*** Scattered.

dis·per·sion ***n.*** A solvent that has been dispersed within a solute.

dis·play ***n.*** A presentation of data on a video screen.

dis·pro·por·tion·ate ***adj.*** Out of proportion.

dis·sect ***v.*** **1.** To cut open in order to examine. **2.** Figuratively, to analyze.

dis·si·pate ***v.*** **1.** To break up and cause to disappear. **2.** To use up.

dis·so·ci·ate *See* disassociate.

dis·solve ***v.*** **1.** To go into solution. **2.** To cause to go into solution.

dis·till ***v.*** To purify or separate constituents by boiling and condensing.

dis·til·late ***n.*** The product resulting from distillation.

dis·til·la·tion ***n.*** The act of distilling.

dis·trib·ute ***v.*** To spread.

dis·tri·bu·tion ***n.*** The manner in which data are spread, in time or in space.

div·i·dend ***n.*** A number that is to be divided.

di·vid·ers ***n.*** An instrument used in drafting consisting of two arms connected by a hinge.

di·vi·sor ***n.*** The number by which another number (the dividend) is to be divided.

DNA ***n.*** Deoxyribonucleic acid, the genetic material contained in all living cells.

do·dec·a·he·dron ***n.*** A 12-sided figure.

do•main *n.* A magnetic region in a crystal.

do•nor *n.* An impurity added to a semiconductor to increase the number of free electrons.

dope *v.* To add an impurity to a semiconductor. *n.* dopant.

Dopp•ler *adj.* Pertaining to a frequency shift. *See* Doppler effect.

Dopp•ler ef•fect *n.* The apparent change in frequency that takes place as an emitter moves relative to a receiver. An example is the change in pitch of a train whistle as the train approaches and then goes by.

dor•mant *adj.* Inactive but capable of resuming activity.

do•sim•e•ter *n.* An instrument for measuring x-ray or other radiation exposure.

dot–ma•trix print•er *n.* A computer peripheral device. A printer that produces characters composed of tiny dots arranged to resemble letters of the alphabet, numbers, or symbols.

doub•ler *n.* An electronic circuit that doubles the input frequency.

doub•let *n.* Two electrons that are shared by two atoms.

dow•el *n.* A pin that connects two parts.

down•time *n.* The period of time a computer (or other machine) is inoperative because of a malfunction.

DPDT *adj.* *D*ouble-*p*ole *d*ouble-*t*hrow; an electric switch.

DPST *adj.* *D*ouble-*p*ole *s*ingle-*t*hrow; an electric switch.

draft *n.* The angle of clearance between close-fitting parts.

dregs *n.* Sediment or sludge.

drone *n.* A pilotless aircraft that is either preprogrammed or operated by remote control.

dry–bulb tem•per•a•ture *n.* Air temperature taken with an ordinary thermometer. *See* wet-bulb temperature.

du•al *adj.* Double.

duc•tile *adj.* Referring to a material that is easily drawn out and elongated. *n.* duc•til•i•ty.

dumb ter•mi•nal *n.* A teleprinter that has no microprocessor control; thus it simply records output. *See* intelligent terminal.

du•plex *adj.* Referring to two parts working together.

dy•nam•ic *adj.* Active, moving, changing state.

dy•nam•ics *n.* That branch of mechanics that deals with motion.

dy•na•mo *n.* A generator of electric power.

dy•na•mom•e•ter *n.* **1.** An instrument that measures mechanical force. **2.** An instrument that measures current, voltage, or power.

dy•na•tron *n.* A type of vacuum tube in which the anode current decreases as the anode voltage increases, resulting in negative resistance.

dyne *n.* A unit of force. *Abbr.* dyn

E

e *See* natural base.

EBCDIC *n.* Extended Binary Coded Decimal Interchange Code. An 8-bit code used in IBM data-communications equipment. *Pronounced* ′ib-suh-dick. *See* ASCII.

ec•cen•tric *adj.* Off-center.

ec•cen•tric gear *n.* A gear whose axis is not at the center.

Ec•cles Jor•dan *n.* A "flip-flop" switch used in the earliest vacuum-tube computers. *See* flip-flop.

echo *n.* A reflected wave.

eclipse *n.* The obscuring of one object by another.

ed•dy *n.* A rotating electric current induced by a nonuniform magnetic field.

ed•it *v.* To rearrange, correct, and corroborate input data.

ed•i•tor *n.* A computer program used to review text materials and program instructions.

EDP *n.* *E*lectronic *d*ata *p*rocessing.

EEPROM *n.* *E*lectrically *e*rasable *p*rogrammable *r*ead-*o*nly *m*emory chips.

ef•fect *n.* The result of an occurrence. ***v.*** To cause to occur.

ef•fi•cien•cy *n.* A measure of useful output from total input, expressed as a ratio or a percentage.

ef•flo•resce *v.* To lose water to the degree that the substance crumbles into powder. ***n.*** ef•flo•res•cence

elab•o•rate *v.* To extend a treatment in greater detail.

elas•tic *adj.* Able to return to a previous shape and size after being deformed.

elas•tic•i•ty *n.* The ability to recover entirely a previous size and shape after being subjected to deformation.

elas•tic lim•it *n.* The highest force to which a material can be subjected and still return to its original dimensions on the removal of that force.

elas•to•mer *n.* Synthetic rubber or plastic.

el•bow *n.* A right-angled pipe joint.

elec•trode *n.* An element that conducts electricity into or out of a medium, such as a gas or vacuum.

elec·tro·form *v.* To build up a structure by electroplating.

elec·trol·y·sis *n.* The production of chemical change in an ionic solution by passing an electric charge through it.

elec·tro·lyte *n.* The chemical solution used in electrolysis.

elec·tro·mag·net *n.* A coil structure which generates magnetic force when an electric current passes through it.

elec·trom·e·ter *n.* An instrument for measuring voltages while drawing very little current.

elec·tro·mo·tive force *n.* The difference in electric potential between two unlike electrodes (positive and negative) that are bathed in the same electrolyte solution. This difference permits electrochemical reactions to take place, such as metal plating. *Abbr.* emf

elec·tron *n.* A subatomic, negatively charged particle.

elec·tron cloud *n.* A "cloud" of negative electricity (electrons) surrounding the nucleus of an atom.

elec·tron con·fig·u·ra·tion *n.* The arrangement of an atom's electrons in spatial shells about the nucleus. The first seven shells are designated *K*, *L*, *M*, *N*, *O*, *P*, and *Q*.

elec·tron mi·cros·co·py *n.* The use of electrons to provide a high-magnification image in an electron microscope.

elec·tron spin *n.* The spinning of an electron about its axis; this gives rise to its characteristic angular momentum.

elec·tro·pho·re·sis *n.* The movement of particles as a result of applied voltage forces.

elec·tro·plate *v.* To deposit metal on an electrode from an electrolyte by the action of electricity.

elec·tro·stat·ics *n.* The study of electric charges at rest.

elec·tro·ther·mal *adj.* Producing heat by electrical means.

el·e·ment *n.* **1.** One of the basic constituents of matter. **2.** An electrical component.

el·e·va·tion *n.* A drawing of the vertical plane of an object, for example, the front, back, or one side of a building.

el·lipse *n.* A curve. *See* conic section.

elon·ga·tion *n.* The increase in length as a function of applied force, expressed as a percentage.

elu·tri·ate *v.* To remove by washing.

em·a·na·tion *n.* That which is emitted. In nuclear physics, in a cloud of radioactive gas.

em·boss *v.* To produce a pattern of indentations and projections.

em·bryo *n.* An early state of a system or living thing.

em•ery *n.* An abrasive powder.

emf *See* electromotive force.

emis•sion *n.* The emanation of radiation or particles.

emis•siv•i•ty *n.* The measure of efficiency of emitting radiation.

emit•ter *n.* A structure or a surface from which waves or particles emanate.

em•pir•i•cal *adj.* Based on data.

emul•si•fy *v.* To break up into tiny particles within another liquid.

emul•sion *n.* A suspension of tiny particles.

en•am•el *n.* A glassy layer deposited on a substrate for decorative or protective purposes. *v.* To deposit an enamel layer.

en•cap•su•late *v.* To enclose in a package.

en•code *v.* To convert data into a code.

en•do•ther•mic *adj.* Referring to a heat-absorbing chemical reaction. *See* exothermic.

en•er•gy *n.* The capability of doing work.

en•er•gy band *n.* For each atom, regions of allowed levels of energy of its associated electrons.

en•er•gy gap *n.* The forbidden energy levels of electrons in an atom which separate allowed levels (energy bands).

en•thal•py *n.* The amount of heat associated with any chemical or physical state. *Symbol* H

en•thal•py change *n.* *See* enthalpy. *Symbol* ΔH

en•ti•ty *n.* A being or substance complete in itself.

en•tro•py *n.* A thermodynamic measure of disorder in any chemical or physical state. *Symbol* S

en•tro•py change *n.* *See* entropy. *Symbol* ΔS

en•ve•lope *n.* An enclosing structure.

en•vi•ron•ment *n.* All the materials and conditions that surround a material or a process.

en•zyme *n.* A protein molecule which catalyzes chemical reactions in living organisms.

epi– *prefix* Outer; after.

ep•i•sode *n.* An occurrence.

ep•i•taxy *n.* Growth of one crystal upon another. Growth of the second crystal is oriented by the lattice structure of the substrate. *adj.* ep•i•tax•i•al

epit•o•me *n.* The ideal.

ep•oxy ***n.*** A resin, used as an adhesive.

EPROM ***n.*** *E*rasable *p*rogrammable *r*ead-*o*nly *m*emory chips.

equate ***v.*** To declare the identity of two or more terms, or the parts of a chemical reaction.

equi– ***prefix*** Equal.

equi•an•gu•lar ***adj.*** Having equal angles.

equi•axed ***adj.*** Having similar lengths and widths (used to describe crystals).

equi•lat•er•al ***adj.*** Having equal sides.

equi•lib•ri•um ***n.*** A stable condition.

equi•po•ten•tial ***adj.*** Having equal voltages.

equiv•a•lent ***adj.*** Virtually the same and with the same significance; equal.

erad•i•cate ***v.*** To remove entirely.

erg ***n.*** A unit of work or energy.

er•go ***conj.*** Therefore *(Latin).*

erode ***v.*** To wear away.

ero•sive ***adj.*** Wearing away.

er•ra•ta ***n. pl.*** A list of errors. *sing.* er•ra•tum.

er•rat•ic ***adj.*** Not predictable.

es•ter ***n.*** A chemical formed by combining an alcohol and an organic acid.

etch ***v.*** To dissolve a surface chemically.

etch•ant ***n.*** The chemical used in the etching process.

eu•clid•e•an ***adj.*** Referring to traditional, two- and three-dimensional geometry.

eu•tec•tic ***adj.*** Having the lowest melting-point composition of an alloy.

evac•u•ate ***v.*** To create a vacuum; to empty a vessel.

ex•ci•ta•tion ***n.*** **1.** The act of achieving a higher energy state. **2.** The application of signal voltage in an electron tube.

ex•e•cute ***v.*** In a computer program, to carry out a set of instructions.

ex•fo•li•ate ***v.*** To flake off or peel.

ex•haust ***v.*** **1.** To evacuate. **2.** To empty. ***n.*** A gas given off in a process.

ex•o•sphere ***n.*** The uppermost level of Earth's atmosphere, 250 miles and higher, grading off into space. In the exosphere, a molecule of air has a 50 percent chance of either hitting another air molecule or escaping into space.

ex·o·ther·mic *adj.* Referring to a chemical reaction that gives off heat. *See* endothermic.

ex·pend *v.* To consume.

ex·po·nent *n.* A superscript number showing a power to which a number is raised; thus x^3 is x times x times x. *adj.* ex·po·nen·tial

ex·po·nen·tial growth func·tion *n.* An equation that is used to calculate population growth, radioactive decay of an element, compound interest, and many similar cases.

ex·pres·sion *n.* **1.** A mathematical statement. **2.** The separation of a liquid from a solid, under pressure.

extra– *prefix* Outside.

ex·tract *v.* To separate a desired constituent out of a mixture by a chemical or a physical process.

ex·trap·o·late *v.* **1.** To extend a line on a graph. **2.** To extend an idea beyond limits imposed by available data; to make a prediction based on those data.

ex·trem·i·ty *n.* The farthest point.

ex·trin·sic *adj.* Not part of; external to.

ex·trude *v.* To squeeze a semisoft material through a small aperture.

F

fac·et *n.* A flat surface, natural or made by cutting, on a crystal. *v.* To create a facet.

fac·tor *n.* One algebraic term multiplied by another. *v.* To separate a term into its components; thus the factors of $3a$ are 3 and a.

fac·to·ri·al *n.* For a number n, which is any given number, the product of all the whole numbers (integers) from 1 to n. *Symbol* ! (The notation for factorial n is $n!$).

Fahr·en·heit *adj.* Referring to the temperature scale in which 32° is the freezing point of water and 212° the boiling point. *Abbr.* F

far·ad *n.* The unit of capacitance. *Abbr.* F

far·a·day *n.* The electric charge required to deposit or dissolve a gram equivalent of a substance by electrolysis.

fa·tigue *n.* Failure in a material caused by repeated flexings at one point.

fat·ty ac·id *n.* An organic acid. The fatty acids are important as lubricants, and in detergents.

fault *n.* **1.** A defect in a crystal or a circuit. **2.** A fracture in rock.

fea·si·bil·i·ty *n.* The likelihood that a procedure will achieve its goal.

feed•back ***n.*** **1.** The return of a portion of the output in a circuit to the input. **2.** A type of circuit.

femto– ***prefix*** One-million-billionth. *Symbol* f

fer•men•ta•tion ***n.*** A chemical reaction that breaks down large molecules in the absence of oxygen.

ferri–, ferro– ***word root*** Relating to iron. (From Latin *ferrum*—iron.)

fer•ric ***adj.*** Relating to a compound of trivalent (Fe^{3+}) iron.

fer•rite ***n.*** A magnetic material, of high electrical resistivity, used for magnetic cores, magnetic memories, switches, and many other applications.

fer•ro•mag•net•ic ***adj.*** A highly magnetic material.

fer•rous ***adj.*** Relating to a compound of divalent (Fe^{2+}) iron.

fer•rule ***n.*** A metal ring clamped around a shaft to strengthen it.

FET ***n.*** Field effect transistor.

fi•ber ***n.*** A threadlike material, natural or manufactured. ***adj.*** fi•brous

fi•ber op•tics ***n.*** The technique of transmitting light signals through long fibers of glass or plastic; used for communications.

field ***n.*** **1.** A region across which an electric force acts. **2.** In data processing, an area of a card or computer memory set aside for certain information.

fil•a•ment ***n.*** A fine wire or thread. ***adj.*** fil•a•men•ta•ry.

fi•lar ***adj.*** Filamentary.

file ***n.*** A set of related records treated as a unit in data processing.

fil•let ***n.*** A flat, narrow strip of wood or metal.

fil•ter ***n.*** **1.** A porous material for separating solids from liquids. ***v.*** To separate solids from liquids by passing through a porous material. **2.** An electronic circuit that selectively enhances desired input signals.

fil•trate ***n.*** The liquid transmitted by a filter.

fin ***n.*** A flat plate projecting from a surface and perpendicular to it.

fi•nite ***adj.*** Limited or bounded.

firm•ware ***n.*** Programs used so often that they are stored in the ROM. *See* ROM.

fis•sile ***adj.*** Capable of being split.

fis•sion ***n.*** Division of an atomic nucleus to release energy. ***adj.*** fis•sion•a•ble.

flint ***n.*** **1.** A variety of chalcedony. **2.** An alloy used to make the "flint" in a cigarette lighter.

flint glass ***n.*** A brilliant, high-quality glass containing lead oxide that is used in optical systems.

flip–flop *n.* A binary counter, stable in either of two states until it receives a signal to switch to its alternate state. *See* Eccles-Jordan.

floc *n.* A cluster of particles formed in a fluid.

floc•cu•la•tion *n.* The formation of a cluster of particles.

flop•py disk *n.* A flexible magnetic disk used in computer inputs and outputs and auxiliary storage. *See* hard disk.

flow•chart *n.* A diagram showing the progression of a process.

flu•id *n.* A gas or a liquid.

flu•id bed *n.* Fine powders which act in a fluidlike manner in a rising gas stream. Used in petroleum cracking.

flu•id•ics *n.* The technology of fluids.

flu•id•i•ty *n.* A measure of the ability of a material to flow.

flu•o•res•cence *n.* Light emission from a material as a result of incident radiation. Thus, certain minerals fluoresce when their atoms are excited by ultraviolet light—known as the "black light" phenomenon.

flux *v.* The rate of transfer of energy across a given space; used to describe the strength of electric and magnetic fields.

FM *n.* Frequency modulation.

fo•cal point *n.* The location of light-ray convergence as a result of lens refraction.

fo•cus *n.* The point at which light rays converge. ***adj.*** fo•cal.

foot–pound *n.* A unit of energy or work. *Abbr.* ft·lb

forge *v.* To form hot metal by pressing or hammering.

for•mat *n.* An arrangement of data for use in a computer.

FOR•TRAN *n.* Formula Translator, a high-level computer language.

Fou•ri•er trans•form *n.* A formula that expresses the energy relationship between a momentary oscillation of voltage or current and the adjacent frequency spectrum. *Also* Fou•ri•er se•ries.

frac•tion *n.* A portion of a whole; a quotient of two numbers expressed with a bar line.

frac•tion•ate *v.* To separate a material into its components.

fray *v.* To abrade.

free en•er•gy *n.* Internal energy of a system − (temperature × entropy) = free energy. *Also called* Helmholtz free energy. *Abbr.* G, F

free fall *n.* The rate of fall of an object influenced by gravity only.

french curve *n.* A templatelike drafting device used as an aid to drawing curves.

Fren•kel de•fect *n.* A crystal defect consisting of a vacancy and an interstitial, which arises when an atom is plucked from its normal position and forced into another point in the lattice.

fre·quen·cy ***n.*** The number of oscillations completed by a repeating wave form per second. Formerly known as cycles per second *(abbr.* cps), now called the hertz. *Abbr.* Hz

fri·a·ble ***adj.*** Easily crumbled.

fringe ***n.*** In optics, a dark or light band produced by interference or diffraction.

frit ***n.*** Powdered ceramic mixture used in enameling.

ful·crum ***n.*** The point on which a lever is supported and around which it rotates.

fume ***n.*** Smoke or vapor, usually noxious. ***v.*** To give off smoke or vapor.

func·tion ***n.*** **1.** A mathematical formula that relates variables to each other. Thus profit can be expressed as a function of the number of units sold by a business. **2.** In FORTRAN, a subroutine that returns a computational value.

func·tion·al ***adj.*** Able to be used; working.

fuse ***v.*** **1.** To melt **2.** To join two materials by heating. ***adj.*** fus·i·ble.

fu·se·lage ***n.*** The main body of an aircraft.

fuze ***n.*** An electronic mechanism designed to detonate the charge in a missile.

G

gage *See* gauge.

gain ***n.*** Power increase in an electronic signal produced by an amplifier; generally expressed in decibels. *See* attenuate.

ga·le·na ***n.*** An ore of lead.

gal·van·ic ***adj.*** Referring to direct current electricity.

gal·va·nom·e·ter ***n.*** An instrument for measuring small voltages.

gan·try ***n.*** A movable scaffold used in positioning rocket vehicles.

Gantt chart ***n.*** A chart which plots target events against predicted dates.

gar·net ***n.*** A mineral used as an abrasive. Some garnets are of gem quality.

gate ***n.*** An electronic circuit in which one signal switches another signal on and off.

gauge ***n.*** **1.** A measure of the thickness of metal sheet or wire. **2.** A device for measuring small distances. ***v.*** To measure small distances. *Also* gage.

gauss ***n.*** The measure of magnetic induction. ***pl.*** gauss. *Abbr.* G

gauss·ian dis·tri·bu·tion *n.* In statistics, a symmetrical distribution of data. *Also known as* a bell-shaped curve; a normal distribution.

gear *n.* A toothed wheel in a machine that transmits motion between two shafts.

gear up *v.* To arrange gears so that the driven shaft rotates faster than the driver.

Gei·ger–Mül·ler count·er *n.* An instrument for measuring radioactivity. Also known as Geiger counter.

gel *n.* A jellylike material. *v.* To cause to become jellylike.

gene *n.* The unit of inheritance which carries codes that determine qualities of a living organism.

gen·er·a·tor *n.* A machine that produces electric energy from mechanical energy.

ge·ner·ic *adj.* Pertaining to a type.

ge·ol·o·gy *n.* The study of the history and structure of the Earth; now extended to all solid astronomical bodies.

geo·phys·ics *n.* The physics of the Earth, including such fields as oceanography and seismology.

germ *n.* **1.** The beginning of an idea. **2.** A microorganism. **3.** An egg or a sperm cell.

get·ter *n.* A material that absorbs gases in a vacuum system. Also known as a degasser. *v.* To absorb gases in a vacuum system.

giga– *prefix* Billion (thousand million). Symbol G

GIGO *n.* Computer proverb: garbage in, garbage out.

gil·bert *n.* A unit of magnetomotive force. *abbr.* Gb

gim·bal *n.* A support which permits free movement along two axes. Thus, aboard a ship, a compass or table swung in gimbals will always present a level surface to the user.

gim·mick *n.* **1.** A clever feature of a device or a device itself. **2.** A feature or a device that is attention-getting but poorly based in scientific research. *v.* To lash together crudely.

gird·er *n.* A large horizontal beam, metal or concrete.

gla·cial *adj.* Referring to pure acetic acid.

glob·u·lar *adj.* Spherical.

glob·ule *n.* A small spherical object.

glu·cose *n.* The most common sugar.

glyc·er·ine *n.* An alcohol used in antifreeze, medicine, cosmetics, and many other substances.

glyc·er·ol *See* glycerine.

gneiss *n.* A metamorphic rock usually containing bands of mica.

-gon- *word root* Angle.

go·ni·om·e·ter *n.* An instrument that measures the angle between two crystal faces.

grad *n.* A vector function.

gra·da·tion *n.* A series.

grade *v.* **1.** To smooth. **2.** To sort by grades.

gra·di·ent *n.* **1.** A rate of change. **2.** A slope, expressed as feet per mile, in a ratio of the horizontal to the vertical distance.

grad·u·a·ted *adj.* Marked off in units.

grain *n.* **1.** A crystal. **2.** Surface texture.

gram *n.* A unit of weight ($\frac{1}{454}$ pound). *Abbr.* g

gran·ule *n.* A particle.

graph *n.* A representation that plots variables as points. *v.* To draw a graph.

graph·ite *n.* A soft, noncrystalline form of carbon, used as a lubricant.

grat·ing *See* diffraction grating.

grav·i·met·ric *adj.* Relating to the weight of a compound, in analytical chemistry.

grav·i·ta·tion *n.* The attraction that exists between large bodies.

grav·i·ty *n.* The force of gravitation at the surface of the Earth or other large body. *Abbr.* G

grid *n.* **1.** A perforated or mesh structure. **2.** The electrode in a vacuum tube that controls electron flow from cathode to anode.

ground *n.* The connecting path between an electrical system and the Earth. *v.* To make such a connection.

gyro- *word root* Pertaining to rotation. (From Greek *gyros*—rounded.)

gy·ro·scope *n.* A spinning disk balanced between two low-friction supports that keeps its balance even in a "tilted" environment; used for stabilization in moving vehicles, for example, ships and spacecraft. *Abbr.* gyro

H

hab·it *n.* The form of a crystal.

ha·la·tion *n.* A region of light surrounding an object or image.

half-life *n.* A measure of the persistence of radiation.

hal·ide *n.* A salt having atoms of any of the halogen family: chlorine, fluorine, iodine, bromine, or astatine.

ha•lo *n. See* halation.

hal•o•gen *n.* The chemical group comprising fluorine, chlorine, bromine, iodine, and astatine.

hard copy *n.* Any copy that is written or printed on paper.

hard disk *n.* A nonflexible disk used in conjunction with a computer for data storage and whose storage capacity usually exceeds that of a floppy disk. *See* floppy disk.

hard•ware *n.* The physical components of a computer system, as contrasted with the software (the programs).

har•mon•ic *adj.* Referring to waves. *n.* A musical overtone with a frequency that is a multiple of the fundamental tone. Simple harmonic motions include waves in the open sea, the movement of an atom in a sound wave, the swing of a pendulum, and the stroke of a piston in an engine. Harmonic motion is described and graphed by means of trigonometric ratios. *See* trigonometric ratios.

Hart•ley prin•ci•ple *n.* The number of bits that can be transmitted on a channel, equal to the product of bandwidth and transmission time.

H beam *n.* A structural member shaped like the letter H.

heat•er *n.* The hot wire that heats the cathode of a vacuum tube.

hecto– *prefix* Hundred. *Symbol* h

heli–, helio– *word root* Relating to the sun. (From Greek *helios*—sun.)

helic–, helico– *word root* Relating to a spiral. (From Greek *helix*—spiral.)

hel•i•cal *adj.* Shaped like a spiral; an example is a screw thread.

he•lix *n.* A spiral.

hem•i•sphere *n.* Half a sphere.

hen•ry *n.* A unit of induction. *pl.* hen•rys. *Abbr.* H

her•met•ic *adj.* Gastight.

hertz *n.* A unit of frequency, 1 cycle per second. *Abbr.* Hz

hetero– *prefix* Different.

het•er•o•dox *adj.* Disagreeing with accepted belief.

het•er•o•dyne *v.* To mix two different frequencies in order to produce new frequencies, which are the sum and difference of the original frequencies. Heterodyning has many applications in radio and television, code transmission, and analysis of audio signals.

heu•ris•tic meth•od *n.* An approach to problem solving in which various paths are tried and the results are evaluated after each trial.

heu•ris•tic rou•tine *n.* A program in which the computer tries various solutions to a problem and evaluates its own progress after each attempt.

hex·ag·o·nal *adj.* Having six angles. *See* crystal systems.

hex·a·dec·i·mal *adj.* Pertaining to a number in base 16.

hex·a·dec·i·mal no·ta·tion *n.* A notation used to count in base 16. The characters are the digits 0 to 9 plus letters A to F.

hi·a·tus *n.* A break, pause, or interruption.

hi·er·ar·chy *n.* An ordering from lowest to highest value.

hi–fi *n.* and *adj.* High fidelity; high-quality sound reproduction.

high–level lan·guage *n.* A programming language (such as BASIC, COBOL, FORTRAN, or Pascal) that is intelligible to human users but that must be compiled or interpreted into machine language in order for the computer to carry out the instructions. *See* machine language.

his·to·gram *n.* A bar chart in which the width of the bar is proportional to the width of the class, and the height is proportional to the frequency within each class.

hole *n.* In a semiconductor, the absence of an electron; a hole functions as an electric carrier.

ho·log·ra·phy *n.* A photographic laser technique for creating extremely realistic three-dimensional images.

homo–, homeo– *word root* Same.

ho·mol·o·gous *adj.* Having the same or a corresponding function.

hop·per *n.* A box for holding parts or materials.

ho·rol·o·gy *n.* The science of time measurement.

horse·power *n.* A unit of power. 1 horsepower equals approximately 746 watts. *Abbr.* hp

hot work·ing *n.* The shaping of metal at high temperature.

hub *n.* The center of a wheel.

hue *n.* The name of a color.

hunt *v.* In radio, to oscillate around a given position before settling into that position.

hy·brid *n.* A system made up of two or more dissimilar elements.

hy·brid·ize *v.* To produce hybrids.

hy·drate *n.* A chemical compound with associated water molecules.

hy·drau·lics *n.* The study of the mechanics of fluids, particularly liquids. *See* fluid.

hy·dride *n.* A chemical compound containing hydrogen and another element.

hydro– *word root* Water-related. (From Greek *hydor*—water.)

hy·dro·car·bon *n.* A chemical compound of hydrogen and carbon only.

hy•dro•dy•nam•ics *n.* The study of fluids in motion.

hy•dro•foil *n.* A winglike fin that lifts the hull of a hydrofoil boat out of the water at high speeds.

hy•dro•gen, iso•topes of *n.* The isotopes of hydrogen are protium (atomic mass 1), the most common isotope; deuterium (atomic mass 2), known as "heavy water" and formerly used in nuclear reactors; and tritium (atomic mass 3), used as a radioactive tracer in chemical research.

hy•drol•y•sis *n.* The decomposition or alteration of a chemical substance by a process that involves splitting a bond and adding the elements of water. *v.* hy•dro•lyze

hy•dro•phil•ic *adj.* Strongly attracted to water; likely to absorb water.

hy•dro•pho•bic *adj.* Strongly repelled by water; likely not to be wetted by water.

hy•dro•stat•ic *adj.* Pertaining to fluids "at rest" and to the forces and pressures they transmit.

hy•dro•stat•ics *n.* The study of fluids in equilibrium.

hy•drox•ide *n.* A chemical compound having a hydroxyl (OH^-) radical.

hy•drox•yl *n.* The chemical radical OH^-.

hygro– *word root* Relating to humidity. (From Greek *hygros*—wet.)

hy•grom•e•ter *n.* An instrument for measuring humidity.

hyper– *prefix* Excessive.

hy•per•bo•la *n.* The twin curves, called branches, that would be obtained if an hourglass were sliced, top and bottom, along a plane. *adj.* hy•per•bol•ic *See* conic section.

hy•per•bol•ic func•tions *n.* Six functions related to the hyperbola as the trigonometric functions are related to the circle. They are: hyperbolic cosecant (*abbr.* csch), hyperbolic cosine (*abbr.* cosh), hyperbolic cotangent (*abbr.* coth), hyperbolic secant (*abbr.* sech), hyperbolic sine (*abbr.* sinh), and hyperbolic tangent (*abbr.* tanh). Uses of the hyperbolic functions include plotting the position of a ship at sea given two radio beacons at different locations; and calculating the curvatures of the main mirror and the auxiliary mirror in a reflecting telescope. *See* hyperbola; trigonometric ratios.

hypo– *prefix* Chemical prefix. The lowest oxidation state of an acid.

hy•pot•e•nuse *n.* The side of a right triangle oppposite the right angle.

hy•poth•e•sis *n.* A proposed explanation to be proved.

hy•po•thet•i•cal *adj.* Conjectural, conditional, not yet proved.

hys•ter•e•sis *n.* A phenomenon of lagging during oscillations that consumes energy, gives off heat, and lessens efficiency.

I

I beam *n.* A structural member shaped like the letter I.

–ic *suffix* Chemical suffix. The highest oxidation state of an acid. *Compare* –ous.

–ide *suffix* Chemical suffix indicating a binary acid compound.

idler *n.* A wheel or gear that transfers motion or acts as a guide.

IF *n.* Intermediate frequency.

ig•ne•ous *adj.* Pertaining to rock that has hardened from magma. *See* magma.

ig•ni•tron *n.* An electron tube that contains an ignition electrode. Used in heavy-duty switches.

–ite *suffix* **1.** In chemistry, the second-highest oxidation state of a salt. **2.** Indicating a rock or a mineral.

il•lu•so•ry *adj.* Not real.

im•age *n.* An optical representation of an object. *v.* To represent an object optically.

imag•i•nar•y num•ber *n.* A number which is theoretically impossible, such as the square root of −1 (two negative numbers multiplied together will produce a positive number).

im•mi•nent *adj.* About to occur.

im•mis•ci•ble *adj.* Pertaining to two liquids unable to dissolve into each other.

im•mu•ta•ble *adj.* Not changeable.

im•pact *n.* A sudden collision.

IM•PATT di•ode *n.* A junction diode. Derived from *imp*act *a*valanche and *t*ransit *t*ime diode.

im•ped•ance *n.* Electric resistance.

im•pel•ler *n.* An engine part that pushes a fluid, used in turbines, fans, mixers, and the like.

im•per•me•a•ble *adj.* Not permitting fluid to pass through, as of a membrane. *Also known as* impervious.

im•per•vi•ous *adj.* Not able to be penetrated or damaged by the environment. *See* impermeable.

im•pe•tus *n.* A stimulus to movement.

im•pinge *v.* To strike.

im•plant *v.* To place within.

im•ple•ment *n.* A tool. *v.* To carry through to a conclusion.

im•plic•it *adj.* Understood on the basis of previous information.

im•plode *v.* To burst inward. *n.* Im•plo•sion.

im•preg•nate *v.* To permeate a porous solid with a liquid.

im•press *v.* To indent with a tool under pressure.

im•pres•sion *n.* The pattern on a surface made by a tool under pressure.

im•pulse *n.* A sudden force.

im•pu•ri•ty gra•di•ent *n.* A change of distribution of impurities from one point to another.

in•board *adv.* Toward the center of a ship or aircraft.

in•can•des•cence *n.* The light emitted by a hot surface.

in•cep•tion *n.* The beginning.

in•ci•dence *n.* **1.** The rate of occurrence. **2.** The striking of radiation on a surface.

in•cip•i•ent *adj.* About to happen.

in•cise *v.* To cut into.

in•cli•na•tion *n.* The angle of deviation from the horizontal or the vertical.

in•clu•sion *n.* A foreign body, for example, a crystal or gas pocket imbedded in another crystal.

in•co•her•ent *adj.* Disordered, scattered.

in•com•men•su•ra•ble *adj.* Without a common base of measurement.

in•com•pat•i•ble *adj.* Not able to function together.

in•con•tro•vert•i•ble *adj.* Without any doubt.

in•cor•po•rate *v.* To blend in; to include.

in•cre•ment *n.* A small change, either increase or decrease. *Symbol* Δ

in•cu•bate *v.* To maintain in conditions favorable to development.

in•dent *v.* To produce a hollow in a surface by means of pressure.

in•de•pen•dent *adj.* Uninfluenced by other events.

in•de•ter•mi•nate *adj.* Uncertain, unclear.

in•dex *n.* A pointer on a scale. *v.* To move down one position in a grid.

in•dex of re•frac•tion *n.* The ratio of the speed of light in a vacuum to its speed in a particular medium. This change in speed is accompanied by a change of direction as the ray passes into the medium. The index of refraction is given in terms of the medium. Thus, for water, it is 1.332; for air, 1.000293; and for diamond, 2.4173. *Symbol* α

in•dex word *n.* *See* modifier.

in•di•cate *v.* To suggest, to point out.

in·di·ca·tor *n.* **1.** A result that signals an effect. **2.** A pointer in a measuring device.

in·duce *v.* To cause.

in·duc·tance *n.* A coil in which electromotive force is generated by a varying magnetic flux.

in·duc·tion *n.* The generation of electric energy in one circuit by another.

in·duc·tive *adj.* Pertaining to or induction or inductance.

in·duc·tor *n. See* inductance.

in·e·las·tic *adj.* Referring to the inability of a material to be deformed and then return to its original size and shape.

in·ert *adj.* Nonreactive chemically.

in·er·tia *n.* The tendency of moving or stationary bodies to remain in motion or at rest.

in·fil·trate *v.* To penetrate.

in·fi·nite *adj.* Without limit.

in·fin·i·tes·i·mal *adj.* Extremely small, having values very close to zero.

infra– *prefix* Below; less than.

in·fra·red ra·di·a·tion *n.* Electromagnetic waves invisible to the eye that can be felt as heat; in the wavelength range 0.75–1000 micrometers (between visible red light and the lowest microwaves).

in·fra·son·ic *adj.* Lower than audible sound waves, about 15 hertz.

in·fuse *v.* To steep a substance in a liquid.

in·got *n.* A bar of solid metal.

in·her·ent *adj.* Essential to a structure or a process.

in·hib·i·tor *n.* A substance used in low concentrations to reduce chemical activity.

in·oc·u·late *v.* To introduce a small amount of one material into another.

in·or·gan·ic *adj.* Referring to chemical compounds that do not contain carbon in the form of plant or animal materials.

in·put *n.* Data or instructions given to a computer.

in·put im·ped·ance *n.* The resistance offered by an electric circuit when its output terminals are shorted.

in·put/out·put *n.* The process of entering data into or printing it out from a computer. *adj.* Pertaining to such data. *Abbr.* I/O

in si·tu *adv.* In place (*Latin*).

in·sol·u·ble *adj.* Not dissolvable.

instruction, in•struc•tion set *n.* A command or set of commands used in computer programs.

in•su•la•tor *n.* A nonconductor of electricity.

in•te•gers *n.* The natural numbers 1, 2, 3, . . . , plus the negative numbers . . . −3, −2, −1. Zero is also an integer.

in•te•gral *n.* The result of an integration, and the inverse of differentiation. The expression "integral of the rate" means the total of the rate. *Symbol* ∫ *See* differential; integration.

in•te•grate *v.* **1.** To perform an integration. **2.** To incorporate.

in•te•grat•ed cir•cuit *n.* A miniature electronic circuit produced within and upon a single semiconductor. Used in computer memories, microcomputer central processing units, calculators, watches, and so forth. *Abbr.* IC

in•te•gra•tion *n.* An operation in calculus that asks, "Given the rate of change of a quantity and the value of the quantity at some certain instant, what is the value of the quantity at any instant?" *See* differentiation.

in•te•gra•tor *n.* A circuit whose output is a function of its input.

in•tel•li•gent ter•mi•nal *n.* A terminal that contains a microcomputer. *See* dumb terminal.

inter– ***prefix*** Between, as in *interconnect* (to establish a link between).

in•ter•e•lec•trode *adj.* Referring to forces between electrodes.

in•ter•face *n.* **1.** The place where two surfaces meet. **2.** In computer technology, the connection between two different units, such as a cathode-ray tube and the central processing unit, that makes their interaction possible.

in•ter•fer•ence *n.* **1.** The effect of wave energies on each other. **2.** Undesired signals that obscure radio reception.

in•ter•fer•ence fringe *n.* In optics, a dark line caused by the interaction of light waves of different frequency.

in•ter•fer•om•e•ter *n.* An optical instrument for measuring light interference. Used in measuring wavelengths, very small distances, and in astronomy to study spectral lines and diameters of stars.

in•te•gral *n.* The results of the integration operation in calculus. ***adj.*** Essential for completeness.

in•ter•im *n.* The time between.

in•ter•po•late *v.* To insert.

in•ter•pret•er *n.* A computer program that translates a high-level language into machine instructions.

in•ter•sperse *v.* To place into a system.

in•ter•stic•es *n.* The open spaces between closely packed units, especially the voids around atoms in crystals. ***adj.*** In•ter•sti•tial.

in•ter•val *n.* The difference frequency between two frequencies.

intra- *prefix* Within, as in *intramural* (occurring within a community or organization).

in•trin•sic *adj.* **1.** Inherent; fundamental; basic. **2.** Referring to a region in a semiconductor in which *n*-type and *p*-type doping are in balance.

in va•cu•o *adv.* In a vacuum.

in•verse *adj.* Opposite; negative. In an inverse relation, x increases as y decreases.

in•ver•sion *n.* A reversal of order.

in•vert•er *n.* A device that converts direct current into alternating current.

in•vest•ment casting *n.* A very accurate method used to make small castings.

in vi•tro *adj.* Pertaining to experiments performed "in glass" (in a Petri dish or some other piece of laboratory equipment). *Compare* in vivo.

in vi•vo *adj.* Pertaining to experiments performed on a living organism. *Compare* in vitro.

in•vo•lute *adj.* Rolled in at the edge.

I/O *See* input/output.

ion *n.* A chemical radical or an atom having a positive or a negative charge. *adj.* ion•ic.

ion im•plan•ta•tion *n.* The process of driving ions into the surface layer of a material; used in semiconductor fabrication.

ion•ize *v.* To impart a positive or negative charge to an atom or a radical.

ion•o•lu•mi•nes•cence *n.* Light emission due to ion impact.

ion•o•sphere *n.* The layer in the Earth's atmosphere that is ionized by solar ultraviolet radiation so that radio waves are affected. Extends from about 25 miles to about 250 miles above the Earth's surface.

ip•so fac•to *adv.* By the nature of the thing; necessarily.

ir•i•des•cence *n.* A rainbow effect produced by interference in soap bubbles and by diffraction in birds' feathers. *adj.* ir•i•des•cent

iris *n.* A diaphragm with a small hole that admits light. Used in cameras and other optical systems.

ir•ra•di•ate *v.* To expose to x-rays, gamma rays, or other ionizing radiation.

ir•ra•tion•al num•ber *n.* A number with an infinite number of digits after its decimal point in which no pattern of digits repeats itself; a number that cannot be produced by multiplying two integers. Examples include the square root of 2 and the ratio pi (3.14159265 . . .).

ir·ri·gate *v.* To wash with flowing water.

is–, iso– ***word root*** The same. (From Greek *isos*—equal.)

iso·bar *n.* A contour line indicating equal pressure on either side, as on a meteorological chart.

iso·mer *n.* A chemical compound with the same composition as another compound but with a different structure.

iso·met·ric drawing *n.* In drafting, a nonperspective method that shows three perpendicular axes all equally foreshortened.

iso·mor·phic *adj.* Identical in shape, form, or structure.

isos·ce·les *adj.* Referring to a triangle with two equal sides.

iso·ther·mal *adj.* At a constant temperature.

iso·tope *n.* One of two or more atoms with the same atomic number but different mass numbers. Thus, carbon-12 is the most stable and most common isotope of carbon, which also exists as carbon-13 and carbon-14.

iso·trop·ic *adj.* Having equal tendency for growth in all directions. *n.* isot·ro·py.

is·sue *n.* The result.

it·er·a·tive meth·od *n.* To repeat a series of approximations, each one coming closer to the desired result. *Also known as* iteration.

izod test *n.* An impact test of the strength of a material.

J

jam *v.* To interfere with radiated signals.

jet *n.* A high-velocity stream of gas or liquid.

jig *n.* A structure to aid in positioning parts for machining. *v.* To design and construct such a structure.

joint *n.* The point where two pieces of a structure are attached.

joist *n.* A structural member that directly supports a floor.

joule *n.* A unit of measurement of energy, work, or heat. The work performed in 1 second by 1 ampere of current flowing against 1 ohm of resistance. *Abbr.* J

Joule's equiv·a·lent *n.* The amount of work that will raise the temperature of 1 pound of water by 1 degree Fahrenheit, when that work is converted into heat.

junc·tion *n.* The transition region between the *n*-type and the *p*-type zones of a semiconductor.

K

K *n.* **1.** In computer technology, 1024 bytes, but often discussed as if it equaled 1000; a 64K memory refers to one that stores approximately 64,000 bytes. **2.** *See* Kelvin.

ka·o·lin *n.* A fine-grained clay used in ceramics, filters, and papermaking.

kar·at *n.* A measure of the proportion of pure gold in an alloy. Thus, 24-karat gold is pure gold; 18-karat gold is alloyed with 75 percent of another metal, usually copper. *See* carat. *Abbr.* k *or* kt

Kel·vin *adj.* A temperature scale that uses Celsius units. The Kelvin scale is used when it is necessary to express very high or very low temperatures. On this scale, water freezes at 273.15 K. Named after the English physicist, Lord Kelvin. *Note* that the degree sign is not used. *Abbr.* K

kerf *n.* The thickness of the cut made by a saw blade.

ker·nel *n.* **1.** The core. **2.** An atom that has been stripped of its valence electrons.

ke·tone *n.* A class of organic chemical compounds, important as solvents and as intermediates in the synthesis of other organic compounds. An example is acetone (used in nail-polish remover).

key *n.* A unique identification for a record.

key·board *n.* A set of push-button switches used to input information.

kiln *n.* An oven used for the firing of ore or ceramics.

kilo– *prefix* Thousand. *Symbol* k

kil·o·cal·o·rie *n.* A heat unit equal to 1000 calories. *Abbr.* kcal

kil·o·cy·cle *n.* A unit of frequency equal to 1000 cycles. *Abbr.* kc

ki·lo·gram cal·o·rie *n.* The amount of heat required to raise the temperature of one kilogram of water one degree Celsius. *Abbr.* kg-cal *Also called* large calorie. *Abbr.* Cal

kilo·hertz *n.* A unit of frequency equal to 1000 hertz. *Abbr.* kHz

ki·lo·me·ter *n.* A measure of distance; 1000 meters. *Abbr.* km

kil·o·volt *n.* A unit of potential difference equal to 1000 volts. *Abbr.* kV

kil·o·watt–hour *n.* A unit of electric power equal to 1000 watts per hour. *Also* kilowatthour. *Abbr.* kWh or kW·h or kWhr

kin·e·mat·ics *n.* The study of moving objects.

kit *n.* A set of parts for a system.

klys·tron *n.* A microwave electron tube used as an amplifier or as an oscillator.

knurl *n.* Projections on a surface, designed to provide a firm grip.

K shell *See* electron configuration.

kVa *n.* Kilovolt-ampere.

kWh or kW·h *n.* Kilowatt-hour.

L

lab·y·rinth *n.* A complex structure with many paths. ***adj.*** lab·y·rin·thine.

lam·bert *n.* A unit of luminance. *Abbr.* L

la·mel·la *n.* A thin layer. ***pl.*** la·mel·lae. ***adj.*** la·mel·lar.

lam·i·nate *v.* To bond layers together.

lan·guage *n.* A set of rules used to convey information.

lan·guage lev·el *n.* The complexity of a computer language and its closeness to English. *See* high-level language; machine language.

lap *v.* To polish.

lap joint *n.* An overlapping joint.

large–scale in·te·gra·tion *n.* The method of producing multiple electronic components on a single chip of semiconductor material. *Abbr.* LSI

la·ser *n.* A light-emitting device that operates in the optical or infrared regions, with applications in many fields, from medicine to communications. Derived from Light Amplification by Stimulated Emission of Radiation.

la·ser di·ode *n.* A semiconductor component that emits laser light.

la·ser disk *n.* A rotating platter for the recording of data by means of a laser beam.

la·ser gy·ro *n.* A device to measure rotation with great accuracy.

la·tent *adj.* Inactive.

lat·tice *n.* A regular crystal structure.

Laue pic·ture *n.* An x-ray pattern describing a crystal lattice.

leach *v.* To dissolve one substance out of another by a percolating liquid.

lens *n.* A nearly spherical transparent medium, usually glass, used in light refraction.

lev·i·tate *v.* To cause to rise.

life·time *n.* The time a carrier of electricity exists.

lift *n.* The upward force on a flying body.

li·gand *n.* An atomic group bound to a central atom in an organic compund.

light–emit·ting di·ode *n.* A device that converts electric energy into luminescence. Used in indicator lights. *Abbr.* LED

light pen *n.* A device that permits direct input to a video display.

lim·it·er *n.* An electronic circuit that limits the shape of a waveform to a specified level.

line print·er *n.* A printer that can print a complete line of characters simultaneously and at high speed.

lin·ear equa·tion *n.* An equation whose slope is a straight line.

liq·uid *adj.* Pertaining to the state of matter intermediate between the solid and the gaseous state. A substance in the liquid state will flow, whereas most solids will not. A gas will expand to fill any container, whereas a liquid will lie on the bottom. Also a gas can be greatly compressed, whereas a liquid cannot. *Compare* fluid.

liq·ui·dus *See* solidus/liquidus.

li·ter *n.* A measure of volume; 1000 cubic centimeters. *Also* litre. *abbr.* *l* or *L*

li·thog·ra·phy *n.* A printing process used in the fabrication of semiconductor devices.

lit·mus *n.* An acid-base indicator, derived from powdered lichens. Turns red at pH 4.5 (acid) and blue at pH 8.3 (alkaline).

load *n.* The amount of electric power drawn from a power line or other source.

lo·cant *n.* A number or letter that designates the position of an atom or radical in an organic compound.

lock·ing *v.* Applying a signal of constant frequency to a circuit in order to control the frequency of an operator.

lo·cus *n.* A collection of points that satisfy an equation. *Pl.* lo·ci.

lode·stone *n.* A naturally magnetic mineral, magnetite.

log *See* logarithm.

log·a·rithm *n.* The exponent giving the power to which the base must be raised to render it equal to a given number. In the expression $a^x = b$, a is the base and x is the logarithm of b. *Abbr.* log

log·ic *n.* The flow of a computer program.

log·ic di·a·gram *n.* *See* flowchart.

lo·gis·tics *n.* The procurement and assignment of military materiel and personnel; hence detailed plans for the management of any operation.

loop *n.* **1.** A portion of an electric circuit. **2.** In computer programming, a portion of the program that repeats by means of an end statement referring back to an opening statement.

lo•ran *n.* A system of long-range electronic navigation; derived from *lon*g *ran*ge.

loss *n.* The energy consumed in a system.

L shell See electron configuration.

LSI *See* large-scale integration.

lug *n.* A structural projection or head on a nut, to which attachment may be made.

lu•men *n.* A unit of light intensity. *Abbr.* lm

lu•mi•nes•cence *n.* Light emission due to sources other than heat, such as radiation and electromagnetic fields.

lu•mi•nous *adj.* Emitting visible radiation.

M

ma•chine lan•guage *n.* A low-level computer language whose commands are written in binary code. This is the language that the computer hardware "understands" in its circuits. It specifies an operation code, the number of characters to be processed, and the location of the data.

ma•chine–read•a•ble *adj.* Able to be read, or understood, by a computer or other mechanical device.

Mach num•ber *n.* The ratio of the speed of a body to the speed of sound, in air. Mach 1 represents the speed of sound (1,086 feet per second); lower numbers represent subsonic speeds, higher numbers supersonic speeds.

macro– ***word root*** Large; long. (From Greek *makros*—long.)

mac•ro•im•age *n.* An image larger than life.

mac•ro•mol•e•cule *n.* A large molecule containing many simple structural units of several atoms each.

mac•ro•scop•ic *adj.* Able to be seen by the unaided eye.

mag•ma *n.* Molten rock, as it issues from volcanoes or volcanic fissures.

mag•ne•sia *n.* Magnesium oxide, MgO. Used in cement, medicine, and other substances.

mag•net *n.* A material that exerts magnetic force by creating a magnetic field in its environment.

mag•net•ic am•pli•fi•er *n.* A device that modulates the flow of ac electric power in response to a dc input signal.

mag·net·ic dec·li·na·tion *n.* The angle between true (geographic) north and magnetic north; measured by a declinometer. *Also known as* compass variation.

mag·net·ic do·main *n.* The smallest portion of a crystal that is magnetic.

mag·net·ic flux *n.* A distribution of magnetic induction in a magnetic field.

mag·net·ic tape *n.* A high-speed device for recording and storing information.

mag·ne·to *n.* An electric generator that uses permanent magnets to create its field. Used for ignition in engines.

mag·ne·to·hy·dro·dy·nam·ics *n.* The study of the motion of ionized gas interacting with a magnetic field. *Abbr.* MHD

mag·ne·tom·e·ter *n.* An instrument for measuring magnetic intensity.

mag·ne·to·mo·tive force *n.* **1.** The agent which produces magnetic lines of force. **2.** The strength of such an agent. *Abbr.* mmf

mag·ne·to·stric·tion *n.* A dimensional change due to magnetic-field influence.

mag·ne·tron *n.* A microwave electron tube. Used in radar and in microwave ovens.

main·frame *n.* **1.** A central processing unit for a large computer. **2.** Such a computer, with large memory.

makeup *n.* The way in which a system is put together.

mal·le·a·ble *adj.* Able to be distorted, especially by hammering. *n.* mal·le·a·bil·i·ty.

man·drel *n.* A form on which material may be wound or wrapped, for support during machining.

man·i·fest *adj.* Apparent. *v.* To show the visible signs of.

man·i·fes·ta·tion *n.* A demonstration or appearance.

man·i·fold *n.* A branching structure of pipes.

ma·nom·e·ter *n.* An instrument used to measure the difference in pressure between two fluids.

mar·gin *n.* A boundary or limit.

mar·gin·al *adj.* Extra.

mar·tem·per·ing *n.* Quenching austenitic steel to the martensite temperature range, then letting it cool to room temperature.

mar·tens·ite *n.* A crystal phase of carbon in steel.

ma·ser *n.* A device that emits a coherent beam of microwave energy. Derived from *m*icrowave *a*mplification by *s*timulated *e*mission of *r*adiation. *See also* laser.

mass *n.* The amount of matter in an object, as measured by its resistance to change in speed and direction. *Compare* weight. *Abbr. m*

mass spec·tro·graph *n.* An instrument for detecting and measuring on a photographic plate the distribution of atom species.

mask *n.* A perforated sheet used to produce patterns in semiconductor manufacture.

mass num·ber *n.* The sum of the numbers of neutrons and protons in an atom; different for every isotope of every element. *See* isotope.

mas·ter file *n.* A computer file containing authoritative data.

ma·trix *n.* **1.** An array of numbers. **2.** A fixed pattern. **3.** A parent structure.

matte *n.* **1.** A dull surface finish. *Also* mat or matt. **2.** A sulfide slag.

max·well *n.* A unit of magnetic flux. *Abbr.* Mx

Max·well e·qua·tions *n.* A set of equations describing electric and magnetic fields, the basis for the theory of electromagnetic waves.

maze *n.* A complex network.

mean *n.* The average of a set of values. *See* median.

mean free path *n.* The average path length a particle may travel without colliding with another particle. *Abbr.* mfp

mech·a·nism *n.* **1.** A device. **2.** The basic principle on which a process depends.

mech·a·nize *v.* To convert from hand to machine labor.

me·di·an *n.* The midpoint in a range of values, so that 50% of the items lie below the median and 50% above it. *See* mean.

me·di·um *n.* The substance through which particles or radiation may be transmitted.

mega– *prefix* Million. *Symbol* M

me·ga·hertz *n.* One million cycles per second. *Abbr.* MHz

me·ga·ton *n.* One million metric tons. Used in measuring the intensity of a nuclear explosion.

me·ga·volt *n.* One million volts. *Abbr.* MV

me·ga·watt *n.* One million watts. *Abbr.* MW

Meg·ger *n.* Brand name for an ohmmeter capable of measuring very high values of electric resistance.

meg·ohm *n.* One million ohms. *Abbr.* MΩ

mei·o·sis *n.* A type of cell division that reduces the number of chromosomes by one-half.

mel·a·mine *n.* A resin.

mem·ber *n.* A discrete part of a system.

mem•brane *n.* A thin flexible sheet.

mem•o•ry *n.* A subsystem in a computer in which information is stored and from which it can be called up.

me•nis•cus *n.* The curved surface of liquid contained in a vessel. The shape of the meniscus helps identify the liquid.

men•sur•a•ble *adj.* Measurable.

menu *n.* A list of files or stored programs in a computer.

mer•cap•tan *n.* A group of chemical compounds, usually with disagreeable odors, found in crude petroleum.

Mer•ca•tor *adj.* Relating to a projection system used in mapping. Useful for navigation, since it preserves the correct latitude-longitude relationship, but on a world map it causes serious vertical distortion at both poles.

me•rid•i•an *n.* A circle on the surface of a sphere that passes through its poles. On maps, a north-south reference line.

me•sa *n.* A truncated cone.

me•sa tran•sis•tor *n.* A transistor in the shape of a mesa.

mesh *n.* **1.** A perforated or woven screen. **2.** An electric network. *v.* To engage gears.

me•son *n.* A subatomic particle.

meta– *prefix* Later; more highly organized; transformed. ***Chemical prefix*** **1.** The lowest stage of hydration. **2.** Characterized by two positions on the benzene ring that are separated by one carbon atom.

me•tab•o•lism *n.* The chemical reactions in a living cell.

met•al•lo•graph *n.* An instrument for observing and photographing metal crystals under magnification.

met•al•loid *n.* A nonmetallic element such as carbon that combines with a metal to form an alloy.

met•a•mor•phic *adj.* Pertaining to rock that has been altered in the Earth's depths under enormous heat and pressure. Marble, for example, is metamorphized limestone.

met•a•sta•ble *adj.* Not stable.

me•ter *n.* **1.** A measuring device. **2.** A unit of length in the metric system. *Also* metre *Abbr.* m *v.* To measure.

meth•od•ol•o•gy *n.* A plan for conducting an investigation.

met•ric *adj.* Referring to one of several decimal systems of measurement.

met•ric sys•tem *n.* A decimal system of measurement. It includes the centimeter-gram-second (cgs) system, the meter-kilogram-second (mks) system, and the International System of Units (SI system).

MHD *See* Magnetohydrodynamics

mho *n.* The unit of conductance. *Also known as* sie•mens. *See* conductance; siemens (preferred). *Symbol* ℧

mi•ca *n.* A mineral that occurs in thin, easily split sheets. Used in electrical components.

mi•cro *n.* Colloquial for microcomputer.

micro– *prefix* One-millionth. *Symbol* μ *word root* Small. (From Greek *mikros*—small.)

mi•cro•com•pu•ter *n.* A small computer; typically a personal computer with 4–64K of memory.

mi•cro•e•lec•tron•ics *n.* The technique of constructing electric circuits in very small packages.

mi•cro•fiche *n.* A sheet of film, about 4 by 6 inches, used to store large quantities of data.

mi•cro•film *n.* A roll of film containing a photographic image, greatly reduced, of printed matter.

mi•cro•gram *n.* One-millionth of a gram. *Abbr.* μg

mi•crom•e•ter *n.* A device for measuring small distances.

mi•cro•me•ter *n.* A millionth of a meter. *Abbr.* μm

mi•cron *n.* A millionth of a meter. *Abbr.* μ For most uses, **micrometer** is preferred.

mi•cro•pho•to•graph *n.* A photograph made at high magnification.

mi•cro•proc•es•sor *n.* A chip that can perform basic logic, arithmetic, and storage.

mi•cro•pro•gram *n.* A sequence of pseudoinstructions to be translated by a logic subsystem in the central processing unit into machine subcommands.

mi•cro•ra•di•o•graph *n.* An x-ray photograph made at high magnification.

mi•cro•struc•ture *n.* Patterns observed through a microscope.

mi•cro•wave *n.* A high-frequency electromagnetic wave, at frequencies of 1–100 gigahertz. *adj.* Referring to such waves.

mi•cro•wave plumb•ing *n.* Rectangular hollow pipes used as conductors of microwave electric energy.

mil *n.* One-thousandth of an inch.

mi•lieu *n.* Environment.

mill *v.* To cut into a surface with a rotating tool.

Mil•ler in•di•ces *n.* A number system used to identify crystal faces.

milli– *prefix* One-thousandth. *Symbol* m

min•er•al *n.* An inorganic chemical compound which occurs naturally, usually crystalline.

mini ***n.*** Colloquial for minicomputer.

min•i•com•pu•tor ***n.*** A computer that stands between the microcomputer or personal computer, and the large, mainframe; typically with 64 kilobytes to 1 megabyte of memory.

mi•ni•fi•ca•tion ***n.*** The act of making an image smaller optically.

mis•ci•ble ***adj.*** Mixable.

mis•sile•ry ***n.*** Missile technology.

mi•ter ***n.*** A 90° corner formed by pieces beveled at a 45° angle. *Also* mi•tre.

mi•ti•gate ***v.*** To moderate.

mi•to•sis ***n.*** The chromosome duplication process that yields two daughter nuclei carrying an arrangement of chromosomes identical to the parent cell's.

mix•er ***n.*** An electronic circuit that combines separate audio and video signals to create an output signal.

mne•mon•ic ***n.*** **1.** An aid to memory. **2.** An abbreviated command to a computer, such as SA (save) or CO (copy).

mo•bil•i•ty ***n.*** The ease with which electrons or holes move in a crystal lattice.

mock–up ***n.*** A full-sized nonoperating model.

mo•dal *See* mode.

mode ***n.*** **1.** A method of operation of a system. Thus, a microcomputer may operate in a word-processing mode, a graphics mode, or other condition. **2.** In statistics, the value that appears most often in a series of observations. ***adj.*** mo•dal.

mod•el ***n.*** A representation, either physical or theoretical. ***v.*** To shape.

mo•dem ***n.*** A device that modulates and demodulates signals transmitted over telephone lines so that they may be understood by a sending or receiving computer.

mod•er•a•tor ***n.*** A material that slows a chemical or physical reaction.

mod•i•fi•er ***n.*** A quantity that alters the address of an operand in a computer. *Also known as* index word.

mod•u•late ***v.*** To impose a signal wave over a carrier wave.

mod•ule ***n.*** A subsection of a structure or a system. ***adj.*** mod•u•lar.

mod•u•lus ***n.*** A constant or coefficient.

mod•u•lus of elas•tic•i•ty ***n.*** The ratio of stress to strain.

mo•dus ope•ran•di ***n.*** The way something functions.

Mohs' scale ***n.*** A hardness scale for minerals. Diamond is 10 (hardest), talc is 1 (softest), with other minerals ranged between them.

mo•lal•i•ty ***n.*** The concentration of a solution in moles per 1000 grams of solvent. ***adj.*** mo•lal.

mo·lar·i·ty ***n.*** The concentration of a solution by weight of the compound dissolved in moles per liter. ***adj.*** mo·lar. *Abbr. M*

mold ***v.*** To cast. ***n.*** The form into which a casting material is poured.

mole ***n.*** The molecular weight of a substance in grams. *Abbr.* mol

mo·lec·u·lar weight ***n.*** The weight of an element or compound equal to the sum of all the atomic weights of all the atoms that constitute it.

mol·e·cule ***n.*** The smallest unit of matter which still retains all the qualities of a chemical compound. ***adj.*** mo·lec·u·lar.

mol·ten ***adj.*** Melted.

mo·ment of in·er·tia ***n.*** A measure of a body's ability to resist angular acceleration.

mon·a·to·mic ***adj.*** Composed of one atom.

mon·au·ral ***adj.*** Referring to a sound system with one sound source.

Mo·nel Me·tal ***n.*** A trade name for a copper-nickel alloy.

mon·i·tor ***v.*** To observe a process. ***n.*** An instrument or a person that monitors.

mon·o·chro·mat·ic ***adj.*** Of one color.

mon·o·clin·ic *See* crystal systems.

mon·o·lay·er ***n.*** A film one molecule thick.

mo·no·lith·ic ***adj.*** **1.** Undivided **2.** Referring to a single silicon substrate that contains an integrated circuit.

mon·o·mer ***n.*** A structural unit within a polymer.

mor·dant ***n.*** An agent that fixes a dye. Alum and aniline are examples.

morph– or –morph ***prefix or suffix*** Pertaining to form.

mor·phol·o·gy ***n.*** The structure and form of an organism.

mo·sa·ic ***n.*** In aerial photography, overlapping photographs which make a whole image.

MOS tran·sis·tor ***n.*** A *m*etal *o*xide *s*emiconductor device in which a metallic oxide acts as an insulating layer.

mount ***v.*** To set up, as of a job on a machine.

mouse ***n.*** A hand-held structure that acts as a cursor for a computer.

M shell *See* electron configuration.

mul·ti·far·i·ous ***adj.*** Put together from many parts.

mul·ti·form ***adj.*** Having more than a single shape.

mul·ti·lat·er·al ***adj.*** Having many sides.

mul·ti·plex ***v.*** To transmit many signals in a single channel. Used in communications networks.

mul·ti·pli·cand ***n.*** The number to be multiplied by another.

mul·ti·pli·er *n.* An electric circuit used to increase the voltage range.

mul·ti·vi·bra·tor *n.* An oscillator circuit.

mu·ri·a·tic acid *n.* Hydrochloric acid.

mu·ta·ble *adj.* Able to be changed.

mu·tant *n.* The product of genetic mutation.

mu·ta·tion *n.* A sudden change in genetic material. *v.* mu·tate.

mute *v.* To soften a sound.

N

n *n.* An unspecified sequence number in a series. Thus, for the counting numbers 0, 1, 2, 3, . . . , n, n represents a number at the far end of the sequence. *adj.* nth

na·bla *n.* A vector operation symbol, ∇.

NAND *n.* A computer logic operation, derived from "not-and."

nano– *prefix* One-billionth. *Symbol* n

na·no·sec·ond *n.* One billionth of a second. *Abbr.* nsec

nas·cent *adj.* Newly formed.

nas·cent ox·y·gen *n.* Oxygen released by a chemical reaction. It is more active (more likely to form compounds) than when in its ordinary state of O_2 molecules, because it exists as temporarily uncombined O atoms.

nat·u·ral base *n.* The base of the natural logarithms, approximately equal to 2.17828. *Symbol* e, an important constant in formulas that describe continuous growth.

nat·u·ral num·bers *n.* The range of positive whole numbers: 1, 2, 3, . . .

neo– *prefix* New.

neb·u·la *n.* An interstellar cloud of gases and dust, for example, the Horsehead Nebula in the constellation Orion.

ne·gate *v.* To render ineffective.

ne·o·prene *n.* A synthetic rubber.

net·work *n.* **1.** A collection of electronic components that interconnect to form an operating electrical system. **2.** An interconnection of computer systems.

neu·ron *n.* A nerve cell.

neu·tral *adj.* **1.** Neither acidic nor basic. **2.** Neither positive nor negative.

neu·tral·ize *v.* **1.** To render ineffective. **2.** To make a solution neutral, neither acidic nor basic.

neu•tri•no *n.* An uncharged subatomic particle.

neu•tron *n.* An uncharged particle in the nucleus of an atom.

new•ton *n.* A unit of force. *Abbr.* N

New•ton•ian *adj.* A modifier applied to many scientific laws and systems developed by Isaac Newton.

New•ton•ian me•chan•ics *n.* Classical mechanics, based on Newton's laws of motion. *See* relativistic mechanics.

Ni•col prism *n.* An optical system for producing polarized light.

nil *n.* Nothing.

ni•tride *v.* To diffuse nitrogen into steel in order to harden it.

no•ble me•tal *n.* A metal such as gold, platinum, or silver that is resistant to corrosion.

node *n.* A nonmoving location in a vibrating structure.

nod•u•lar *adj.* Having the form of small rounded lumps. *n.* no•dule.

noise *n.* Random sound, with no regular wave pattern.

no•men•cla•ture *n.* The method of naming the elements in any system.

nom•i•nal val•ue *n.* The expected, as opposed to the actual, value.

no•mo•gram *See* nomograph.

no•mo•graph *n.* A chart that represents an equation in three variables.

non•fer•rous *adj.* Not containing iron.

NOR *n.* A computer logic operation, derived from "not-or."

nor•mal dis•tri•bu•tion *See* gaussian distribution.

nor•mal•ize *v.* To bring into conformity with a norm, or standard, so that a comparison can be made.

nose cone *n.* The front or leading section of a spacecraft or missile, usually protected against the heat generated by reentry into Earth's atmosphere.

no•ta•tion *n.* A system of symbols used to describe a system.

notch sen•si•tiv•i•ty *n.* The tendency of a material to fracture under stress at a notched point.

nox•ious *adj.* Unpleasant, even poisonous, as of chemical fumes.

noz•zle *n.* A spout for emission of jets of gas or liquid.

npn *adj.* Pertaining to a transistor with an *n*-type emitter and collector and a *p*-type base.

N shell *See* electron configuration.

n–type *adj.* Referring to a semiconductor doped with an impurity of the donor type.

nu•cle•ar *adj.* **1.** Pertaining to the nucleus of an atom. **2.** Pertaining to nuclear energy.

nu•cle•ate *v.* To initiate crystal growth at a seed site in a solution.

nu•cle•on *n.* A proton or neutron; the main constituents of the atomic nucleus.

nu•cle•us *n.* The positively charged center of an atom.

nu•clide *n.* An atom specified numerically by its atomic number and mass number, for example, $^{6}_{3}Li$.

null *n.* A zero or negligible quantity

null set *n.* In mathematics, the empty set.

nu•mer•a•tor *n.* The expression above the line in a fraction.

nu•ta•tion *n.* An eccentric (off-center) rotation.

O

ob– *prefix* Against.

ob•jec•tive *n.* The lens farthest from the eyepiece in an optical system, and the one through which light first passes.

ob•ject pro•gram *n.* A program in machine language, prepared by the interpreter or the compiler from the high-level language. *See* high-level language; machine language.

ob•late spher•oid *adj.* Having the shape of a football.

ob•lique *adj.* At an angle.

ob•scure *adj.* Not well-defined; dark.

ob•ser•va•tion *n.* The record of a data point.

ob•tuse an•gle *n.* An angle between 90 and 180 degrees.

ob•vi•ate *v.* To render invalid.

oc•clude *v.* To adsorb a gas onto a surface.

oc•tal *adj.* **1.** Pertaining to a number in base 8. **2.** Pertaining to the base of a vacuum tube with eight pins.

oc•tane num•ber *n.* A rating of the antiknock quality of gasoline.

oc•tave *n.* The interval between any two frequencies having a ratio of 2:1.

oc•u•lar *n.* The eyepiece of microscope or a telescope.

OD *n.* Outside diameter.

–ode *suffix* A path as it pertains to an electronic element or component.

o•dom•e•ter *n.* A distance-measuring device.

o•dor•ant *n.* A substance added to an odorless gas to give it a distinctive odor.

oer•sted *n.* A unit of magnetic-field strength. *Abbr.* Oe

off–line *adj.* Referring to data or equipment not in direct communication with the central processing unit.

ohm *n.* A unit of electric resistance. *Symbol* Ω

ohm•ic *adj.* Resistive.

ohm•me•ter *n.* A meter for measuring resistance.

Ohm's law *n.* The mathematical relation between voltage and electric current across a given resistor.

–oic *suffix* Indicates a carboxyl group.

–oid *suffix* Similar to.

on–line *adj.* Pertaining to equipment directly connected to a computer.

o•pac•i•ty *n.* A measure of the portion of light transmitted through a partially transparent medium.

o•pal•es•cent *adj.* Iridescent, from the play of color within an opal.

op code *n.* Operation code.

op•er•and *n.* The data or equipment operated on.

op•er•a•tion *n.* The outcome of a single computer instruction.

op•er•a•tion code *n.* The instruction code specifying the operations to be performed by a computer.

op•er•a•tor *n.* That which designates an action to be performed.

op•ti•cal char•ac•ter rec•og•ni•tion *n.* A means of reading hard copy by a photosensitive device. *Abbr.* OCR

op•ti•cal read•er *n.* A device able to read hard copy by means of optical character recognition.

op•ti•cal scan•ner *n.* Optical reader.

op•ti•cal flat *n.* A disk of quartz glass ground to extreme flatness and a high polish.

OR *n.* A computer logic operation.

or•bit *n.* Any closed path, such as the orbit of the Earth around the Sun.

or•bit•al *n.* **1. Atomic:** The energy level (distance from the nucleus) at which an orbiting electron is to be found. For convenience's sake, the electron's orbital around the nucleus is pictured as shaped like a planet's orbit around the sun, but in fact an electron's motions are more complex than those of a planet and in some ways even resemble the behavior of waves. **2. Molecular:** The path of an electron around the nuclei of atoms bonded together as a molecule.

or•der *n.* **1.** A way of designating the level of a mathematical term according to its highest exponent. The order of a differential equation is the number of times a function has been differentiated to give a certain derivative. The order of a curve is the number of points at which the curve is intersected by a straight line. **2.** A classification of a chemical reaction in terms of the number of molecules that enter into the reaction. **3.** The order of a crystal is the regular and repeated arrangement of the atoms in a configuration that identifies a particular mineral. **4.** Any ranking.

or·dered *adj.* Put into an order, by quantity or other relationship.

or·di·nal num·bers *n.* The counting numbers, . . . , −3, −2, −1, 0, 1, 2, 3. . . .

or·di·nate *n.* The vertical coordinate of a point on a graph.

ord·nance *n.* Military equipment.

or·gan·ic *adj.* Pertaining to the chemistry of carbon compounds that also contain hydrogen, and possibly oxygen, nitrogen, and other elements.

or·thi·con *n.* A tube used in a television camera.

orth-, ortho- ***word root*** Straight, upright. (From Greek *orthos*—straight.) ***chemical prefix*** **1.** In the fullest possible state of hydration. **2.** Characterized by two neighboring positions on the benzene ring.

or·thog·o·nal *adj.* Perpendicular.

ortho·rhom·bic *See* crystal systems.

os·cil·la·tor *n.* A device or circuit for converting direct current to alternating current.

os·cil·lo·scope *n.* A cathode-ray tube which displays electrical values and waveforms. Used as a test instrument.

O shell *See* electron configuration.

os·mo·sis *n.* The movement of a solute through a semipermeable membrane that separates two solutions of different concentrations.

-otic ***suffix*** Affected by.

-ous ***suffix*** Chemical suffix. The second-highest oxidation state of an acid.

out·crop *n.* A rock formation that is exposed at ground level.

out·gas *v.* To remove gases from materials or a vacuum system by heat and evacuation.

out-of-phase *adj.* Referring to two waves of the same frequency, of which one lags behind the other in time.

out·put *n.* The information obtained from processing by the central processing unit.

over·lay *n.* A layer of one substance over another.

over·tone *n.* A frequency that is a whole-number multiple of a base frequency.

ovoid *adj.* Egg-shaped.

ozone *n.* A trivalent oxygen molecule O_3. Used as a powerful industrial bleach. Ozone molecules are created in the air when lightning strikes.

ozone lay•er *n.* The layer of the upper atmosphere that maintains a balance between heat radiated from Earth and ultraviolet radiation received from space.

P

pack•age *v.* To encapsulate.

pad *n.* A region in a thick-film or thin-film circuit to which a connection is made.

pal•pa•ble *adj.* Able to be felt by touch.

pan•to•graph *n.* A tool for duplicating a drawing.

para– *prefix* Beside; closely related to. ***chemical prefix*** **1.** Closely related to. **2.** Characterized by two opposite positions on the benzene ring.

pa•rab•o•la *n.* A curve. *See* conic section.

par•al•lax *n.* The difference between the observed position of an object (such as the pointer on an instrument dial) and its true position; occurs when the observer's eye is not directly perpendicular to the plane of the object.

par•al•lel•e•pi•ped *n* A solid, all of whose faces are parallelograms.

par•al•lel•o•gram *n.* A four-sided figure whose opposite sides are parallel.

par•a•mag•net•ic *adj.* Referring to a material (such as aluminum or platinum) that can be partially magnetized and whose ability to be magnetized is independent of the strength of the magnetic field acting on it.

pa•ram•e•ter *n.* A quantity that is constant under one set of conditions and whose change under another set of conditions can be measured.

par•a•met•ric amp•li•fi•er *n.* **1.** An ultra-high frequency microwave amplifier. **2.** In laser technology, a crystal that amplifies an optical or an infrared beam. *Abbr.* paramp

par•a•sit•ic cur•rent *n.* In machinery, an eddy current that causes power loss.

par•a•site *n.* an organism that lives on or in another organism, of a different species.

par•i•ty *n.* A self-checking code in a computer.

par•tial de•riv•a•tive *n.* Calculus equations that solve problems dealing with two or more dependent variables. Examples are the calculation of a profit (which depends on sales, overhead, cost of materials, and other factors) or the efficiency of a solar collector (which depends, among other things, on the size of the panels, the number of sunny days, and the angle at which the sun's rays strike it). *Symbol* ∂.

par•tial dif•fer•en•tial *n.* The same as partial derivative.

par•ti•cle *n.* An object smaller than an atom.

par•tic•u•lates *n.* Fine solid particles.

Pas•cal *n.* A high-level computer programming language. Named for the French mathematician Blaise Pascal (1623–1662), inventor of the first calculating machine and the originator of probability theory. Since Pascal is a proper name and not an acronym, the language named after him is not commonly written in all-capital letters.

pas•siv•ate *v.* To render impervious to chemical attack.

patch *n.* An expedient computer program modification. *v.* To plug a device into a circuit, for example, a portable telephone.

pat•i•na *n.* A surface film on metal which is the result of slow oxidation.

pat•tern rec•og•ni•tion *n.* The ability of a computer to recognize shapes and symbols.

peak–to–peak *adj.* Referring to the measurement of a wave from positive peak to negative peak.

pearl•ite *n.* Ferrite and cementite layered together in high-carbon steel and cast iron.

peen *v.* To hammer a surface in order to harden it.

pen•tode *n.* A five-electrode vacuum tube.

pe•num•bra *n.* A partially illuminated shadow at the edge of a deep shadow.

pep•tide bond *n.* A bond between protein molecules.

per– *prefix* Thoroughly. ***chemical prefix*** Containing an element in its highest oxidation state.

per•cen•tile *n.* A value in a percentage distribution of data. It separates the data range into two groups, with a given percentage lying below this value. Thus, a test score higher than 90 percent of all test scores in the sample is said to be in the ninetieth percentile.

per•co•late *v.* To seep through a permeable substance. An example is the percolation of rainwater down through soil and porous rock until it reaches the water table.

per•i•he•li•on *n.* The closest approach of an orbiting object.

pe•ri•od *n.* The time of one cycle of a wave.

pe•ri•od•ic *adj.* Repeating regularly. *n.* pe•ri•o•dic•i•ty

pe•riph•er•al de•vice *n.* Equipment that works with a computer but is not part of it. Examples are a cathode-ray terminal, a printer, and a modem.

pe•riph•ery *n.* The boundary of a figure, object, or system.

per•i•tec•tic *adj.* Referring to a reaction in which a liquid phase reacts with a solid phase during cooling to produce another solid phase.

per•lite *n.* A glassy mineral that breaks into small, porous spheres when heated. Used to lighten potting soil.

perm *n.* A unit that measures the permeability of a porous material to the passage of fluids.

Perm•al•loy *n.* Trade name for magnetically permeable alloy of iron and nickel.

per•me•a•bil•i•ty *n.* **1.** The ratio of magnetic induction to the magnetizing force that produces it. Describes how strongly or how weakly a material can be magnetized. **2.** The quality of being permeable.

per•me•ance *n.* The conductivity of a magnetic circuit. *Compare* reluctance.

Per•men•dur *n.* Trade name for an iron-cobalt magnetic alloy, of high permeance.

per•mit•tiv•i•ty *n.* The ratio of electric flux displaced across a given area to the electric force at the same point. *Symbol* ϵ

per•mu•ta•tion *n.* A transformation.

PERT *n.* Program Evaluation Review Technique; a management-control system used for monitoring goal-directed programs.

pf *See* Power factor.

pH *n.* A logarithmic measure of acidity; 1 is the most acidic, 14 is the least acidic (most basic), and 7 is neutral.

phase *n.* **1.** A crystal structure. **2.** A state of matter (gas, liquid, or solid). **3.** A condition of a wave with respect to another with regard to whether they are coincident (in-phase) or not (out-of-phase). ***v.*** To merge.

phase di•a•gram *n.* A graph of the equilibrium relationships between phases, such as vapor-liquid or liquid-solid.

phase in•vert•er *n.* A circuit that changes the phase of a signal by 180 degrees.

phase shift•er *n.* An electronic circuit that changes the phase relation between two alternating-current values.

phe•no•lic plas•tic *n.* A thermosetting plastic. Used in adhesives, for molded parts, and in home water softeners.

phe•nol•phthal•ein *n.* A chemical indicator that shows the degree of acidity by changing color—bright red to bases, colorless to acids.

phil–, philo– ***word root*** Having an affinity for. (From Greek *philos*—dear, friendly.)

-phobe ***word root*** Having a fear of or an aversion for. (From Greek *phobos*—fearing.)

phon- ***word root*** Referring to sound, speech.

pho·non ***n.*** A quantum of heat energy vibrating in a crystal lattice.

phos·phor ***n.*** A material that gives off light when impacted with particles or waves. Used in television tubes and phosphorescent paint.

phot ***n.*** A unit of illumination.

phot-, photo- ***word root*** Relating to light. (From Greek *phos*—light.)

pho·to·cell ***n.*** A solid-state device that provides an electric signal when subjected to light. *Also known as* an electric eye.

pho·to·chem·is·try ***n.*** The study of the chemical reactions activated by light.

pho·to·di·ode ***n.*** *See* photocell.

pho·to·e·lec·tric ***adj.*** Pertaining to electric responses to light.

pho·to·en·grave ***v.*** To etch a pattern on masked photosensitive material.

pho·tom·e·ter ***n.*** An instrument used for measuring light.

pho·to·mic·ro·graph ***n.*** A photograph of a microscope image.

pho·ton ***n.*** A quantum of light.

pho·to·re·sist ***n.*** A light-sensitive material used in the manufacture of semiconductor devices.

pho·to·syn·the·sis ***n.*** Chemical reactions in living matter, particularly in plants, as a result of light activation.

pho·to·vol·ta·ic ***adj.*** Referring to the electromotive force generated by the exposure of the junction point of a metal and a semiconductor to radiation.

pico- ***prefix*** One-trillionth. *Symbol* p

pic·o·sec·ond ***n.*** One trillionth of a second. *Abbr.* psec or ps (preferred)

pier ***n.*** A vertical support for a concentrated load from an arch or bridge.

pi·e·zo·e·lec·tric·i·ty ***n.*** Electric signals generated by mechanical pressure, and vice versa.

pi·las·ter ***n.*** A vertical member that is structurally a pier and architecturally a column.

pile (atomic) ***n.*** A nuclear reactor.

PIN di·ode ***n.*** A semiconductor device with *p*-type, intrinsic, and *n*-type regions.

pip ***n.*** A sharp, bright pulse visible on a video display.

pi•pette *n.* A calibrated glass tube used in chemical analysis. *Also* pipet.

pitch *n.* **1.** The frequency, high or low, of a sound wave. **2.** The inclination of a slope.

Pi•tot tube *n.* An instrument that measures the pressure of a flowing fluid.

PL/1 *n.* Programming Language number 1, a high-level business-oriented language.

pla•nar *adj.* Lying in one plane.

pla•nar tran•sis•tor *n.* A transistor whose electrodes lie in parallel planes.

plane po•lar•i•za•tion *n.* The quality of light rays that have passed through a polarizing medium and emerged with their vibrations all moving in one direction. *See* polarize, 2.

pla•nim•e•ter *n.* A device for measuring the area of a flat surface by tracing its boundary.

plas•ma *n.* An ionized flowing gas at a very high temperature.

plas•tic *adj.* Able to be permanently deformed by an applied force. *n.* A material which exhibits plastic properties.

plas•tic•i•ty *n.* Capacity for being molded or altered in shape.

plate *n.* **1.** The anode of an electron tube. **2.** A flat layer. *v.* To deposit metal on a cathode (the object to be plated) in an electrolyte solution.

plat•en *n.* A flat metal die against which materials are pressed, as when molding plastic parts.

ple•num *n.* A chamber capable of holding higher than atmospheric pressure.

plot *v.* To draw points on a graph; to locate a curve on a graph using plotted points.

plot•ter *n.* In computer graphics, the device that transfers the operator's instructions onto paper.

plumb *adj.* Exactly vertical. *v.* To explore in depth.

ply *n.* One of many layers, as of plywood.

pnp *adj.* Referring to a type of transistor with emitter and collector and N-type base.

Pois•son's ra•tio *n.* The ratio of the transverse contracting strain to the longitudinal elongating strain.

po•lar *adj.* Pertaining to points that display opposite qualities, such as the north and south poles of a magnet. *n.* po•lar•i•ty.

po•lar•ize *v.* **1.** To create poles of opposite (positive-negative) charge, as in an electromagnet. **2.** In optics, to pass light through a plane-polarizing filter, such as a Nicol prism. ***n.*** po•lar•i•za•tion. *See* plane polarization.

poles ***n.*** The positively or negatively charged ends of a polarized object.

poly– ***prefix*** Many.

pol•y•es•ter ***n.*** A type of organic polymer, capable of being drawn into fibers or molded.

pol•y•mer ***n.*** A large compound made up of chains of smaller molecules (monomers). ***adj.*** pol•y•mer•ic.

po•lym•er•i•za•tion ***n.*** The process of bonding monomers to make up a much larger (giant) molecule.

po•ly•mor•phism ***n.*** The condition of occurring in many different shapes.

pol•y•no•mi•al ***n.*** A mathematical expression consisting of two or more terms.

pol•y•phase ***adj.*** Referring to a system of two or more power-line voltages displaced in phase relative to each other.

pop•u•la•tion ***n.*** In statistics, a defined universe which is examined for distribution of data. *Symbol N*

po•ros•i•ty ***n.*** The ability of a structure such as a filter to absorb or let pass fluid through extremely small holes.

port ***n.*** An entrance or exit in a system.

pos•i•tron ***n.*** A positively charged subatomic particle.

post ***v.*** To list data.

pot ***v.*** **1.** To encapsulate in a solid medium. **2.** To embed (as electronic components) in an insulating or protective container. ***n.*** Colloquial for potentiometer.

po•ten•tial ***n.*** A quality of electric circuits, ionization, and radiation: the energy required to bring a unit positive charge to a particular point within the system from some reference point outside the system (such as the surface of the Earth or some other large conductor). Measured in volts. ***adj.*** Latent; ready to occur.

po•ten•tial dif•fer•ence ***n.*** A voltage difference between two locations; the electromotive force required to move the unit charge from one location to the other.

po•ten•ti•om•e•ter ***n.*** A variable resistor, used to measure potential difference.

pow•der met•al•lur•gy ***n.*** The technology of pressing metal powders into required shapes, then firing them to create a rigid body.

pow•er ***n.*** **1.** The energy expended in a unit time. **2.** The exponent to which a base is raised; 2^4 equals two to the fourth power, or 16.

pow•er fac•tor *n.* A measure of the efficiency of an electrical system using alternating current power. *Abbr.* pf

prac•tice *n.* An accepted procedure.

pre•cess *v. See* precession.

pre•ces•sion *n.* The deviation from a given axis of a rotating body.

pre•cip•i•tate *n.* The solid substance that appears in a solution when its solubility limit is exceeded. *v.* To separate, or cause to separate, a solid substance from solution.

pre•cip•i•ta•tion hard•en•ing *n.* The process of hardening an alloy by a low-temperature heat treatment that causes certain components of the alloy to precipitate from solution.

pre•ferred ori•en•ta•tion *n.* A nonrandom arrangement of crystals or fibers.

pre•form *n.* A small punched-out piece of foil of a shape to match that of parts being assembled.

pre•mix *n.* A mixture of resin, extenders, and other ingredients for use in a plastics molding process.

pre•preg *n.* A woven structure of fibers used in the construction of reinforced plastic composites; combined with the resin before molding takes place.

pres•sure *n.* A measure of force exerted per unit area.

pri•ma fa•cie *adj.* On the basis of initial observation.

pri•ma•ry stor•age *n.* The main computer memory.

prin•ci•pal *adj.* Most important. *See* principle.

prin•ci•ple *n.* A rule or law. *See* principal.

print•out *n.* The output of a computer's printer on a sheet of paper. *Also known as* hard copy.

pri•or art *n.* That which has already been learned and is known.

prism *n.* An optical system used for the refraction of light into a spectrum. *adj.* pris•ma•tic.

pro•cess stream *n.* A system of continuous actions, as in an oil-refining process.

pro•file *n.* An outline. *v.* To describe.

pro•gram *n.* A set of instructions directing a computer to solve a problem. *v.* To write a computer program.

pro•gram•ma•ble read–on•ly mem•o•ry *n.* A read-only chip that can be programmed after manufacture by the user. *Abbr.* PROM

pro•ject *n.* A formal activity with defined goals. *v.* To perform a prediction.

pro·jec·tion *n.* **1.** A representation of a three-dimensional object on a flat plane. **2.** A system for representing the sphere of the Earth on a flat map. *See* Mercator.

pro·lif·er·ate *v.* To expand.

PROM *n. See* programmable read-only memory.

prop·a·ga·tion *n.* Of a wave, to move through a medium.

pro·pel·lant *n.* A fuel that supplies ejection thrust to a projectile.

pro·por·tion·al *adj.* Having a constant ratio of values.

pro·pri·e·tar·y *adj.* Referring to information owned by an organization or an individual under a patent or a trademark.

pro·pul·sion *n.* The force that moves a body.

pro·spec·tus *n.* A detailed plan for a new project, especially a commercial venture.

pro·tein *n.* A class of large organic molecules in living matter composed of amino acids linked by peptide bonds. There are up to 100,000 different proteins in the human body.

pro·ti·um *See* hydrogen, isotopes of.

proto– *prefix* First.

pro·ton *n.* An elementary particle, positively charged.

pro·to·type *n.* A model on which subsequent production of a device is to be based.

pro·vi·so *n.* A stipulated precondition.

pseu·do·code *n.* A programming code used by the programmer for analysis but not machine-readable.

P shell *See* electron configuration.

p–type *adj.* Pertaining to the acceptor-type dopant used in a semiconductor, rendering the material hole-conductive.

pulse *n.* A short, sudden deviation from a regular pattern.

pulse width *n.* The duration of a pulse.

punched tape sys·tem *n.* A computer configuration whereby paper tape with punched holes is the means of recording and storing data.

purge *v.* **1.** To clean forcefully and thoroughly. **2.** To replace the atmosphere in a vessel with an inert gas to prevent the creation of an explosive mixture.

push–pull *adj.* Pertaining to a circuit with two elements operating in phase to reinforce the desired wave and cancel undesirable effects.

pyro– ***root word*** Relating to fire or heat. ***chemical prefix*** Made from ortho compounds by means of heat. *See* ortho-.

py•ro•ce•ram *n.* A ceramic of high strength and resistance to thermal shock, formed from glass by recrystallization. Used for electrical parts and coatings.

py•rol•y•sis *n.* The breaking down of a complex chemical compound by heat action. ***adj.*** py•ro•lyt•ic.

py•rom•e•ter *n.* Any of several instruments used for measuring temperature.

Q

Q *n.* **1.** A symbol of the "quality" of an electric circuit, namely its ability to store energy. **2.** The electronic symbol for transistor.

Q shell *See* electron configuration.

quad *n.* **1.** A combination of four transistors. **2.** Any linking of four components.

quad•rant *n.* One-fourth of a circle.

quad•rat•ic equa•tion *n.* An equation containing a second-degree polynomial; that is, x^2 but no higher power of x occurs.

quad•ri•lat•er•al *adj.* Having four sides.

quad•ri•va•lent *adj.* Having a valence of 4.

quad•ru•pole *adj.* Having two electric or magnetic dipoles.

qual•i•ta•tive anal•y•sis *n.* Chemical analysis of a gas, liquid, or solid sample to determine its components. *Compare* quantitative analysis.

qual•i•ty as•sur•ance *n.* An analysis that ensures that the product is defect-free.

qual•i•ty con•trol *n.* Inspection during production to ensure that each processing step is being performed according to specification.

quan•ti•fy *v.* To describe numerically.

quan•ti•ta•tive *adj.* Referring to continuously varying factors.

quan•ti•ta•tive anal•y•sis *n.* Chemical analysis to determine the amounts or proportions of the components in a sample. *Compare* qualitative analysis.

quan•tum *n.* One of several possible values ascribed to a wave; an energy packet. *pl.* quan•ta.

quan•tum me•chan•ics *n.* The postnewtonian theory of matter and of the interaction of matter with energy; studies the structure and activity of atoms.

quarks *n.* Elementary particles, theorized as the building blocks from which subatomic particles are made.

quar•ter wave *adj.* Having an electrical length equal to a quarter of a wavelength.

quartz–crys•tal os•cil•la•tor *n.* An electronic device that resonates at a single fixed frequency, the natural frequency of the quartz crystal.

quasi– *prefix* Similar to; almost.

qua•ter•nary *adj.* Made of four parts.

quench *v.* **1.** To cool rapidly by plunging into a liquid. **2.** To extinguish a flame or spark.

quick•sil•ver *n.* Mercury.

qui•es•cent *adj.* Dormant.

quo•tient *n.* The result of dividing one number by another.

R

rab•bet *n.* In engineering, a groove cut into a part for the purpose of joining it to another part. *v.* To cut a groove.

race *n.* A channel to hold ball bearings.

rack *n.* A bar having gear teeth which will engage a rotating gear for lateral movement.

ra•dar *n.* An electronic system used to detect objects at a distance by means of a reflected beam of electromagnetic energy.

ra•di•al *adj.* Referring to straight-line paths starting from a single point and pointing in any direction.

ra•di•an *n.* A unit of angle measurement. The radian system of measurement is based on the fact that any angle can be considered to be the central angle in a real or imaginary circle. In the radian system, a 180-degree angle equals pi, a 360-degree angle equals 2 times pi, and a 90-degree angle equals ½ pi. This system is widely used in scientific and technical work because of its convenience in performing calculations. *Abbr.* rad

ra•di•ate *v.* To emit energy in the form of waves or particles. ***adj.*** ra•di•ant.

ra•di•a•tion *n.* Energy which is radiated. Heat, light, and x-rays are examples of radiant energy.

ra•di•a•tor *n.* A body that emits radiant energy.

rad•i•cal *n.* **1.** A group of atoms that behave like a single ion, for example, the hydroxyl radical OH^-. **2.** In mathematics, the root of a quantity; thus the square root of 4 is 2, and the fourth root of 27 is 3. *Symbol* $\sqrt{}$

radio– ***prefix*** Referring to radiation.

ra•di•o•ac•tiv•i•ty ***n.*** Radiation emitted by a radioactive substance, such as radium or uranium.

ra•di•o•chem•is•try ***n.*** The branch of chemistry that deals with radioactive phenomena and reactions.

ra•di•o fre•quen•cy ***n.*** The frequency range from about 10 kilohertz to about 100 gigahertz, used for communication. *Abbr.* rf

ra•di•o•gen•ic ***adj.*** Referring to material produced by radioactive decay, for example, lead produced by the decay of uranium.

ra•di•o•graph ***n.*** A photograph taken with x-ray radiation.

ra•di•og•ra•phy ***n.*** The procedure of taking x-ray pictures.

ra•di•om•e•ter ***n.*** Any one of several instruments for measuring radiant energy.

ra•di•o•nu•clide ***n.*** A radioactive nuclide.

ra•di•o•sonde. ***n.*** A weather balloon.

ra•di•us ***n.*** The line connecting the center of a circle to its circumference, also the length of such a line.

ra•di•us of cur•va•ture ***n.*** A geometric means of measuring approximate degree of curvature, for example, in the curve of a railroad track between two points.

ra•dome ***n.*** **1.** The nose cone of a missile. **2.** The cover of a radar system or of another emitter of electromagnetic energy which is transparent to that energy.

RAM *See* random-access memory.

Ra•man ef•fect ***n.*** The scattering of light by a medium through which it is being transmitted.

ramjet ***n.*** A type of jet engine in which the air required for combustion is compressed by the forward motion of the aircraft.

ran•dom–ac•cess mem•o•ry ***n.*** A computer memory whose files may be accessed in any desired order. *Abbr.* RAM.

ran•dom•ize ***v.*** To distribute the members of a sample so that they present no order or pattern.

Ran•kine ***adj.*** Referring to an absolute temperature scale that uses Fahrenheit degrees. *Abbr.* R

Ran•kine cy•cle ***n.*** An ideal thermodynamic cycle, the performance standard for heat engines.

rare-earth elements ***n.*** A group of metals, atomic numbers 58 to 71.

rar•e•fac•tion ***n.*** The reduction of the density of a gas.

rasp ***n.*** A woodworking tool with a rough surface of small points.

ras•ter *n.* A closely set pattern of scanning lines on a video screen.

ratch•et *n.* A toothed wheel, prevented from slipping back by a pawl, or catch.

ra•tio *n.* The proportion or quotient of two quantities.

ra•tion•al *adj.* Referring to the set of all numbers that are the quotient of two integral numbers, such as ½, ¾; the fractions.

Ray•leigh scat•ter•ing *n.* The scattering of light waves by small particles; the cause of colors in the sky.

RC *adj.* Referring to a resistor-capacitor.

re•act *v.* Of substances, to interact chemically, forming new compounds.

re•ac•tance *n.* That part of the impedance which does not convert electric power into heat. Expressed in ohms. *Symbol X*

re•ac•tion *n.* **1.** A chemical interaction. **2.** The consequence of an action.

re•ac•tor *n.* A vessel or system in which reactions may take place.

read *v.* To acquire information, especially from a computer.

read•er *n.* A device that converts information from one storage medium to another.

read–on•ly mem•o•ry *n.* Memory programmed by the manufacturer and which cannot be written to by the user. *Abbr.* ROM

read•out *n.* The presentation of information from a computer in a variety of human-readable forms.

read/write head *n.* A magnetic head that senses and records data.

re•a•gent *n.* A chemical that tests for the presence of other chemicals.

real num•ber *n.* The largest class of numbers, taking in the irrationals, rationals, integers, and natural numbers; a nonterminating decimal.

real time *n. See* real-time, adj.

real–time *adj.* Referring to information processing which is fast enough so that the computer response appears to be instantaneous, even though the computer may be managing several other processes at the same time.

ream *v.* To enlarge a hole.

Re•au•mur *adj.* Referring to a temperature scale that sets the freezing point of water at zero degrees and its boiling point at 80 degrees. *Abbr.* R

re•cip•ro•cal *n.* The quotient of 1 divided by a number. Thus, the reciprocal of x is $1/x$.

re•cip•ro•ca•ting en•gine *n.* A piston engine.

rec•i•proc•i•ty *n.* The quality of a system in which input and output can be switched without changing its response to a stimulus.

re•coil *n.* A reaction to the application of a force, as of a gun that is fired.

rec•on•cile *v.* To justify, to bring into balance.

re•cov•er•y *n.* The percentage yield of a desired material in a process.

re•crys•tal•li•za•tion *n.* The seeding and repeated growth of new crystals to replace a previous crystal array.

Rec•ten•na *n.* Trademark; an antenna which converts incident microwave radiation to direct current power.

rec•ti•fi•er *n.* An electric component or circuit that allows more current to flow in one direction than in another, thus effectively converting it to direct current.

rec•ti•lin•e•ar *adj.* Bounded by straight lines.

re•cu•per•a•tor *n.* A device which extracts heat from a flowing fluid.

re•cy•cle *v.* To reprocess for further use, for example, soft-drink cans for their aluminum content.

re•duc•tion *n.* A chemical reaction in which oxygen is removed from a compound or electrons are added.

re•dun•dan•cy *n.* A deliberate duplication of circuitry or information to prevent loss or breakdown in case of emergency.

reel *n.* A rotating structure for the support of spools of magnetic tape or of photographic film.

re•en•trant *adj.* Pertaining to shapes projecting inward.

re•en•try *n.* The return of a space vehicle to Earth.

re•fine *v.* To free from impurities.

re•flec•tance *n.* A measure of the reflected brightness of light.

re•flex *adj.* Reflected. For example, a reflex camera contains a mirror that reflects the image onto a ground glass.

re•flux *n.* The flowing back of a chemical to a chemical reaction so that reaction can continue.

re•frac•tion *n.* The bending of light rays as they pass from one medium to another.

re•frac•to•ry *n.* A material with a high melting point. *adj.* Referring to the ability of a material to resist heat.

re•frig•er•ant *n.* A fluid that in going through changes of state causes the lowering of temperature in its environment.

re•gen•er•ate *v.* To restore by cleansing from impurities.

re•gen•er•a•tive am•pli•fi•er *n.* An amplifier circuit that returns part of

the power to the input, as positive feedback, to reinforce the strength of the signal.

re•gime *n.* A pattern through which a process may go. *Also known as* reg•i•men.

reg•is•ter *n.* An area of computer storage.

re•gres•sion *n.* **1.** The return to an earlier state of development. **2.** In statistics, the tendency to move from an extreme (high or low) to an average value. Thus, when very tall and very short plants are cross-bred, the majority offsprings' heights tend to be statistically somewhere in between the heights of the parents, neither very tall nor very short.

reg•u•la•tion *n.* The automatic compensation for variation in input conditions.

re•it•er•ate *v.* To repeat.

re•la•tion *n.* A correspondence or a connection; for example, the direct relation that exists between temperature and expansion or contraction of concrete pavement; as temperature rises, the pavement expands. A rubbery substance that forms a joint between two paving blocks absorbs these movements and prevents the blocks from cracking.

rel•a•ti•vis•tic me•chan•ics *n.* The system of mechanics that was developed from Albert Einstein's special and general theories of relativity. *See* newtonian mechanics.

rel•a•tiv•i•ty *n.* The theory that recognizes the universal interdependence of the speed of light, space, time, and the position of the observer. ***adj.*** rel•a•ti•vis•tic.

re•lax•a•tion *n.* The relief of stress in a material.

re•lax•a•tion os•cil•la•tor *n.* An oscillation circuit that produces saw-tooth or rectangular wave forms.

re•lay *n.* An electrical switch. ***v.*** To pass on an electric signal after it has been amplified, especially in a communications network.

re•lease a•gent *n.* A lubricant that keeps surfaces from adhering to each other.

re•lief *n.* **1.** Clearance provided around the cutting edge of a tool by removal of tool material. **2.** The topography of a surface.

re•luc•tance *n.* Resistance to a magnetic force. *See* permeance.

rep•li•ca•tion *n.* **1.** The duplicating of a structure. **2.** The duplication of nucleic acid in living organisms by copying a molecular template. *See* RNA.

re•pul•sion *n.* A force which acts to increase the distance between objects.

rep rate *n.* Colloquial for repetition rate.

res•er•voir *n.* **1.** A stored fluid. **2.** A place of storage.

re•set *n.* To clear an instrument and return it to a previous condition.

re•sil•ience *n.* The ability of a deformed body to return to its original shape with the removal of stress.

res•in *n.* **1.** Natural substances like rosin and shellac, secreted by some plants and insects. **2.** The synthetic resins: any plastic produced by polymerization, with properties resembling a natural resin. They are based on formaldehyde, casein, urea, vinyl derivatives, and other organic compounds.

re•sis•tance *n.* The opposition to the flow of electric current offered by a circuit component known as a resistor. *Unit* ohm. *Symbol* Ω

re•sis•tance weld *n.* A weld made by the heat generated by a pulse of electric current.

re•sis•tiv•i•ty *n.* The resistance in ohms of a unit volume of a material. *Symbol* ρ

re•sis•tor *n.* A circuit component that provides a given amount of electric resistance to limit current flow.

res•o•lu•tion *n.* In television, a measure of the distinctness of an optical image, expressed as the number of lines per inch that can be distinguished from a distance equal to the height of the screen; usually 350–400. *Also known as* re•solv•ing pow•er.

res•o•nance *n.* An amplitude peak based on a combination of physical conditions and the radiation frequency.

re•sul•tant *n.* The average of two or more vectors.

re•tard•ing field os•cil•la•tor *n.* An electric device that maintains a positive charge on the grid of the triode. The field around the grid retards the passage of electrons in either direction.

re•ten•tiv•i•ty *n.* The residual magnetic flux in a material after removal of the magnetizing field.

ret•i•cle *n.* Fine lines engraved on a lens.

re•tort *n.* A vessel for high-temperature reactions.

re•triev•al *n.* The obtaining of desired information from a computer memory.

retro– *prefix* Back or backward.

ret•ro•fit *v.* To modify equipment made earlier to make it conform with later models.

ret•ro•grade *adj.* Pertaining to moving backward. *n.* ret•ro•gres•sion.

rev•o•lu•tions per min•ute *n.* The number of times in a minute a body rotates a full 360 degrees and returns to its starting position. *Abbr.* r/min (preferred) *or* rpm

rev·o·lu·tions per sec·ond *n.* The number of times in a second a body rotates a full 360 degrees and returns to its starting position. *Abbr.* r/s (preferred) *or* rps

re·write *v.* To restore a memory device to its state prior to reading.

Reyn·olds num·ber *n.* A number used to calculate the viscosity of a fluid. *Symbol* N_{Re}

rf *See* Radio frequency.

rhe·ol·o·gy *n.* The study of the flow of materials.

rhe·o·stat *n.* A variable resistor, used in dimmer switches.

right an·gle *n.* An angle of 90 degrees.

right tri·an·gle *n.* A triangle having one right angle.

rms *See* root-mean-square.

RNA *n.* Ribonucleic acid. An organic acid essential for the production of protein in living cells.

Rock·well hard·ness test *n.* A measure of the hardness of a metal. A point, called a brale, is pressed into the test sample and the size and depth of the indentation are measured.

roent·gen *n.* A unit of exposure to x-rays. *Abbr.* R

roll *n.* **1.** An oscillatory motion of an aircraft or seacraft. **2.** A cylindrical device used to flatten metal.

roll·ing mill *n.* A machine that flattens metal sheet by squeezing it between rotating rolls.

ROM *See* Read-only memory.

root *n.* In the equation $x^n = y$, x is a root of y.

root mean square *n.* A kind of average; the square root of the sum of the squares of a series of values. *Abbr.* rms

ros·in–core sol·der *n.* A solder wire which contains a rosin flux core to aid in wetting during soldering.

ro·tam·e·ter *n.* An instrument for measuring fluid flow.

ro·tor *n.* The rotating element of an electrical motor or generator.

round·ing *n.* A method of terminating, for practical purposes, a long or a nonterminating decimal value. If the last digit is 6 or more, 1 is added. If the last digit is 4 or less, it is merely dropped. If the last digit is 5, the digit following the 5 is taken into consideration. Thus, 3.141 rounds to 3.14; 3.14159 rounds to 3.1416 and on further rounding to 3.142. In computers and programmable calculators, it is important to know whether the instrument rounds or truncates (drops digits without rounding). *See* truncate.

round·ing er·ror *n.* The error introduced by rounding.

rou·tine *n.* A set of computer instructions.

RPG *n.* A computer programming language, Report Program Generator.

ru·by la·ser *n.* A solid-state laser which uses a synthetic ruby crystal.

run time *n.* The time required for the completion of a computer program.

ryd·berg *n.* A unit of energy used in atomic physics. *Abbr.* ry

S

sag *n.* The slack in a cable.

sa·line *adj.* Containing salt.

sa·lin·i·ty *n.* Saltiness.

sal·i·nom·e·ter *n.* An instrument for measuring the quantity of salt in a solution.

salt *n.* **1.** The reaction product when a metal displaces the hydrogen of an acid. **2.** The compound sodium chloride, table salt.

salt bath *n.* Molten salts used as an environment for the heat treatment of metals.

SAM *n.* *S*urface-to-*a*ir *m*issile.

sand–cast *v.* To form by pouring molten metal into a sand mold.

sa·pon·i·fi·ca·tion *n.* The process of converting chemicals to soap. *v.* sa·pon·i·fy. *See* soap.

sap·phire *n.* A mineral of aluminum oxide.

sat·el·lite *n.* **1.** A celestial body in orbit around another larger one (example: the moon). **2.** A manufactured object placed in orbit around a celestial body.

sat·u·rate *v.* To cause to absorb to capacity.

sat·u·ra·ble re·ac·tor *n.* An iron-core reactor, used in the construction of a magnetic amplifier.

SAW *See* surface acoustic wave.

sawtooth wave *n.* A waveform with a slow rise time and a short fall, like a series of sawteeth.

S band *n.* A microwave frequency, from 1550 to 5200 megahertz.

sca·lar *n.* A quantity which has only magnitude and not direction. *Compare* vector. *adj.* Referring to a nondirectional quantity.

scale up *v.* To increase in size while retaining all relative proportions, as when a smaller model is used for the design of a product of commercial size.

scalp *v.* To remove surface layers.

scan·ner *n.* A device which performs character recognition and converts characters into electric signals.

scan·ning elec·tron mi·cro·scope *n.* A microscope that scans the specimen with an electron beam and transmits the signal to a cathode-ray tube. It is capable of very high magnification. *Abbr.* SEM

scar·i·fy *v.* To scratch a surface.

scat·ter *n.* The diffusion of beams of electromagnetic radiation in a random manner.

sche·mat·ic *n.* A part-by-part diagram showing the relationship of all elements of a system.

Schott·ky de·fect *n.* A defect in a crystal.

Schott·ky di·ode *n.* A semiconductor that contains a barrier at the junction where a semiconductor layer meets a thin metal coating; this barrier permits electrons ("hot carriers") to flow in one direction only, from the semiconductor to the metal base; this diode is useful for its high forward conductivity and fast switching speeds. *Also known as* carrier diode.

scin·til·la·tion *n.* Random fluctuation in electromagnetic radiation.

scope *See* oscilloscope.

–scope ***word root*** An instrument for visual display of data. (From Greek *skeptesthai*—to watch, to look at.)

SCR *See* silicon controlled rectifier.

scram·bler *n.* An electronic circuit that converts speech signals into unintelligible sounds, for the purpose of sending secret messages, which can be unscrambled only by someone with the proper equipment.

scratch pad stor·age *n.* A memory device used for the temporary storage of data.

screen *n.* **1.** A grid in an electron tube. **2.** The viewing screen of a cathode-ray, radar, x-ray, or oscilloscope image. **3.** *v.* To separate rock, ore, or aggregate by means of a mesh through which only particles below a certain size can pass.

screen·ing *n.* The prevention of interaction between an electronic device and incident radiation by placing a nonconducting material around the device. *Also known as* shielding.

screw dis·lo·ca·tion *n.* A defect in a crystal.

scrib·er *n.* A sharp-pointed tool.

scrub *v.* To cancel a scheduled procedure.

scrub·ber *n.* A device that removes impurities by wet action.

sec *See* secant; second.

se·cant *n.* *See* trigonometric ratios. *Abbr.* sec

sech *See* hyperbolic secant.

sec·ond *n.* **1.** One-sixtieth of a minute. **2.** A unit of plane angle, equal to $\frac{1}{3600}$ degree. *Abbr.* sec *or* s (preferred)

sec·tion *n.* A drawing of an object as it would appear if cut along a plane. *v.* To cut a solid into sections.

sec·tor *n.* **1.** A pie-shaped portion of a circle. **2.** Such a portion on a radar screen. **3.** On a diskette, boundaries (hard in the case of physical holes, soft in the case of software-formatted sectors).

sed·i·men·ta·ry *adj.* Pertaining to rock formed from deposits of sand or clay in the beds of ancient streams, rivers, and oceans.

See·beck ef·fect *n.* The voltage produced in a circuit by temperature differences between two junctions of different metals.

seed *n.* The nucleus, in a solution, on which a crystal may grow. *v.* To add such a nucleus to a solution.

shield·ing *See* screening.

sieze *v.* To jam and lock together, as of close-fitting moving parts.

self–dif·fu·sion *n.* The spontaneous movement of an atom in a crystal lattice to another site in that lattice.

sel·syn mo·tor *n.* A device that transmits and receives angular motion over wires.

SEM *See* scanning electron microscope.

sem·i·con·duc·tor *n.* A material whose electrical conductivity lies between that of metals and that of insulators; used in the fabrication of solid-state devices, usually silicon or gallium arsenide.

sem·i·con·duc·tor stor·age *n.* A memory device that uses semiconductor chips for storage of information.

sen·sor *n.* A device that senses a value or a change in a physical quantity and transforms that response into an input signal. One example is a television camera.

se·quen·tial ac·cess *See* serial access.

se·ri·al *adj.* Pertaining to the processing of data in a prescribed order.

se·ri·al ac·cess *n.* A process utilizing devices in a computer that store data in a given order and that can be accessed only in that order.

se·ries cir·cuit *n.* An electric circuit that has a single path.

ser·vo·mech·a·nism *n.* An automatic feedback control system whose output is mechanical motion.

ser·vo·mo·tor *n.* The final control element in a servomechanism.

set the·o·ry *n.* The mathematical system that assigns numbers to sets, in a matching procedure.

set·up *n.* The preparation of a machine or a system to perform a process.

sheaf *n.* A bundle of fibers.

shear stress *n.* Lateral stress.

sheave *n.* A grooved wheel or pulley.

shell *n.* A set of electron orbital states.

shil·ling frac·tion *n.* A fraction written on a single line with an oblique stroke separating the numerator and the denominator (1/2). *Compare* built-up fraction.

shim *n.* A thin piece of metal used to correct for fit between surfaces. *v.* To correct for fit by insertion of thin metal pieces.

shock re·sis·tance *n.* The ability of a material to resist fracture when struck a sudden blow.

shore *v.* To prop up with timber or other material.

shor·ing *n.* A system of props.

short *See* short circuit.

short cir·cuit *n.* A low-resistance path between two sides of a circuit, usually accidental. *v.* To lead current down a low-resistance path.

short·wave *n.* Electromagnetic frequencies used in radio broadcasting, of 1600–30,000 kilohertz, above the standard broadcast band.

shot *n.* Small spherical steel particles used as cutting agents.

shunt *n.* A resistive parallel path in a circuit. *v.* To place such a path in a circuit.

side·band *n.* A frequency band above or below the frequency of the carrier band.

si·de·re·al *adj.* Referring to time measured relative to the motion of the stars.

sie·mens *n.* A unit of conductance (preferred to mho). *Abbr.* S

sig·moid *adj.* S-shaped.

sig·nal–to–noise ra·tio *n.* A measure of interference with a signal. *Abbr.* S/N

sig·na·ture *n.* A characteristic identifying pattern.

sign on *v.* To address the computer so that it is prepared for any input/output instructions.

sil·i·ca *n.* Silicon dioxide.

sil·i·cone *n.* A group of heat-stable, water-repellant materials containing silicon. Used in adhesives, cosmetics, and synthetic rubber.

sil·i·con con·trolled rec·ti·fi·er *n.* A semiconductor device. *Abbr.* SCR

silk–screen *n.* The process of printing a pattern through a selectively porous mask. Used in thick-film deposition. *See* thick-film.

sim·u·late *v.* To model a process on a computer or by other means.

sin *See* sine.

sine *n. See* trigonometric ratios. *Abbr.* sin

sine wave *n.* A wave whose amplitude varies as the sine of a linear function of time. *See* trigonometric ratios.

sin·gle–pole dou·ble–throw *n.* A three-terminal switch that connects one terminal to either of two other terminals. *Abbr.* SPDT

sin·gle–pole sin·gle–throw *n.* A two-terminal switch that opens or closes one circuit. *Abbr.* SPST

sink *n.* A defined region into which energy (such as heat) may drain.

sin·ter *v.* To bond powders by subjecting them to heat and pressure without melting.

si·nus·oi·dal *adj.* Pertaining to harmonic motion.

size *v.* To surface-treat textiles, paper, or leather.

siz·ing *n.* Material used as a pore filler in surface-treating. *See* size.

skewed *adj.* **1.** Not parallel, distorted. **2.** Referring to a nonsymmetrical frequency distribution in statistics.

skin ef·fect *n.* An effect (greatest at high frequencies) that causes the amount of current on the surface of a conductor to exceed that in the interior.

slag *n.* The molten glassy material that rises to the top in a crucible in which metal is being refined.

slave *n.* A mechanism whose movement is controlled by the movement of another mechanism.

slew *v.* To move an antenna rapidly.

slice *n.* A wafer of semiconductor single-crystal material.

slip *n.* **1.** A suspension of fine clay particles used to cast ceramic ware or to cast a shape. **2.** The movement of intracrystalline planes over each other as a result of applied stress.

slip ring *n.* A wiping contact in a motor or generator.

slope *n.* The path of a curve through a point.

slug *n.* A unit of mass.

slug tun·ing *n.* Changing the inductance of a coil by moving a metal core within the coil.

slur·ry *n.* A semifluid suspension of particles.

small–sig·nal pa·ram·e·ter *n.* A parameter that describes the performance of an electronic device at low input levels.

smart ter·mi·nal *n.* A computer terminal with microprocessor decision-making capability.

smut *n.* Contamination left on the surface of a metal.

S/N *See* signal-to-noise ratio.

soap *n.* **1.** A salt of one of the fatty acids. Thus, lead soap, a lubricant, is made of lead salts saponified with fats. **2.** A detergent.

soft•ware *n.* The programs, or set of instructions, used in a computer.

sol *n.* A colloidal solution or suspension.

so•lar *adj.* Pertaining to the Sun.

so•lar cell *n.* A semiconductor device that converts sunlight or radiant energy to electric energy.

sol•der *v.* To join metal surfaces with an alloy, usually of tin and lead, having a low melting point.

sol•der dip *n.* A process used in the manufacture of printed circuit boards.

so•le•noid switch *n.* A switch whose action is controlled by the magnetic field of a coil.

sol•id–state *adj.* Pertaining to electronic devices using semiconductor materials rather than vacuum tubes.

sol•i•dus *n. See* solidus/liquidus.

sol•i•dus/liq•ui•dus *n.* A graph of the points at which the different constituents of alloys in a melt begin to solidify. It defines the solidification range of alloys, silicate melts, and so on.

sol•u•bil•i•ty *n.* The ability of a substance to dissolve in another substance.

sol•ute *n.* The substance which is dissolved in a solvent.

sol•vent *n.* A substance (usually liquid) in which another substance will dissolve. In the case of substances which can dissolve in each other, the one present in greater quantity is called the solvent.

so•nar *n.* An underwater electronic sound system that locates objects by echoes.

sonde *n.* An instrument that obtains weather data during ascent or descent through the atmosphere.

so•no•buoy *n.* An acoustic receiver and radio transmitter mounted on a buoy.

sorp•tion *n.* The process of absorbtion and adsorbtion.

sort *v.* To arrange data into a desired sequence.

source code *n.* A code used in a source program.

source pro•gram *n.* A program written in any of the basic programming languages, which then must be converted by the compiler into machine language (object program).

space charge *n.* The net voltage that exists within a given volume.

spa•ghet•ti *n.* Insulation over fine wires.

spall *v.* To chip away fine fragments from a surface.

SPDT *See* single-pole double-throw.

SPST *See* single-pole single-throw.

spe•cif•ic grav•i•ty *n.* The measure of the density of a material.

spe•cif•ic heat *n.* The amount of heat needed to heat a unit quantity of a material a unit degree of temperature.

spec•tral *adj.* Referring to a range of frequencies; pertaining to a spectrum.

spec•trom•e•ter *n.* An instrument for measuring wavelengths.

spec•tro•pho•tom•e•ter *n.* An instrument for analyzing color or comparing luminous intensities of two spectra.

spec•tros•co•py *n.* The technique of identifying materials by measuring the radiation they emit or absorb when externally activated.

spec•trum *n.* A range of radiation frequencies. ***pl.*** spectra.

spec•u•lar *adj.* Highly reflective.

spher•i•cal co•or•di•nates *n.* A coordinate system in three-dimensional space whose points lie on the surface of a sphere.

sput•ter•ing *n.* The process of depositing a thin layer of metal onto a substrate, in a vacuum. Used in the manufacture of integrated circuits.

sphe•roid *n.* A round body which is not a perfect sphere.

sphe•roi•dize *v.* To heat-treat steel so as to convert its iron carbide to globular shape.

spline *n.* A groove cut into a shaft in order to receive and lock a corresponding part.

sport *n.* An unexpected genetic result, usually resulting from mutation.

spot *n.* The area immediately illuminated by an electron beam in a cathode-ray tube.

spot test *n.* Analysis of a substance by the addition of a drop of reagent to one or two drops of the substance.

spot weld•ing *n.* Resistance welding which is confined to small, circular areas. *See* resistance welding.

spring•back *n.* The return of a bent material to its original shape after the release of a bending force.

sprue *n.* The channel through which molten metal is poured into the mold during casting.

SPST *See* single-pole single-throw.

square wave *n.* A wave with sudden changes of amplitude to plateau levels, each cycle of the wave being shaped like a square.

stage *n.* A single portion of an electrical system involving an active device such as a vacuum tube or a semiconductor and performing a single function.

stamp *v.* To punch parts out of sheet stock with a die.

stamp•ing *n.* A part which has been stamped out of sheet stock.

stand–a•lone ma•chine *n.* A computer that possesses its own memory, and is not attached to a master computer.

stan•dard de•vi•a•tion *n.* A measure of the variability of data.

stand•ing wave *n.* A wave which takes up a stationary position between two nodes. Two examples are: the movement of a rope with one end tied to a post when its free end is shaken; and a "haystack" (a wave that does not move downstream with the current) in a white-water river.

stand•off *adj.* Referring to a spacer that separates objects.

star net•work *n.* A circuit with three or more branches which connect at a common node.

–stat *word root* Referring to an instrument that steadies, stabilizes, or brings to a halt. (From Greek *states*—one that stops or steadies.) ***physics prefix*** Relating to static electricity. *See* statampere, statcoulomb.

stat•am•pere *n.* The electrostatic unit of current.

stat•cou•lomb *n.* The electrostatic unit of charge.

state *n.* The condition of a system.

state of the art *n.* The level of understanding and the analytical tools presently available for the solution of a problem.

stat•ic charge *n.* The electric charge that is accumulated on an object.

stat•ics *n.* The branch of physics that treats forces or bodies in a state of equilibrium.

sta•tis•ti•cal me•chan•ics *n.* The branch of physics that interprets macroscopic behavior of a system on the basis of the behavior of that system's microscopic constituents.

sta•tor *n.* The nonrotating portion of a motor or generator.

steady state *n.* The condition of a system in which it does not change with time.

step–down trans•form•er *n.* A transformer in which the secondary windings have a lower voltage than the primary windings.

step–up trans•form•er *n.* A transformer in which the secondary windings have a higher voltage than the primary windings.

stereo– *word root* Referring to the illusion of three-dimensionality produced by using two separated sources of sense data. (From Greek *stereos*—solid.)

sto•chas•tic *adj.* **1.** Referring to random variables. **2.** Involving chance or probability.

stoi•chi•o•met•ric *adj.* Referring to the number of atoms participating in a chemical reaction.

stoi•chi•om•e•try *n.* The numerical relationship of the elements and compounds that take part in a chemical reaction.

stop•cock *n.* A small valve, used in chemical apparatuses.

stor•age (data) *n.* The function performed by a computer's memory.

stor•age tube *n.* A cathode-ray tube that stores its image on a screen behind the viewing screen. Images will remain on the viewing screen until the storage screen is erased.

strain *n.* A change in length in one dimension as a result of an applied force.

strain gauge *n.* A sensor that measures strain.

strat•o•sphere *n.* The second layer of the atmosphere surrounding Earth, about 15–25 miles above Earth's surface.

stra•tum *n.* A layer. ***pl.*** strata.

stream•er *n.* A type of computer tape.

stress *n.* Force expressed as units of weight per unit of area.

stress cor•ro•sion *n.* Corrosion accelerated by stress in metal.

stress rais•er *n.* A notch at which stress may concentrate, resulting in possible fracture.

stress re•lief *n.* The removal of retained stress by low-temperature heat treatment.

stress–strain curve *n.* A curve showing the relationship of stress or load on a material to the strain or deformation that results.

strike *n.* A thin initial layer of electroplated film.

string•er *n.* A horizontal support member used in construction.

strip *v.* **1.** To remove the top layer of soil or rock from a quarry or open-pit mine. **2.** To remove insulation from a wire.

strip line *n.* A microwave transmission line.

strobe *n.* A small, bright spot of light in a display.

stro•bo•scope *n.* An instrument that illuminates a moving object with flashes of light so as to make it appear to stand still.

sty•lus *n.* In computer graphics, the device that contacts the record sheet. ***pl.*** styli.

sub•li•ma•tion *n.* The change of state from solid to gas or vice versa, without passing through the liquid state.

sub•rou•tine *n.* A set of computer instructions designed to be used by other routines.

sub•set *n.* Selected members of a given set.

sub•son•ic *adj.* **1.** Pertaining to frequencies below the level of human hearing (about 15 hertz). **2.** Pertaining to speeds below the speed of sound (about 1087 feet per second in air), as in subsonic flight.

sub•strate *n.* The material on which a microcircuit is fabricated. Used primarily for insulation and physical support.

sub•tra•hend *n.* A number which is subtracted from another number.

sump *n.* A pit or tank that receives drainage from a system.

super–, supra– ***prefix*** Over, above, exceeding.

su•per•con•duc•tiv•i•ty *n.* The very high electrical conductivity attained in many compounds at extremely low temperatures.

su•per•cool *v.* To cool a material to a temperature below that at which it would change state, but with that change of state suppressed. For example, to cool a liquid past its freezing point without freezing actually taking place.

su•per•fi•cial *adj.* At or on the surface.

su•per•sat•u•ra•tion *n.* The condition in which more solute is contained in a solution than its normal saturation point. *Also known as* supersolubility.

su•per•sol•u•bil•i•ty *See* supersaturation.

su•per•son•ic *adj.* **1.** Pertaining to speeds above the speed of sound (about 1087 feet per second in air), as in supersonic flight. **2.** In acoustics, the same as ultrasonic, which is the preferred word.

sup•pres•sor *n.* A device that limits noise in an electric circuit.

sur•face acous•tic wave *n.* A sound wave that propagates along the surface of a solid. *Abbr.* SAW

sur•face–bar•ri•er di•ode *n.* A diode using thin-surface layers.

sus•cep•ti•bil•i•ty *n.* The ratio of the magnetization of a material to the magnetic field strength.

sus•pen•sion *n.* A mixture of fine particles of a solid within a liquid or gas.

swage *v.* To forge, squeeze, or hammer a metal tube into a tapered shape or a shape with reduced diameter.

sweep *n.* **1.** The steady movement of an electron beam across a cathode-ray tube. **2.** The steady change in frequency of a signal generator.

switch•ing *n.* The making, breaking, or changing of connections in an electric circuit.

sym•bol•ic ad•dress *n.* A symbol defined by the programmer to locate a particular word in computer memory; the symbol stands for the entire word. *See* word.

syn•chro•nous *adj.* Referring to simultaneous phenomena, such as two circuits that are in phase.

syn•er•gy *n.* Greater results from the combination of factors than from the sum of their individual effects.

syn•tax *n.* Rules for constructing statements in a computer language.

syn•the•sis *n.* Any process or reaction for building up a complex compound by the union of simpler compounds or elements. *v.* syn•the•size

syn•thet•ic *adj.* Produced by chemical or biochemical synthesis; artificial or manufactured.

sys•tem soft•ware *n.* Software that supports the operation of the computer and which was provided by the manufacturer, as distinct from software generated by the user.

tab *n.* A small projection, used as a fastener.

ta•bling *n.* The separation of particulate materials by passing a water suspension over an inclined table.

tab•u•lar *adj.* Referring to data presented in table form.

tab•u•late *v.* To print data in table form.

ta•chom•e•ter *n.* An instrument used to measure revolutions per minute.

tac•tile *adj.* Able to be perceived by touch.

tag•ging *n.* The act of labeling a particular element of a process.

tagged mol•e•cule *n.* A radioactive molecule used in tracer or tagging studies.

tail•ings *n.* The residue of a mineral refinement process.

tail•stock *n.* The part of a lathe that holds the work being machined.

take–up *n.* A tensioning device to take up the slack of loose lines and cords.

tan•gent *adj.* Pertaining to a line that touches a curve at a single point; tangential. *n. See* trigonometric ratios. *Abbr.* tan

tanh *See* hyperbolic tangent

tank cir•cuit *n.* A circuit capable of storing electric energy.

tap *v.* **1.** To machine threads into a hole. **2.** To break the seal of a container. *n.* The tool used to machine threads into a hole.

tape drive *n.* A device that causes tape to be transferred from reel to reel.

tape ed•i•tor *n.* A program used to edit and correct a routine recorded on a magnetic tape.

tap•pet *n.* A lever moved by a cam; used in piston engines.

tare *n.* The weight of an empty container. To obtain the weight of a container's contents, the tare must be subtracted from the weight of the full container.

tar•get *n.* **1.** The electrode bombarded by electrons in a vacuum system. **2.** In sputtering, the source of sputtered ions. **3.** In radar, the object observed. *v.* To point; to set sights on.

tax·on·o·my *n.* The science of classification, usually of living things.

Tay·lor se·ries *n.* In mathematics, a power series representation of a given function.

tec·ton·ics *n.* **1.** The technology of construction with reference to the final use. **2.** A branch of geology that deals with the large structural units of the earth's crust.

tee joint *n.* A joint in which the members meet in a right angle.

tele- *word root* At a distance. (From Greek *tele*—far off.)

te·lem·e·ter *v.* To transmit instrument readings over a distance by wire, radio waves, or other means.

tel·e·scop·ing *adj.* Nesting into each other.

tel·e·tex *n.* A system that transmits data between terminals (such as two word processors) via the telephone network.

tel·e·text *n.* A computer communications service that uses a home television receiver to display material broadcast via radio waves. The user selects information to be screened from a menu of preprogrammed frames of data but cannot interact with the data base. Examples of available data are stock market results, sports results, weather conditions, and listings of hotels and restaurants.

tem·per *n.* A designation of the hardness and strength of a metal. *v.* To soften hardened metals by reheating them.

tem·plate *n.* **1.** A pattern which guides a machining tool. **2.** Any pattern by which one part shapes another, including the cells of living organisms.

ten·sile strength *n.* The maximum stretching force a material can bear without tearing.

ten·sion *n.* Stretching force brought to bear on a material.

ten·sor *n.* The magnitude of a vector. The general theory of vector quantities, called *tensor analysis,* developed by the mathematician Gregorio Ricci in 1887, gave Albert Einstein a way of representing the force of gravity in his general theory of relativity. *See* vector.

tera- *prefix* Million million (trillion). *Symbol* T

term *n.* A signed (plus or minus) component of a mathematical expression. Thus, $+a$, $-b$, and $+c$ are the terms of the expression $a - b + c$.

ter·mi·nal *n.* **1.** A location in an electric circuit where connections may be made. **2.** In computer technology, a device composed of keyboard and screen for input/output of data.

ter·mi·nal sym·bol *n.* An oval-shaped flowchart figure which indicates start and stop points.

ter·na·ry *adj.* Consisting of three.

ter•ti•ary *adj.* Third.

tes•la *n.* The unit of magnetic flux density. *Abbr.* T

tes•sel•lat•ed *adj.* Of a mosaic, having a pattern built up of small stone or glass squares.

test bed *n.* A base or frame on which experimental equipment is mounted.

te•trag•o•nal *See* crystal systems

tet•rode *n.* A four-element electronic component.

tex•ture *n.* The tactile quality of a surface.

the•o•rem *n.* A conclusion demonstrated by a series of deductive steps, that is, by reasoning from principle.

therm *n.* A unit of heat energy.

ther•mal con•duc•tiv•i•ty *n.* A measure of heat flow through a plate of specified thickness.

ther•mal flux *n.* The flow of heat across a surface.

ther•mal gra•di•ent *n.* The distribution of temperature in a material in terms of temperature difference from point to point.

therm•i•on•ic *adj.* Referring to the emission of electrons from a hot surface.

therm•is•tor *n.* A resistive circuit element for sensing temperature; as temperature increases, resistance in the thermistor decreases.

ther•mo•com•pres•sion bond•ing *n.* The process of assembly of semiconductor components using a combination of heat and pressure. *Abbr.* TC bonding

ther•mo•cou•ple *n.* A pair of dissimilar conductors which generate an electric signal at their junction, indicating temperature.

ther•mo•dy•nam•ics *n.* The study of the physics of heat and energy.

ther•mo•e•lec•tric•i•ty *n.* The conversion of heat into energy or vice versa by means of thermocouple elements.

ther•mog•ra•phy *n.* **1.** The process of mapping surface temperatures by observing color changes in luminescent materials. **2.** A photocopy process using infrared rays.

ther•mo•pile *n.* An array of thermocouples used to generate electricity from heat.

ther•mo•plas•tic *adj.* Able to be shaped when heated.

ther•mo•set•ting *adj.* Becoming rigid when heated.

ther•mo•son•ic bond•ing *n.* The process of joining fine wires to regions on chips or hybrid circuit substrates by means of vibrating pressure combined with heat.

ther·mo·stat·ic bath *n.* An apparatus for maintaining a constant temperature.

thick–film *adj.* Referring to a film deposited by silk-screening on a substrate and fired, for use in microcircuits.

thin–film *adj.* Referring to a film deposited by evaporation or sputtering on a substrate, for use in integrated circuits.

thio– *word root* Containing sulfur. (From Greek *theion*—sulfur.)

thread *n.* A continuous spiraling rib.

thresh·old *n.* The least value of a quantity that produces the minimal detectable response.

thy·ra·tron *n.* A hot-cathode gas-filled electron tube.

thy·ris·tor *n.* A semiconductor device that performs like a thyratron.

time con·stant *n.* The time required for a physical quantity to reach a defined level.

time–shar·ing *n.* The apparently simultaneous use of a computer system by different users.

ti·tra·tion *n.* The process of analyzing a chemical substance by adding measured amounts of a liquid that reacts with the chemical in a known way.

tog·gle switch *n.* An electric switch operated by a small finger-activated lever.

to·ler·ance *n.* The permissible plus-or-minus variation in a structure or system.

tongue and groove *n.* A joint in which a projection of one member fits into a groove of the other member.

tooth *n.* A projection on the edge or face of a gear.

to·pog·ra·phy *n.* A surface's configuration, including its relief.

to·roi·dal *adj.* Doughnut-shaped.

torque *n.* A force that causes torsion and rotation.

torr *n.* A unit of gas pressure.

tor·sion *n.* A twisting deformation of a solid body about an axis.

touch plate. *n.* A panel on a device which performs switching when touched by the finger.

trace *n.* **1.** A tiny, almost imperceptible, amount. **2.** The track of a moving spot on a video screen.

trac·er *n.* A radioactive additive used to monitor the distribution or location of a given substance.

trans·ceiv·er *n.* An electronic system that combines the functions of transmitting and receiving signals.

trans•duc•er *n.* A device that converts one form of energy into another.

trans•form *n.* The conjugate of a complex number, itself complex.

trans•form•er *n.* An electric component that transfers electric energy from one circuit to another by magnetic induction.

tran•sient *n.* A short sudden pulse that occurs in a system before it reaches a steady state condition.

tran•sis•tor *n.* An electronic device made of semiconducting material and capable of performing many functions in a circuit.

tran•sit time *n.* The time required for an electron to travel between two electrodes in an electron tube or transistor.

trans•late *v.* **1.** To convert computer information from one language to another. **2.** To "slide" a plane so that points (x, y) move to a new position $(x + h, y + k)$, while remaining in the same relationship to one another.

trans•mu•ta•tion *n.* The changing of the nuclides of one element to another; a continuous process in radioactive elements. Thus, over a period of time, radium, thorium, and uranium will be reduced to lead.

tran•son•ic *adj.* Located at the boundary between sonic and supersonic speeds, about Mach 0.8–1.2. *See* Mach number. *Also* trans-son•ic.

tran•spi•ra•tion *n.* The passage of a gas or vapor through a membrane.

tran•spond•er *n.* An electronic device that automatically receives and responds to an incoming signal.

trans•po•si•tion *n.* The interchanging of locations.

trans•u•ran•ic elements *n.* Elements having atomic numbers larger than 92, that of uranium.

trap *n.* Vacancy in a semiconductor in which an electron can be caught and trapped.

trap•e•zoid *n.* A four-sided figure with two sides parallel.

trav•el•ing–wave tube *n.* A microwave electron tube used as an amplifier or oscillator. *Abbr.* TWT

tri•ad *n.* A group of three.

tri•an•gu•late *v.* To divide an area into triangles for the purpose of surveying a piece of land or, in navigation, finding the position of a ship.

tri•clin•ic *See* crystal systems.

trig•ger *v.* To initiate an action.

tri•go•nal *See* crystal systems.

trig•o•no•met•ric ra•tios *n.* In a right triangle ABC, the six possible ratios (sine, cosine, tangent; secant, cosecant, cotangent) of the hypotenuse (AB), the side adjacent to angle A (AC), and the side opposite to angle A (BC). These ratios were calculated by the ancient Greeks in the solution of surveying and engineering problems, and in the study of the heavenly bodies, their sizes, and the distances between them. The earliest tables of these ratios were computed by Hipparchus, about 150 B.C. Since the seventeenth century, trigonometric ratios were extended to solving problems for any variable quantity that involves measuring an angle, such as the movement of a body down an inclined plane or the swing of a pendulum. In modern times, these ratios have been used to graph and describe mathematically the shape of water waves, sound waves, electric currents, and radio waves. *Abbrs.* sine = sin; cosine = cos; tangent = tan; secant = sec; cosecant = csc; cotangent = cot.

tri·no·mi·al *n.* A mathematical expression consisting of three terms, such as $2a + 3b - c$.

tri·ple–pur·pose me·dia *n.* Computer media used for three purposes.

tri·va·lent *adj.* Having a valence of 3.

-tron *suffix* Referring to an instrument; relating to a vacuum tube.

tro·po·pause *n.* The boundary between the troposphere and the stratosphere.

tro·po·sphere *n.* The lowest 15 miles of the Earth's atmosphere (below the stratosphere).

trun·cate *v.* To drop digits at the end of a decimal number without rounding. Sometimes a source of error in calculations.

trunk line *n.* A path over which information is transmitted in a computer.

tune *v.* To adjust an electronic circuit for resonance at a desired frequency.

tun·nel di·ode *n.* A junction diode semiconductor device.

twin *n.* A compound crystal whose parts form a mirror image of each other.

TWT *See* traveling-wave tube.

U

UHF *See* ultrahigh frequency.

-ular *suffix* Relating to.

-ulent *suffix* Abounding in.

ul·lage *n.* The amount by which a container lacks being full.

ul·tra- *prefix* Beyond.

ul·tra·high fre·quen·cy *n.* A frequency range used in radio, from 300 to 3000 megahertz. *Abbr.* UHF

ul·tra·son·ic *adj.* Pertaining to frequencies above the level of human hearing (about 20,000 hertz).

ul·tra·son·ic bond·ing *n.* The bonding of two metals by vibrating them together under pressure; used in the manufacture of semiconductor devices.

ul·tra·sound *n.* Ultrasonic-frequency sound waves.

ul·tra·vi·o·let ra·di·a·tion *n.* Electromagnetic radiation in the wavelength range 4–400 nanometers, which starts just beyond the violet end of the spectrum of visible light. *Abbr.* UV

um·bra *n.* The darkest portion of a shaded area. *See* penumbra.

un·couple *v.* To disconnect.

u·ni·di·rec·tion·al *adj.* Flowing or radiating in only one direction.

un·ion *n.* A screw or flange connection around a joint.

u·ni·po·lar *adj.* Having only one electric or magnetic pole or direction.

u·nit cell *n.* The basic atomic arrangement in a crystal, which is repeated throughout the crystal.

u·ni·tary pow·er fac·tor *n.* An electric circuit with a power factor of 1.

u·ni·va·lent *adj.* Having a valence of 1.

u·ni·ver·sal gas con·stant *n.* The pressure of a gas times its molar volume divided by its temperature, used in calculating the expansion of gases. *See* molar.

u·ni·verse *n.* The defined system in which an experiment or a process takes place.

un·sat·u·rat·ed com·pound *n.* Any compound having bonds available to participate in new reactions, especially organic compounds containing double or triple bonds.

un·sat·u·ra·tion *See* unsaturated compound.

un·sta·ble *adj.* Capable of undergoing spontaneous change.

un·tuned *adj.* Not resonant to any available frequency.

up·date *v.* To modify a computer instruction.

up·draft *n.* An upward-moving gas.

up·set *n.* An increase in cross section caused by application of pressure. *v.* To cause such an increase in cross section.

up·stream *n.* That part of a process stream that has not yet entered the system.

UV *See* ultraviolet.

V

va·can·cy *n.* An unoccupied lattice position in a crystal; a defect.

vac·u·um *n.* **1.** Theoretically, a space in which there is no matter. **2.** A space in which the pressure is well below normal atmospheric pressure.

vac·u·um dep·o·si·tion *n.* The process of evaporating a film of a material onto a substrate in low-pressure chamber.

vac·u·um evap·o·ra·tion *n. See* vacuum deposition.

vac·u·um fir·ing *n.* Heating in a vacuum to achieve degassing.

vac·u·um gauge *n.* An instrument to measure the absolute gas pressure in a vacuum.

vac·u·um pump *n.* A pump which exhausts air from a chamber.

vac·u·um switch *n.* A switch whose contacts are in a vacuum envelope, to minimize sparking.

vac·u·um tube *n.* An electron tube.

va·lence *n.* A positive number which indicates the combining power of an element with other elements.

val·ue en·gi·neer·ing *n.* The discipline which studies production of a product or service for the optimum cost.

valve *n.* **1.** A mechanism that regulates flow of gases or fluids. **2.** A vacuum tube *(British usage). See* vacuum tube.

Van Al·len ra·di·a·tion belt *n.* A region in space surrounding Earth formed of charged particles that are trapped by Earth's magnetic field.

Van de Graff gen·er·a·tor *n.* A high-voltage generator.

van der Waals force *n.* An attractive force between two atoms.

vane *n.* A flat or curved surface exposed to air or fluid flow, as in a propeller or turbine.

va·por *n.* A substance in the gaseous state.

va·por dep·o·si·tion *n. See* vacuum deposition.

va·por pres·sure *n.* The pressure of a vapor in equilibrium with a liquid or solid.

var·ac·tor *n.* A semiconductor device whose capacitance may be varied by varying the applied voltage.

var·ac·tor di·ode *n. See* varactor.

Var·i·ac *n.* A transformer that gives a continuously adjustable voltage.

var·i·cap *See* varactor.

var·i·ance *n.* A measure of the deviation from a norm or from a predicted value.

var•i•om•e•ter *n.* A variable inductance.

va•ris•tor *n.* A semiconductor device exhibiting voltage-dependent resistance.

V belt *n.* A belt with a trapezoidal cross section which runs in a V-shaped groove.

vec•tor *n.* A quantity which has both magnitude and direction. ***antonym*** sca•lar. *adj.* vec•tor•i•al. *Symbol* a boldface capital letter.

ve•hi•cle *n.* **1.** A physical medium (as in paint, where oil is the medium in which the pigment is suspended). **2.** A medium in which chemical processes may take place.

ven•tu•ri tube *n.* A tube containing a constriction used for measuring fluid flow. *See* Table 4-3.

ver•mic•u•lite *n.* A clay mineral that expands when heated; used in insulation and potting soil.

ver•mi•form *adj.* Wormlike.

ver•ni•er di•al *n.* A dial that performs in the same manner as a vernier scale. *See* vernier scale.

ver•ni•er scale *n.* A short, auxiliary linear scale which is used with a larger scale and which has lines showing subdivisions of the units on the main scale.

vers *See* versed sine.

versed sine *n.* The versed sine of A is $1 - \text{cosine } A$. *See* trigonometric ratios.

ver•sine *See* versed sine.

ver•tex *n.* An angle, formed by two intersecting sides.

very high fre•quen•cy *n.* A frequency range used in radio, from 30 to 300 megahertz. *Abbr.* VHF

very low fre•quen•cy *n.* A frequency range used in radio, from 3 to 30 kilohertz. *Abbr.* VLF

VF Voice frequency.

VHF *See* very high frequency.

vi•a•ble *adj.* Able to develop.

vi•al *n.* A small container, for holding liquids.

vi•bra•tion damp•ing *n.* The process of converting mechanical vibration energy into heat energy.

vi•bra•tor *n.* A vibrating need that converts direct current to alternating current.

Vick•ers hard•ness test *n.* A test used to determine the hardness of materials by means of a diamond-pyramid indenter that is forced into the surface of a specimen.

vid•eo ***adj.*** Pertaining to electronic signals that will be converted to an image.

vid•eo dis•play ***n.*** An image of computer readout on a video terminal.

vid•eo ter•mi•nal ***n.*** The cathode-ray tube which displays outputs of computer systems. *Also known as* visual display terminal.

vid•e•o•tex sys•tem ***n.*** Personal computing and communication networks that allow interaction between people and stored data bases.

vid•e•o•tex ***n.*** Any electronic system that makes computer-based information available to a wide range of users via their home television receivers. Subdivided into *broadcast videotex,* or *teletext* (*see* teletext), and *interactive videotex.* In interactive videotex, the information is carried by a cable (usually a telephone line) and some degree of interaction with the data base (such as transfer of funds from a checking to a savings account) is possible.

vid•i•con ***n.*** A camera tube used chiefly in industrial television cameras.

vir•gin me•di•um ***n.*** A medium which will record computer data, but which has not yet been used.

vir•tu•al im•age ***n.*** An optical image which is not located where it seems to be, for example, a mirror image.

vir•tu•al leak ***n.*** A gradual desorption of gases from the walls of a vacuum container giving the appearance of a vacuum leak where there is none.

vir•tual stor•age ***n.*** Apparent computer memory available to a process; may be larger than the actual physical memory.

vis•com•e•ter ***n.*** An instrument for measuring viscosity of a fluid.

vis•cos•i•ty ***n.*** The flow resistance or internal friction of a fluid. ***adj.*** vis•cous.

Vi•si•Calc ***n.*** Trademark. Software that creates an electronic spread-sheet; a tool for business analysis.

vis•u•al dis•play ter•mi•nal. ***n.*** *See* video terminal.

vi•ti•ate ***v.*** To render ineffective.

vit•re•ous ***adj.*** Glassy.

vit•ri•fi•ca•tion ***n.*** The process of rendering a material glassy by heat and fusion.

VLF *See* very low frequency.

VLSI ***n.*** Very large scale integration; a semiconductor chip.

vo•cod•er ***n.*** An electronic system for synthesizing speech.

voice chan•nel ***n.*** A communication channel with a frequency range suitable for transmitting speech.

voice cod•er *n.* An electronic device that converts speech input into digital signals and back into speech at the receiver.

voice fre•quen•cy *n.* An audio frequency in the range essential for transmission of speech of commercial quality, from about 300 to 3400 hertz. *Also called* speech frequency.

vol•a•tile *adj.* Easily evaporated.

vol•a•tile file *n.* Any computer file which is easily modified.

vol•a•tile stor•age *n.* A computer memory that must continually be supplied with energy to prevent its losing its contents.

vol•a•til•i•ty *n.* The property of having a low boiling point, at ordinary pressure.

volt *n.* The unit of potential difference or electromotive force between two points. *Abbr.* V

volt•age *n.* The potential difference expressed in volts.

volt•age di•vid•er *n.* A resistor having different contacts or taps from which a desired fraction of the total voltage can be obtained.

volt•age doub•ler *n.* A rectifier circuit that approximately doubles the output voltage of a conventional vacuum-tube rectifier.

volt•age drop *n.* The voltage developed across a component in a circuit by the flow of current through the resistance of that component.

volt•age gain *n.* The increase in voltage across a circuit stage due to amplification.

volt•age rat•ing *n.* The maximum allowed voltage for a component.

volt•age reg•u•la•tor *n.* A device that maintains the stability of the voltage level in a circuit in both the operating and the no-load condition.

vol•ta•ic *adj.* Pertaining to electric currents that are chemically developed.

volt•am•me•ter *n.* An instrument that measures either voltage or amperage.

volt–am•pere *n.* The unit of power. *Abbr.* VA or V·A

volt–am•pere hour *n.* The product of 1 volt-ampere and 1 hour. Equal to 1 watt-hour. *See* watt-hour.

volt•me•ter *n.* An instrument for measuring voltage.

vol•ume *n.* **1.** The cubic measure of gas, or body, or a space. *Abbr.* vol **2.** The intensity of sound.

vol•u•met•ric *adj.* Pertaining to volume measurement.

vor•tex *n.* A swirl of rotating fluid.

vul•can•i•za•tion *n.* The process of strengthening and hardening rubber by means of sulfur, etc. or other vulcanizing agent.

W

wa•fer *n.* A slice of single-crystal semiconductor material.

walk•ing beam *n.* An oscillating beam; used in oil well pumping.

ware *n.* Laboratory vessels.

wash *n.* The stream of air sent backward by a propeller or a jet engine.

wa•ter of hy•dra•tion *n.* *See* bound water.

watt *n.* The unit of electric power. *Abbr.* W

wat•tage *n.* The amount of electric power expressed in watts.

wat•tage rat•ing *n.* The maximum allowed wattage for a device.

watt–hour *n.* The unit of electric energy equal to energy consumed at the rate of 1 watt per hour. *Abbr.* Wh or W·h

watt•me•ter *n.* An instrument that measures wattage.

watt•sec•ond *n.* The unit of electric energy equal to energy consumed at the rate of 1 watt per second. *Abbr.* Ws or W·s

wave *n.* A disturbance which contains energy and propagates from point to point.

wave•guide *n.* A device for the conduction of electromagnetic energy.

wave•length *n.* The distance from positive peak to positive peak of a single wave cycle. *Abbr.* λ

web *n.* A continuous length of paper or cloth.

we•ber *n.* The unit of magnetic flux. *Abbr.* Wb

wedge *n.* In optics, a filter in which the transmission decreases from one end to the other.

wedge bond *n.* A weld joint used in the fabrication of semiconductor devices.

weight *n.* The force with which an object is attracted toward Earth (or any other celestial body) by gravitation. Weight = mass × local gravitational acceleration. The weight of an object on Earth is always 32 times its mass. *See* mass.

weight•ing *n.* The adjustment of data to compensate for factors that differ between test and actual situations.

weld•ment *n.* A structure which is welded.

well *n.* **1.** A cavity in an apparatus. **2.** A vertical shaft in a building containing stairs or an elevator.

wet *v.* To form a contact angle of 0–90° between a liquid and a solid surface.

wet-bulb tem·per·a·ture *n.* Temperature dependent upon atmospheric humidity, taken with a thermometer having its bulb wrapped in wet muslin. *See* dry-bulb temperature.

wet·ta·bil·i·ty *n.* The ability of a solid surface to be wetted by a liquid.

Wh *See* watt-hour.

whisk·er *See* cat whisker.

white cast iron *n.* A very hard cast iron, containing about 3 percent carbon.

white met·al *n.* One of a group of metals with low melting points, such as tin, lead, and zinc.

white noise *n.* Random noise that maintains a given amplitude level.

wide·band *adj.* Pertaining to the ability of a circuit to pass a large range of frequencies.

Win·ches·ter *adj.* Referring to a kind of disk drive for computers in which many hard disks are stacked and from which any record can be read.

Win·ches·ter disk *n.* A computer storage device with sealed disk packs that permits random access to the data it contains.

wind·ing *n.* A turn of wire on a coil.

win·dow *n.* **1.** A structure on a microwave system that passes microwave energy without attenuation but provides a mechanical closure. **2.** In VisiCalc software, a command that brings up a portion of a large spreadsheet to full video-screen size.

wind tun·nel *n.* An apparatus for the analysis of airflows and their interaction with solid bodies.

wing nut *n.* A fastener with wings, easily turned by hand.

wire gauge *n.* A standard series of wire diameter sizes.

wire-wrap connection *n.* Wire coiled around a post so tightly that a solder joint is not needed to make an effective electric connection.

wob·bu·la·tor *n.* A signal generator that provides a variable frequency output.

word *n.* The fundamental unit of storage capacity for a computer, usually more than 8 bits in length. *Also known as* computer word.

word length *n.* The number of characters, digits, or bits in a computer word.

word proc·ess·or *n.* A typewriter terminal with a video display used for creation, storage, and retrieval of documents. It is designed to permit editing of stored documents on a screen, and is usually accompanied by a printer for the production of hard copy. *See* hard copy.

work *n.* The transference of energy. *v.* To form metal. ***past tense*** wrought.

work·a·bil·i·ty *n.* The ease or difficulty of forming a metal.

work hard·en·ing *n.* The hardening of metal by mechanically forming it.

work·ing stor·age *n.* That area of a computer's memory that is used during the writing and/or execution of a program.

worm *n.* A short revolving screw whose threads gear with the teeth of a worm wheel.

worm gear *n.* The gear of a worm and a worm wheel.

worm wheel *n.* A gear wheel with curved teeth that meshes with a worm.

write *n.* To input data into a computer's internal storage.

wrought *See* work.

wye *n.* A circuit which, when drawn, resembles the letter Y.

X

x ax·is *n.* The horizontal axis in any system of coordinates. *See y* axis; *z* axis.

X band *n.* A radio-frequency band extending from 5200 to 10,900 megahertz.

xe·rog·ra·phy *n.* A printing process in which the image is formed by heat acting on a resinous powder. It was developed by the Xerox Corporation. Derived from the Greek root *xeros* = dry. ***Note:*** Xerox, a trademark of the Xerox Corporation, is always capitalized and should not be used to describe copies produced by other manufacturers' processes. A safer word to use is photocopy.

xl *n.* An abbreviation for crystal.

x-ray *n.* Penetrating electromagnetic radiation in the wavelength range 0.001–100 nanometers.

x-ray crys·tal·log·ra·phy *n.* The technique of crystal structure analysis using x-ray diffraction.

x-ray dif·frac·tion *n.* The scattering of x-rays, especially in crystals to reveal their structure.

x-ray hard·ness *n.* The penetrating ability of x-rays.

x-ray li·thog·ra·phy *n.* The manufacture of thin-film integrated circuits by the exposure of a sensitive masking film to patterns of x-rays.

XY plot·ter *n.* An automated drafting device that creates very precise drawings at high speed. *Also known as* coordinate plotter.

XY re·cord·er *n.* A recorder that simultaneously traces two variables on a chart.

Y

ya·gi an·ten·na *n.* An antenna array.

yaw *n.* The oscillation of an aircraft or seacraft about a vertical axis. *v.* To oscillate about a vertical axis.

y ax·is *n.* The vertical axis in any system of coordinates. *See* x axis; z axis.

yield *n.* The outcome from a reaction or process. *v.* To give way under mechanical stress.

yield point *n.* The lowest point in a stress-strain test at which deformation becomes significant, without increase in stress.

yig fil·ter *n.* An electronic component using a crystal of yttrium-iron-garnet.

yoke *n.* A ferromagnetic structure that surrounds the neck of a cathode-ray tube to deflect electrons. *Also known as* deflection yoke.

Young's mod·u·lus *n.* A measure of material stiffness; the ratio of tension stress to strain that is parallel to the tension.

Z

z ax·is *n.* The axis that is perpendicular to the x and y axes. *See* x axis; y axis.

Ze·ner di·ode *n.* A semiconductor diode that limits the level of the voltage in a circuit.

Ze·ner volt·age *n.* The voltage at which breakdown occurs in a semiconductor diode. *See* breakdown.

ze·ro ad·just·er *n.* A device for adjusting the pointer of a meter to zero at zero input.

ze·ro grav·i·ty *n.* A condition of weightlessness, in which no acceleration of gravity can be detected.

zone bit *n.* A set of bits, representing a numeric or an alphabetic character.

zone melt·ing *n.* A technique for purification of crystalline solids and for growth of single crystals.

zoom lens *n.* A system of lenses having variable magnification, while keeping the image in the same focal plane.

Index

About the Authors

GEORGE FREEDMAN is a consultant in the field of new-product development and is the author of *The Pursuit of Innovation: Managing the People and Processes That Turn New Ideas into Profits* (AMACOM Books, 1987).

He was the Director of the New Products Center of the Raytheon Company of Lexington, MA, from 1969 through 1987. In this position he managed many programs in a wide variety of technical fields. During the 1960s he was Technical Editor of *Solid State Technology* magazine and has been a lecturer in prosthetic materials at the Harvard School of Dental Medicine. For the years 1979–1981, he was Chairman of the Board of Governors, International Microwave Power Institute, and served as Editor in Chief of its publication, *Journal of Microwave Power and Electromagnetic Energy,* from 1989 through 1991.

He has a degree in metallurgy from M.I.T., and holds an M.A. in physics from Boston University. He is also a graduate of the U.S. Navy Materiel School, where he studied electronics.

DEBORAH FREEDMAN is a teacher of English and drama at Ardsley High School in Ardsley, NY, and has also taught English at Hofstra Univerity in Hempstead, NY. Prior to this she was a technical secretary in the departments of Nuclear Engineering and Mechanical Engineering at the Massachusetts Institute of Technology. She has also worked as account executive for the New York State Department of Commerce *I Love New York* campaign while employed at a New York public relations agency. Throughout her teaching and public relations careers she has been a free-lance writer and editor. She is a graduate of Boston University School of Fine Arts and holds master's degrees from both Emerson College and St. John's University.